GeoJournal Library

Urban Perspectives from the Global South

The Urban Perspectives from the Global South brings together a wide variety of urban scholars under one series title and is purposefully multi-disciplinary. The publications in this series are theoretically informed and explore different facets of varying sized urban places. This series addresses the broad developmental issues of urbanization in developing world countries and provides a distinctive African focus on the subject. It examines a variety of themes relating to urban development in the Global South including: city economic development, issues of local governance, urban planning, and the impact of multi-ethnic and multicultural formations in urban affairs. The series aims to extend current international urban debates and offer new insights into the development of urban places in the Global South from a number of disciplines including geography, sociology, political science, economics, as well as urban studies. A special focus of the series is the challenges of urbanization and cities in Africa.

Timothy Nubi • Taibat Lawanson
Basirat Oyalowo
Editors

Transit Oriented Development in West African Cities

Springer

Editors
Timothy Nubi
Centre for Housing and Sustainable Development
University of Lagos
Akoka, Lagos, Nigeria

Taibat Lawanson
Centre for Housing and Sustainable Development
University of Lagos
Akoka, Lagos, Nigeria

Basirat Oyalowo
Centre for Housing and Sustainable Development
University of Lagos
Akoka, Lagos, Nigeria

ISSN 0924-5499 ISSN 2215-0072 (electronic)
GeoJournal Library
ISSN 2511-2171 ISSN 2511-218X (electronic)
Urban Perspectives from the Global South
ISBN 978-3-031-58725-2 ISBN 978-3-031-58726-9 (eBook)
https://doi.org/10.1007/978-3-031-58726-9

This Springer imprint is published by the registered company Springer Nature Switzerland AG
The registered company address is: Gewerbestrasse 11, 6330 Cham, Switzerland

Contents

Contributors

Michael Adeloye Adebamowo Department of Architecture, University of Lagos, Lagos, Nigeria

Olumide Afolarin Adenuga Department of Building, University of Lagos, Lagos, Nigeria

A. S. Afolayan Department of Estate Management, University of Lagos, Lagos, Nigeria

Franca Agama Department of Urban and Regional Planning, Yaba College of Technology, Lagos, Nigeria

Muyiwa Agunbiade Department of Urban and Regional Planning, University of Lagos, Lagos, Nigeria

Olayinka Oluyemi Agunloye Department of Urban and Regional Planning, University of Lagos, Lagos, Nigeria

John Agwu Department of Urban and Regional Planning, University of Lagos, Lagos, Nigeria

Ernest Agyemang Department of Geography and Resource Development, University of Ghana, Legon, Ghana
Centre for Urban Management Studies, University of Ghana, Legon, Ghana

Femi Aiyegbajeje Department of Geography, University of Lagos, Akoka, Lagos, Nigeria

Oluranti Olupolola Ajayi Department of Building, University of Lagos, Lagos, Nigeria

Oluyemi Akindeju Department of Urban and Regional Planning, University of Lagos, Lagos, Nigeria

Ikpeme A. Ankeli Department of Estate Management & Valuation, Federal Polytechnic, Ede, Nigeria

Isaac Arthur Department of Geography and Resource Development, University of Ghana, Legon, Ghana
Centre for Urban Management Studies, University of Ghana, Legon, Ghana

J. B. Babade Department of Estate Management, University of Lagos, Lagos, Nigeria

Adedoyin Babalola Department of Architecture, University of Lagos, Lagos, Nigeria

Aba Obrumah Crentsil Institute of Statistical, Social and Economic Research, University of Ghana, Legon, Ghana
Centre for Urban Management Studies, University of Ghana, Legon, Ghana

Esther Ilori Ebunoluwa Department of Quantity Survey, Obafemi Awolowo University, Ile-Ife, Nigeria

Adenike Egunjobi Department of Economics, University of Lagos, Akoka, Lagos, Nigeria

Peter Elias Department of Geography, University of Lagos, Akoka, Lagos, Nigeria

Olawale Fabiyi Department of Computer Science and Technology, American International University, West Africa, Serekunda, The Gambia

Gbemiga Bolade Faniran Department of Urban and Regional Planning, Obafemi Awolowo University, Ile-Ife, Nigeria

Olajide J. Faremi Department of Building, University of Lagos, Lagos, Nigeria

Oluwatoyin Fatunsin Department of Industrial Chemistry, University of Lagos, Lagos, Nigeria

O. B. A. Idowu Department of Estate Management, University of Lagos, Lagos, Nigeria

Victor Ilechukwu Department of Urban and Regional Planning, University of Lagos, Akoka, Lagos, Nigeria

Iniobong Beauty John Department of Quantity Surveying, University of Lagos, Lagos, Nigeria

H. A. Koleoso Department of Estate Management, University of Lagos, Lagos, Nigeria

Braima Koroma Institute of Geography and Development Studies, Njala University, Freetown, Sierra Leone

Taibat Lawanson Department of Urban and Regional Planning, University of Lagos, Akoka, Lagos, Nigeria
Centre for Housing and Sustainable Development, University of Lagos, Akoka, Lagos, Nigeria

Joseph Mustapha Macarthy Sierra Leone Urban Research Centre, Njala University, Freetown, Sierra Leone

Francis Mfobujong Department of Accounting & Finance, American International University, West Africa, Serekunda, The Gambia

Oluwaseun Muraina Department of Estate Management, University of Lagos, Lagos, Nigeria

Timothy Nubi Department of Estate Management, University of Lagos, Akoka, Lagos, Nigeria
Centre for Housing and Sustainable Development, University of Lagos, Akoka, Lagos, Nigeria

Fatima Bashar Nuhu Niger State Urban Development Board, Minna, Nigeria

Muhammad Bashar Nuhu Department of Estate Management and Valuation, Federal University of Technology, Minna, Nigeria

Immaculata Nwokoro Department of Urban and Regional Planning, University of Lagos, Lagos, Nigeria

Damilola Odekunle Department of Urban and Regional Planning, University of Lagos, Akoka, Lagos, Nigeria

Tinuola Odugbemi Department of Community Health and Primary Care, College of Medicine, University of Lagos, Lagos, Nigeria

Leke Oduwaye Department of Urban and Regional Planning, University of Lagos, Akoka, Lagos, Nigeria

Yetunde Ohiro Department of Estate Management, University of Lagos, Akoka, Lagos, Nigeria

Timothy Tunde Oladokun Department of Estate Management, Obafemi Awolowo University, Ile-Ife, Nigeria

Oluwafemi Olajide Department of Urban and Regional Planning, University of Lagos, Lagos, Nigeria

Damilola Olalekan Department of Urban and Regional Planning, University of Lagos, Akoka, Lagos, Nigeria

Abel Olaleye Department of Estate Management, Obafemi Awolowo University, Ile-Ife, Nigeria
Department of Finance & Investment Management, University of Johannesburg, Johannesburg, South Africa

J. B. Olaleye Department of Surveying & Geo-informatics, University of Lagos, Lagos, Nigeria

Olorunfemi Ayodeji Olojede Department of Urban and Regional Planning, Obafemi Awolowo University, Ile-Ife, Nigeria

M. M. Omirin Department of Estate Management, University of Lagos, Lagos, Nigeria

Ademola Omoegun Department of Architecture, University of Lagos, Akoka, Lagos, Nigeria

Victor Onifade Department of Urban and Regional Planning, University of Lagos, Lagos, Nigeria

George Owusu Institute of Statistical, Social and Economic Research, University of Ghana, Legon, Ghana
Centre for Urban Management Studies, University of Ghana, Legon, Ghana

Basirat Oyalowo Department of Estate Management, University of Lagos, Akoka, Lagos, Nigeria
Centre for Housing and Sustainable Development, University of Lagos, Akoka, Lagos, Nigeria
School of Built Environment, Oxford Brookes University, Oxford, UK

Benjamin Ozirigbo Department of Accounting & Finance, American International University, West Africa, Serekunda, The Gambia

Simeon Dele Rogers Department of Building, University of Lagos, Lagos, Nigeria

Olufemi Saibu Department of Economics, University of Lagos, Lagos, Nigeria

Taofiki Salau Lagos Urban Development Initiative, Lagos, Nigeria

Nasiru Salihu Department of Estate Management, Bayero University Kano, Kano, Nigeria

Umaru Tanimu Saminaka Department of Accounting & Finance, American International University, West Africa, Serekunda, The Gambia

Adegboyega Sunday Sotunbo Department of Building, University of Lagos, Lagos, Nigeria

E. O. Thontteh Department of Estate Management, University of Lagos, Lagos, Nigeria

Olamide Udoma-Ejorh Lagos Urban Development Initiative, Lagos, Nigeria

Richard Unuigboje Department of Urban and Regional Planning, University of Lagos, Lagos, Nigeria

Charlotte Wrigley-Asante Department of Geography and Resource Development, University of Ghana, Legon, Ghana
Centre for Urban Management Studies, University of Ghana, Legon, Ghana

Suleiman Yakubu Department of Estate Management, Baze University, Abuja, Nigeria

Yasmina Yusuf Sierra Leone Urban Research Centre, Njala University, Freetown, Sierra Leone

Kudirat Ibilola Zakariyyah Department of Building, University of Lagos, Lagos, Nigeria

Part I
Introduction

Chapter 1
Introduction

Taibat Lawanson, Basirat Oyalowo, Isaac Arthur, and Timothy Nubi

1.1 Transport in Infrastructure Development of Africa

Transportation infrastructure across Africa from colonial times has to a large extent determined the role and importance of cities. According to Gwilliam (2011), the chief goal of road and rail networks was to link mines, plantations, and other sites for the exploitation and transportation of natural resources to ports, rather than to provide general connectivity within the region. As such, not a lot of attention was paid to ensure that transport infrastructure served the needs of urban dwellers.

Rapid rates of urban growth across the continent have resulted in fast growing megacities and the proliferation of secondary cities—which range from 100,000 to one million people in population. These already houses more than 40% of Africa's urban population and are expected to absorb two-thirds of Africa's future urban growth (OECD, 2020; Riley & Crush, 2023). This urban growth has resulted in many towns growing without physical development plans and the prevalence of urban sprawl. Urban sprawl comes with an urgent requirement for effective

T. Lawanson (✉)
Department of Urban and Regional Planning, University of Lagos, Akoka, Lagos, Nigeria

Centre for Housing and Sustainable Development, University of Lagos, Akoka, Lagos, Nigeria
e-mail: tlawanson@unilag.edu.ng

B. Oyalowo · T. Nubi
Centre for Housing and Sustainable Development, University of Lagos, Akoka, Lagos, Nigeria

Department of Estate Management, University of Lagos, Akoka, Lagos, Nigeria

I. Arthur
Department of Geography and Resource Development, University of Ghana, Legon, Ghana

Centre for Urban Management Studies, University of Ghana, Legon, Ghana

T. Nubi et al. (eds.), *Transit Oriented Development in West African Cities*, Urban Perspectives from the Global South,
https://doi.org/10.1007/978-3-031-58726-9_1

transportation infrastructure that will serve the resource and mobility needs of people within and across cities on the one hand and between urban and rural areas on the other hand.

A society cannot function optimally if it does not have measures in place to facilitate efficient transport. Efficient transportation is vital for overall productivity and sustainable development. It is important as a lifeline for millions of Africans in that it connects them to people, places and economic opportunities. It also builds thriving communities, creates jobs, eases traffic congestion and promotes a cleaner environment. Investment in public transportation spurs both local and national economies. One way of achieving this according to the Institute for Transport and Development Policy (2020) is through a global shift from urban sprawl to inclusive transit-oriented development.

1.2 What Is Transit-Oriented Development?

Transit-oriented development (TOD) is a form of urban development conceived or implemented to integrate people's everyday activities and public space together, with easy walking and cycling connection between them and near-excellent transit service to the rest of the city (Wood, 2022). It means inclusive access for all, to local and citywide opportunities and resources by the most efficient and healthful combination of mobility modes, at the lowest financial and environmental cost and with the highest resilience to disruptive events. Inclusive TOD is a necessary foundation for long-term sustainability, equity, shared prosperity and peace in cities (Institute for Transportation and Development Policy, 2020). However, it is easier conceptualised than implemented as complex and interdependent variables must be brought together. These variables cut across infrastructure, building codes, planning standards, regulators and finance. As such, numerous stakeholders ranging from transport, planning, infrastructure and finance must engage effectively to ensure that local community needs and interests are provided for and protected.

TOD has been practiced with varying levels of success across the world. Many towns in Japan, Sweden, Denmark, the Netherlands and France have many of the characteristics of TOD communities (Cervero et al., 2002). However, Huang et al. (2018) has argued that despite the success stories of TOD in Asia and Europe, the strategy is yet to be practiced widely in Sub-Saharan Africa. According to him, though TOD's concept of leveraging on major transit infrastructure to affect integrated land-use development for greater benefits has gained recognition, there are few examples of successful TOD on the continent with the probable exception of South African cities, mainly Cape Town and Johannesburg. Interestingly, the implementation of TOD in South Africa has been criticised in the academic discourse (see Robinson et al., 2020; Harrison et al., 2019). However, there is limited documented evidence of similar empirical discourse in West Africa, hence the raison d'être of this book.

1.3 The Book

This book is a collection of theoretical and empirical studies, anchored on scoping studies and transdisciplinary engagements. The first section explores the theoretical perspectives of TOD implementation through a geographic focus on Lagos, West Africa's largest city. The collection of reviews highlights the conceptualisation, implementation, challenges and opportunities for transit-oriented development in Lagos. The chapters keep in sight alternative forms of provision and learnings across Africa while focusing strongly on the complementarity of Sustainable Development Goals (SDGs) to a TOD agenda. The thematic foci cut across urban planning and development, housing, land administration, spatial data infrastructure, urban design and infrastructure, construction and building utilities and maintenance, urban health and well-being as well as pro-poor development.

Complementary to these thematic studies are empirical case studies in the second section of the book. They highlight on the state of urban infrastructure in West Africa with reference to introducing and/or better managing of TOD implementation in six west African cities. They include the capital cities of Nigeria (Abuja), Ghana (Accra), The Gambia (Banjul) and Sierra Leone (Freetown), as well as Lagos and Osogbo, Nigeria. They represent a typology of cities across the subregion with Abuja as a planned city, Accra as a fast-growing city in emerging economy, Banjul as an island city, Freetown as a post-conflict city, Lagos as West Africa's megacity and Osogbo as a typology of secondary city. However, we find that everyday realities to a large extent remain the same with dominant informal transport services, foreign funding for transport infrastructure development and reforms anchored on a cost-recovery model that does not take into full account the everyday lives of urban residents of many African cities.

The authors of the chapters cut across various academia and urban development practitioners in the case study cities. Limitation to anglophone West African countries is largely due to the composition of the African Research Network on Urbanisation and Habitable Cities. The network is part of the African Research Universities Alliance Centre of Excellence for Urbanisation and Habitable Cities, funded by the UKRI. The aim of the network is to foster collaboration between African-based urban researchers and other stakeholders in order to catalyse action towards contextualised solutions for addressing Africa's grand challenges. Since 2019, the centre engaged in scoping reviews, stakeholder meetings and webinars on multifaceted African urban dimensions. The various network activities have highlighted the critical role of nuanced understanding of urban planning and urban infrastructure issues and, with the authors being urban citizens also encountering the everyday dimensions of the challenges they study, are better placed to espouse the structural and institutional reforms that are essential to delivering a better urban future for African urban residents. This volume is borne out of a desire to document and disseminate our findings to both academic and policy audiences whose work is at the interface of urban planning, development, finance, infrastructure and governance in Africa.

1.4 Overview of Chapters

Part I: Introduction

Chapter 1 by Lawanson et al. is a brief but very concise introduction to the topic of discourse. It explains the concept of transport in infrastructure development of Africa, and the authors define transit-oriented development. In order to aid readers in understanding the premise upon which this text was written and the authors, there is a subsection on the book and an overview of the various chapters.

Part II: Conceptualising Transit-Oriented Development: A Thematic Study of Lagos

Chapter 2 by Oduwaye et al. examines the effect of urbanisation on transport infrastructure investments and proposes policies for achieving transit-oriented development (TOD). These authors discuss conceptually the linkages between urban growth and transportation and underscore the relevance of TOD for cities experiencing urban growth challenges to achieve the United Nations Sustainable Development Goals 1, 8, 11 and 13. They argue that despite significant transport infrastructure investments and innovation such as BRT, the lack of an efficient intermodal transport system, poor service delivery and management of existing facilities and infrastructure stifle accessibility in Lagos and negatively impact the functionality of the city. Furthermore, the failure to integrate physical development and transport infrastructure priorities holistically contributes to the gaps in TOD implementation.

Moreover, they extend their argument on the relevance of TOD for cities in both the Global North and Global South by discussing case studies of TOD policies implemented in Copenhagen, Curitiba, Hong Kong, Denver and Johannesburg while highlighting key lessons for Lagos. They demonstrate that these policies resulted in increased utilisation of public transit, thus reducing automobile reliance and carbon footprints. As a result, the chapter recommends the reallocation of road space from cars to pedestrian-friendly walkways, bus and cycle lanes and community advocacy and participation through local organisations supporting incremental short-term improvements, while larger-scale changes in transit networks are implemented.

Chapter 3 by Nubi et al. appraised the nexus between housing and TOD in an urban African context. The appraisal is predicated on an attempt to understand ways in which affordable housing can be achieved in TOD projects, the prospects for affordable housing in neighbourhoods that have experienced transportation investments in the absence of TOD as a conscious planning- investment regime and the available solutions for places where housing development has preceded transportation investment. The authors draw on experiences from cities around the world to establish the relationship between transportation challenges and housing in peripheral areas of the city—where land prices are cheap as opposed to high transport costs for residents and how that influences the residential location choices of individuals. They argued for the viability of TOD as a solution for proving sustainable development interventions in African cities, using examples from Johannesburg (mixed-use infrastructure development) and Addis Ababa (light railway transit project) to validate the concept. The authors posit that the two cases provide important lessons, including the necessity for inclusive, culturally sensitive models, the importance of not ignoring the creation of negative externalities as a result of TOD and the critical role of government and private sector participation in TOD projects.

In examining the viability of the idea, it was revealed that while investments in housing and transportation do not seem to be strategically connected, there are no present intentional plans for TOD interventions in Lagos. For workable strategies, the authors proposed that the expansion of existing BRT network as it offers an incentive for TOD implementation, urban regeneration and creating job opportunities using TOD principles, the current developments along existing highway routes and the proposed light rail routes can be made more transit-friendly through additional infrastructure investment following TOD principles that respect and protect the culture of the local inhabitants. Similar to Oduwaye et al. (this book), the authors present recommendations to the government, investors and communities on the best practices for TOD projects. They conclude with the assertion that although TOD will not address all urban issues, it provides a wide range of social, economic and physical development opportunities in peripheral areas.

Omirin et al. in Chap. 4 acknowledge the multidimensional relevance of TOD in ensuring sustainable African cities. However, they bring attention to the critical role of land policy, planning and market-led articulations of land uses to ensure inclusive benefits of TOD interventions. In their submission, they focus on the existing state of affairs of TOD developments and the potential for successful land resource infusion into TOD in a complex land regime like Lagos.

This chapter develops on the importance of transportation to the economy and social development, noting that rapid urbanisation places a compulsion on government investment in transportation infrastructure. The centrality of their argument is that access to land for TOD and infrastructure provision is fraught with challenges which must be clearly understood and mitigated at key stages of project conceptualisation, planning and execution using the case study of Miami-Dade to highlight the importance of effective land management practices. This argument is germane, taken against the backdrop of the social and economic role of transportation, the need for government involvement and the use of private sector funds. They tie this argument into a conceptual linkage with SDG 11 in particular, highlighting value-chain connections between TOD and land provision, land-use planning and value enhancement.

With specific reference to Lagos, issues around informal land markets, acquisition of land titles, equitable and fair compensation on land compulsorily acquired and rapid urbanisation were all highlighted as challenges to implementing transit-oriented development in the city. They recommend multipronged approaches to regional development, deliberate emphasis on compactness and mixed-use development, land readjustment options to encourage land release and better documentation of land rights as key factors if TOD is to be successful in cities like Lagos.

Lack of data harmonisation frameworks for predictions and decision-making for improved urban systems aggravate inadequate transport systems in cities is the key issue established by Agunbiade et al. in Chap. 5. They argue that spatial data infrastructure framework that combines data activation and multimedia annotation technologies to process and integrate multiple sources and heterogeneous, massive, and decentralised data is key to any successful TOD processes. But not only that; this must accommodate both real-time signal processing and scalability, by supporting data analysis on a large scale. They then proceed to provide an analysis of the nature and depth of the prevailing spatial data infrastructure issues in Lagos, noting that lack of a relevant base collection of technologies, policies and institutional arrangements that facilitate the availability of and access to spatial data limits sustainable urban form and mobility.

In this chapter, the conceptual and theoretical linkages between transportation, TOD and spatial data infrastructure are established through the connections of the SDGs, with TOD being presented as a possible route to achieving the New Urban Agenda. From a spatial data infrastructure perspective, the authors appraise the TOD systems in Cape Town and Curitiba and assert that in the case of Lagos, mobility planning needs to be refocused on the people and not on vehicles, and even though this requires an examination of human issues in data sharing, it still provides a means of integrating data into city transportation systems. The chapter calls for extension of information gathering on travel pattern, trip behaviour and other transportation data to the grassroot/local constituencies and the neighbourhood level as a pivot to the success of any policy engagement.

Against the backdrop of poor infrastructure development in African cities and its effects, Adebamowo et al. in Chap. 6 discuss the intermodal transport system of Lagos from the perspective of TOD. They argue that the intermodal transport system can be a panacea for addressing the transport challenges of an African city like Lagos. This is given its ability to provide sustainable mobility, improve the quality of life of the people and provide an adequate enabling environment for businesses to thrive. The authors begin their discussion by highlighting the links between transportation challenges and urban infrastructure in Lagos and similar to Oduwaye et al. in Chap. 2, conclude that poor urban planning, weak multi-modal coordination and a lack of urban public transportation to meet the city's growing need for mobility, urban traffic congestion, security and environmental friendliness all play a role in Lagos' transportation issues. In establishing the conceptual linkage between TOD, Sustainable Development Goals and urban infrastructure, they developed a sustainable transport infrastructure model. The model consists of soft infrastructure (institutional capabilities and policies used to address emerging challenges) and hard infrastructure (bridges, terminals, roads, railways, tunnels, etc). The authors operationalised the model by demonstrating how the Ministry of Transport has provided policies and strategic plans for improving transportation in addition to the provision of BRT, proposed light rail and water transport services, among others.

The discussion on thematic-based issues and priorities for Lagos is climaxed with highlights on the main outputs of specific plans associated with the extension of Lagos' Strategic Transport Master Plan (STMP) defined in the context of freight, non-motorised transport (NMT), road safety and climate change. The authors proposed some recommendations for the government, investors and communities, and these include the adoption of a policy framework for deployment of environmentally friendly multimodal transport systems across the city.

Adenuga et al. in Chap. 7 trace the connection between TOD and the SDGs through the lens of SDGs 7, 9, 11, 13 and 17. As their emphasis is on the building structure and utilities upon which TODs are anchored, the authors developed a narrative of the adequacy of existing building infrastructure in any attempt at TOD projects in Lagos. They provide evidence from stakeholder engagement that suggests that the existing condition of buildings and utilities around areas which could be designated as TOD is poor and thus highlighted the need for interventions in the areas of facility management and maintenance management, innovations in the designs for new and urban upgrade projects and sustainable practices in the design, construction and post-construction operations of motor parks and garages. Finally, the authors call for investment in other modes of transportation to ensure the interconnection of systems across Lagos.

Transportation and health nexus is the focus of Nwokoro et al. in Chap. 8. Specifically, the chapter explored the relationship between land use, transportation infrastructure, the environment and health in Africa and the significance of these relationships in achieving the African Urban Agenda and the SDGs given the significance of regional connectivity and cooperation required. In essence, it was meant to develop a framework through which sustainable transport, improved urban health and liveability can be achieved in Africa. They argue that through this framework, the implementation of TOD will serve as a link between SDGs and transportation.

Using the case of Johannesburg, South Africa, the authors analysed the condition of transport infrastructure in Johannesburg from a TOD perspective and determine that apartheid and modernist planning to a large extent influenced land-use activities in Johannesburg (Wood, 2022). According to them, post-apartheid interventions were instituted to reverse racial spatial segregation by constructing transit arterial connections to link apartheid-isolated areas to other parts of the city, and the effects were that of inadequate equitable transport accessibility resulting in over-reliance of automobiles and long-distance shuttling with attendant health implications. To broaden the discussion on health issues associated with transportation, the authors highlighted the transportation system in Lagos detailing the modes of transport and statistics of their usage and their operational, technical and design challenges. They argued that these challenges have huge health impacts ranging from physical health and mental health to environmental health problems. The chapter proposed that a feasibility and viability study is required to identify locations that need social and spatial restructuring using TOD; the outcome of the feasibility and viability study should serve as benchmark for city managers to collaborate with relevant stakeholders to develop a sustainable TOD strategy that will control urban development and reverse transport-related health issues, urban sprawl and low-density urban forms in African cities.

In the final chapter of this section (Chap. 9), Lawanson et al. bring to attention the critical role of the informal sector in urban transportation in Africa. The chapter explores the connections between TOD and urban poverty in African cities with a review of transportation, pro-poor development and urban governance in Africa. It highlights how rapid population growth and congestion is hampering the movement of people and goods in many cities, with attendant environmental, health and economic concerns. It focuses on the specific transportation challenges faced by the urban poor and how recent TOD initiatives and transport innovations impact them. Thereafter, conceptual linkages between TOD, the SDGs and pro-poor development are explored. Some case studies of TOD are reviewed with key issues and priorities for the implementation of TOD in other cities elaborated. These include the transit-oriented development used in Addis Ababa, Dar es Salaam, Johannesburg, Curitiba, Ahmedabad and Lagos. The final part of the paper provides key recommendations for improving transportation in Lagos with implications for other African cities. The chapter concludes by calling for creative alternatives of integrating formal and informal transport systems to address mobility challenges in African cities.

The various chapters addressing these thematic issues highlight the linkages between TOD, the SDGs and the creation of inclusive and authentic African city processes using Lagos as a case study.

Part III: Deep Dives of TOD in West African Cities

This section is a review of TOD implementation in six West African cities, highlighting the processes, challenges and potentials of TOD for engendering a sustainable urban future.

In Chap. 10, Arthur et al. describe the Greater Accra Metropolitan Area (GAMA) as being in a process of intense urbanisation and thus the focal point of expansive peri-urbanisation sprawl driven by in-country, regional and international migration. They note that global capital and the presence of international non-governmental organisations (NGOs), multinational corporations (MNCs), expatriates and international returned migrants all add to the economy of the city. The major infrastructural challenge for Accra as far as transportation is concerned is the lack of coordination between road transport investments and urban planning, coupled with the weak capacities of the municipal and metropolitan governments that have contributed to the intensification of urban sprawl and gaps in transport, utility service provision, sanitation and waste management and housing.

Accra's transportation infrastructure is reliant on international donor organisations and financial institutions. In arguing that the transport infrastructure has not been at pace with the urbanisation trends of the city, the authors posit that bureaucratic processes rather than management are a major challenge. Here, they maintain that the application of TOD would be severally called to question, while Accra adopted the National Transport Policy that had the objective of making Accra, and other major Ghanaian urban centres, transit-oriented cities. Arising from this, attempts at introducing mass transit buses in 2003 have failed due to operational problems with the Bus Rapid Transport system picking up only a small proportion of the travelling public. The authors urge adoption of TOD interventions that are culturally focused, especially taking account of the predominant informality in the transport system, the investment by private sector and the involvement of multiple stakeholders.

Abuja, Nigeria's capital city, is the focus of Nuhu et al. in Chap. 11. Abuja is described as a planned city in that it emerged from a 1976 government decree designating it Nigeria's capital city. The intentional creation of a country's administrative capital required a master plan design, and this was vested in the Federal Capital Development Authority (FCDA). The city's design included provision of an Abuja Rail Mass Transit and several ten-lane super multiple carriage highways. Over 40 years later, the rail mass transit is not functional, and the transportation sector is dominated by mostly informal providers. The authors recommended the formulation of policies that would enable informal transit system to play complementary role within an integrated modal transport system, but with regulations to streamline operations and align with formal transport services.

Chapter 12 by Fabiyi et al. presented the Banjul case study. They describe Banjul as The Gambia's port town, city capital and seat of power. In the overview of Banjul's transportation infrastructure, the authors identify the broad challenges facing road transportation, drawing attention to the limitations posed by undefined hierarchical structure of roads within the Greater Banjul Area and unclear institutional structures for the governance of air and water transportation.

The chapter assesses the state of road transportation in Banjul as being one to be concerned about, noting the persistent congestion during rush hours and the poor condition of roads. They argue that the introduction of technological enhancements in road management would be more efficient than road expansion, showing that since 1998, investment in road network provision, with the National Transport Policy (1998–2006), was concentrating on investment in road infrastructure and management in the Greater Banjul areas. They also argue that Banjul's development and investment in road networks requires a sector-wide approach that recognises road classification and hierarchies. Thus, they call for a more clearly defined responsibility of road infrastructure, funding of urban road maintenance, traffic regulation and enforcement and the introduction of traffic demand management measures and signalisation at intersections.

Chapter 13 focuses on Osogbo, a secondary city and UNESCO World Heritage site in South-West Nigeria. Here, Olaleye et al. trace a broad history of the city that bears the heritage of the Yoruba people and note that Osogbo does not have sufficient infrastructure that promotes an integrated basic service delivery. The author contends that the growth of the city from a small commercial centre along a number of radial exit and intra-city road networks has affected its growth, form and land-use pattern.

The authors found that Osogbo is technically a mono-modal city, as road is the only functional transport mode. The moribund railway does not provide commuting services for the residents, and the only river is non-navigable. Even then, road transportation is characterised by predominantly commercial motorcycles and micro-minibuses. Plans exist to convert an aerodrome to an airport but progress is slow. They posit that the transportation challenges in Oshogbo reflect both infrastructure (lack of maintenance, drainage, technologically driven management) and attitudinal deficits (on-street parking, loading and offloading of goods).

However, there currently exists no urban transport policy for any Nigerian city; hence, regulation is done through government ministries in an ad hoc manner. The state government is the most significant funder of transportation investment in the state. Investment itself has been focused on the road infrastructure. The authors argue that the most urgent need for Osogbo today is for a transport policy and master plan that will guide transportation planning in the city. Only within this context can opportunities for transit-oriented development occur.

In Chap. 14, Koroma et al. present a city profile of Freetown, Sierra Leone. The chapter described a city characterised by difficult topography, but incredible urban sprawl arising from its rapidly growing population. An overview of Freetown's transportation network also identifies the primacy of road transport and general problems such as lack of proper hierarchy of roads and poor connectivity, inadequate vehicle capacity of roads and poor pedestrian facilities. In terms of governance framework for the city's transportation sector, Koroma et al. also show that despite the multiplicity of government ministries and departments in planning, financing, running and enforcement of the sector, the delivery of infrastructure and services is complex and inadequate. They highlight the activities of an active informal transportation sector, dominated by minibuses (poda-poda), three-wheel motorcycles/tuk-tuks (kekeh), motorcycles (okada) and shared taxis. In addition to this, the authors identify the lack of mass transit services which inadvertently increases the use of private vehicles. Unplanned growth on land availability, transport and infrastructure development/expansion in the city all add pressure to the system. The complexity of navigating these existing problematic factors may be additional reasons why transit-oriented development is not likely to be adopted as a major policy decision-making process by both national and city authorities. They therefore challenge the government to build resilient, sustainable transport infrastructure that is inclusive, less dependent on foreign aid and more embedded in urban planning frameworks.

The final chapter of the book (Chap. 15) is a profile of Lagos by Oyalowo et al. This chapter identifies the broad infrastructural challenges facing Lagos, West Africa's largest economy and Nigeria's most urbanised city. Like Freetown, it is a primate city.

In an overview of its transportation infrastructure, the authors highlight the dependency on road transportation in the city. Current challenges in the transportation system include traffic congestion occasioned by the overconcentration of industries and businesses in specific areas of the city, an increase in private car ownership, the lack of an efficient public transport system, poor town and highway planning, small roads and poor driving behaviour. The authors also draw attention to transport governance and investment framework of the city, identifying the agency of the informal sector, as areas of synergy, overlap and conflict among actors. The informal sector is presented as an intensely powerful actor imposing road taxes and levies without concomitant infrastructure or even welfare investment. They highlight the massive investment in the Bus Rapid Transit system and its continual expansion though fraught with extensive maintenance gaps. Lagos is unique for the existence of ride-hailing cabs and taxis from different operators.

Rail transportation is under construction to serve intra-city movement, as opposed to the currently existing inter-city rail network operated by the federal government. While investment has occurred in the water transportation sector, the inadequacy of boats and selected route to high-density areas leaves other communities underserved. The authors recommend more intentional synergies between transportation investment and real estate development if TODs are to serve any purpose for the city. They call for the repositioning of urban transportation as a service rather than an infrastructure and require that informal sector operators and their unions be integrated into new projects with clear parameters for engagement.

1.5 Conclusion

The rationale for the book was to better understand how city development processes across West Africa are being shaped by principles of transit-oriented development. The aim was to leverage this understanding to advance pragmatic approaches to the complexities of urban infrastructure deployment and the intricate mix of stakeholder considerations that are required for the implementation of both a transport and planning and management system that enhances inclusion and well-being for urban citizens, thus fostering sustainable development.

With lessons from six West African cities, this edited volume shared conceptual and empirical approaches to navigating urban infrastructure challenges, highlighting what worked, what did not work and why. We hope this volume will offer key learnings beyond the six case study cities and contribute to the discourse and action required to create a better urban future for Africans.

References

Cervero, R., Ferrell, C., & Murphy, S. (2002). *Transit-oriented development and joint development in the United States: A literature review, research results digest* (No. 52). Transit Cooperative Research Programme.

Gwilliam, K. (2011). *Africa's transport infrastructure: Mainstreaming maintenance and management*. World Bank. https://doi.org/10.1596/978-0-8213-8456-5

Harrison, P., Rubin, M., Appelbaum, A., & Dittgen, R. (2019). Corridors of freedom: Analyzing Johannesburg's ambitious inclusionary transit-oriented development. *Journal of Planning Education and Research, 39*(4), 456–468.

Huang, R., Grigolon, A., Madureira, M., & Brussel, M. (2018). Measuring transit-oriented development (TOD) network complementarity based on TOD node typology. *Journal of Transport and Land Use, 11*(1). https://doi.org/10.5198/jtlu.2018.1110

Institute for Transport and Development Policy. (2020). *The TOD standard*. https://www.itdp.org/library/standards-and-guides/tod3-0/about-the-tod-standard/

OECD. (2020). *Africa's urbanisation dynamics 2020: Africapolis, mapping a new urban geography | en | OECD*.

Riley, L., & Crush, J. (2023). Introduction: African secondary city food systems in context. In L. Riley & J. Crush (Eds.), *Transforming urban food systems in secondary cities in Africa*. Palgrave Macmillan. https://doi.org/10.1007/978-3-030-93072-1_1

Robinson, J., Harrison, P., Shen, J., & Wu, F. (2020). Financing urban development, three business models: Johannesburg, Shanghai and London. *Progress in Planning, 154*, 100513.

Wood, A. (2022). Problematising concepts of transit-oriented development in South African cities. *Urban Studies, 59*(12), 2451–2467. https://doi.org/10.1177/00420980211033725

Part II
Thematic Study

Chapter 2
Transit-Oriented Development and Urbanisation in African Cities

Leke Oduwaye, Victor Onifade, Victor Ilechukwu, Peter Elias, Damilola Odekunle, and Damilola Olalekan

2.1 Introduction

The urban population in Africa has been growing at a very high rate, from 27% in 1950 to 40% in 2015 and is projected to reach 60% by 2050 (UN-DESA, 2014). As this population continues to grow, an increasing number of people are moving to cities with the hope of securing better living, higher quality education, good health care and greater economic opportunities. One of the major drivers of urbanisation in Africa is the growth of towns and intermediate cities (Rico, 2016), and this growth is expected to be linked and facilitated by transportation. While urbanisation can be a boost to the national economy, it can also present many infrastructure and humanitarian challenges. The emergence of traffic congestion, inadequate facilities and services, unemployment, poverty and environmental deterioration are common consequences of urbanisation. Currently, Lagos is the largest city in Africa, with a population of 14.4 million at a growth rate of 3.26% (*worldpopulationreview.com*), out of which two-thirds of Lagos residents are experiencing these urbanisation challenges.

In most cities in Africa, traffic congestion could be attributed to the types and qualities of transport infrastructure. Roads dominate the transport sector in most African countries, carrying 80–90% of passenger and freight traffic (World Bank, 2011), followed by railway, air and marine transports. Moreover, the condition of the road system is poor by international standards and the usage is also inefficient. Besides, the road transport system faces challenges of traffic congestion and parking difficulties, longer commuting time, the inadequacy of public transport,

L. Oduwaye · V. Onifade (✉) · V. Ilechukwu · D. Odekunle · D. Olalekan
Department of Urban and Regional Planning, University of Lagos, Akoka, Lagos, Nigeria
e-mail: vonifade@unilag.edu.ng

P. Elias
Department of Geography, University of Lagos, Akoka, Lagos, Nigeria

T. Nubi et al. (eds.), *Transit Oriented Development in West African Cities*, Urban Perspectives from the Global South,
https://doi.org/10.1007/978-3-031-58726-9_2

difficulties for non-motorised transport, loss of public space, high infrastructure maintenance costs, environmental impacts, energy consumption, accidents and safety, among others (Berg et al., 2017).

Transport infrastructure investments can be very costly and need to accompany the fast pace of urbanisation currently occurring in developing countries, especially in Africa. While the provision of transport is potentially crucial for development, there is often a risk that transport investments are not cost-effective and do not produce the range of expected outcomes due to trend or pattern of urbanisation as well as backlogs of transport infrastructure investment, inadequate regulations in the transport sector and rising social costs in terms of congestion, pollution and accidents (Berg et al., 2017).

There is an obvious relationship between the consequences of urbanisation and the level of investment in transport infrastructure. For instance, there is underinvestment in the roads of most suburban areas of African cities where the poor live (Blimpo et al., 2013). Also, poor physical connection between jobs and residences leads to high commuting costs that are detrimental to the workers getting a job (Gobillion & Selod, 2014). Furthermore, transport costs have an impact on educational opportunities and choices that can increase or decrease enrolment rates of students in schools (Khandker et al., 2009). Also, there are potential food and health benefits from better transportation (Blimpo et al., 2013). This underlines the relevance of transit-oriented development which drives smart growth and addresses intimidating urbanisation problems such as traffic congestion, inadequate housing, sprawling, decaying inner-city and disinvestment. Integrating the goals of sustainable transportation, inclusive cities and local economic development is an important consideration in transit-oriented development. The principles of transit-oriented development are derived from the need to create balanced land-use planning and development which accommodates pedestrian footpaths and segregates cycle tracks, a connection that improves access to goods, services and public transportation characterised by frequent, fast and reliable high-capacity rapid transport (ITPD, https://tod.itdp.org/what-is-tod/eight-principles-of-tod.html).

In light of the above discussion, this paper examines the influence of urbanisation on transport infrastructure investment and proposes policies for achieving transit-oriented development in African cities.

2.2 Transportation Challenges and Urbanisation

Sustainable transportation development demands a balance between economic, social and environmental concerns. Urban sustainability development involves several aspects, including populations (socioecology), land use and urban structure and mobility behaviour (transportation) [Alberti, 2008]. The dynamics of urban development is reflected in the interaction between humans and the environment in spatial and temporal contexts. With that, such a complex system, spatial structure strategy alone or partial strategies, such as transportation infrastructure policies, cannot be sufficiently relied upon for achieving sustainability but requires integrated land use, transportation and environmental strategies. Land-use management system has good potential in ensuring a good urban environment, particularly if the land-use system could manage the people's mobility. The spatial distribution of settlement areas, working places, commercial centres and other activities contributed to the trip origin-destination patterns in urban transportation and impacted the trip distance, energy consumption and pollutant emission. As the statement suggests, the urban spatial structure could decrease the trip distance and simultaneously increase urban environmental quality (Petersen, 2010).

Jing and Zhentao (2007) emphasised that the rapid development of urbanisation has resulted in a prominent aggregation of people and buildings among metropolises, creating new urban growth poles or sub-centres. Urban structure evolution, from a monocentric to a polycentric model, always emerges along with economic development and the growth of urban areas at different scales. Uncontrolled development gives rise to urban issues, most of which derive from the imbalances between supply and demand. Urban elements such as land and transportation belong to the static supply side, and human behaviour always belongs to the demand side. The limitation of urban boundaries restricts urban land supply, thus enhancing urban densities. The lack of an efficient public transportation system results in congestion and pollution, especially in rapidly developing cities. Meanwhile, unaffordable housing prices and rents in downtown areas force people to relocate to the undeveloped suburbs, generating crowded and lengthy commutes. Owing to the economic bubble and financial crisis, the unemployment rate is high, resulting in a vacancy problem in both residential houses and commercial houses (Bentley et al., 2016). All of these phenomena reflect urban disequilibrium. To achieve balanced development as a result of the high speed of urbanisation, it is necessary to build a compact, balanced and sustainable city. The relation and interaction of different urban elements and their comprehensive impact on the urban structure are very crucial; these include transportation, road networks, distribution of buildings, land-use patterns, economic performance and human behaviours (Yingqun et al., 2019).

Cities can be described as an engine for economic growth; many cities in the developing world increasingly fail to deliver what is expected of them. Many cities are unable to cope appropriately with rapid population growth and are facing such challenges as lack of urban infrastructure, a deteriorating living environment, growing vulnerability to natural disasters and widening economic disparities. In developing countries, the poor state of transportation infrastructure, including roads, railways, ports and airports, has impeded the attainment of both economic growth and poverty alleviation. The development of transportation infrastructure is therefore indispensable to achieve sustainable economic growth, facilitating the efficient movement of people and goods. The demand for transportation infrastructure is high worldwide. The need to maintain, repair and upgrade aging structures is ever-growing (JICA Annual Report, 2016). There is a reciprocal relationship between transportation and urban growth. Transport plays a vital role in urban development. Transport systems provide essential mobility options for people and goods and influence patterns of growth and the level of economic activity through the accessibility it provides to land (Meyer & Miller, 2001). Urban growth and transport are strongly related issues. On the one hand, transport infrastructure attracts urban development; on the other hand, urban growth and population cause an increase in travel demand and thereby an increase in the requirement for transport infrastructure.

2.3 Transit-Oriented Development, the Relevant SDGs and Urbanisation

Transportation routes open the access of the city to the countryside and are responsible for urban development. Roads are the major catalyst in the process of regional development. Transportation as established by Meyer and Miller (2001) plays a crucial role in urban development and growth; it provides mobility for people and goods and influences patterns of growth as well as the level of economic activity through accessibility. There is a strong relationship between urban growth and the transportation system in a city. There is no urban development and growth without a high increase in population; as the population increases as a result of the influx of people, the existing transport facilities become inadequate, thereby resulting in traffic delay, jams and an increase in man-hour rate on the roads. In the space of just a few decades, urban areas across the world, in both developed and developing countries, have become increasingly automobile-dominated and less sustainable. In developing countries, in particular, cities have experienced rapid growth in transport-related challenges, including pollution, congestion, accidents, public transport decline, environmental degradation, climate change, energy depletion, visual intrusion and lack of accessibility for the urban poor (Dorina & Dominic, 2015).

As reported by the United Nations (2011), nearly half of the world's 3.9 billion urban dwellers reside in relatively urban settlements with fewer than 500,000 inhabitants, while only around one in eight lives in the 28 megacities of 10 million inhabitants or more. Overall, the urban population in developing countries is set to double from 2010 to 2050 while remaining stable in developed countries (United Nations, 2015). At the moment, cities with less than 100,000 inhabitants represent a third of the world's urban population, a figure which is predicted to grow to 40% in 2050. The fastest-growing urban agglomerations are medium-sized cities with less than 1 million inhabitants, located in Asia and Africa (United Nations, 2015). This requires adequate and urgent intervention in order to synergise transit-oriented development and city growth.

Transit-oriented development (TOD) as described by Still (2002) and Bernick and Cervero (1997) is a mixed-use community that encourages people to live near transit services and to decrease dependence on their driving and also the practice of developing or intensive residential land use near rail stations and housing, along with complementary public uses, jobs, retail and services, which are concentrated in mixed-use developments at strategic points along with the regional transit systems

(Calthorpe, 1993; Salvensen, 1996; Boarnet & Crane, 1998). TOD is a new concept focusing on efficient modes of transportation other than the automobile (which is highly polluting with the exception of expensive hybrid or green vehicles). TOD is a growing model of mobility in cities which connects to the UN Sustainable Development Goals (Fig. 2.1). It can contribute towards the creation of a healthy environment (free from air pollution and traffic jams) which links to SDG 8 leading to stress-free, vibrant and sustainable communities in a liveable city and a productive economy (Goal 8). TOD necessitates the creation of compact, walkable, mixed-use communities centred around high-quality train systems (Ngai, 2016).

As many African cities grow with a high population and to achieve sustainable transportation in line with SDG 11, emphasis should be on making cities and human settlements inclusive, safe, resilient and sustainable. It will be necessary for urbanists and local stakeholders worldwide to put urbanisation and territorial development at the heart of sustainable development. This goal and targets require action from subnational urban governments to address a wide range of unique urban challenges such as public transit, systems, planning and governance, cultural heritage, disaster management, air quality, waste management and public green spaces which will enhance the achievement of SDG 1. In response to SDGs, many cities in different parts of the world are using innovative approaches to provide sustainable transport such as in Copenhagen, Curitiba, Hong Kong and Johannesburg as discussed in the following section.

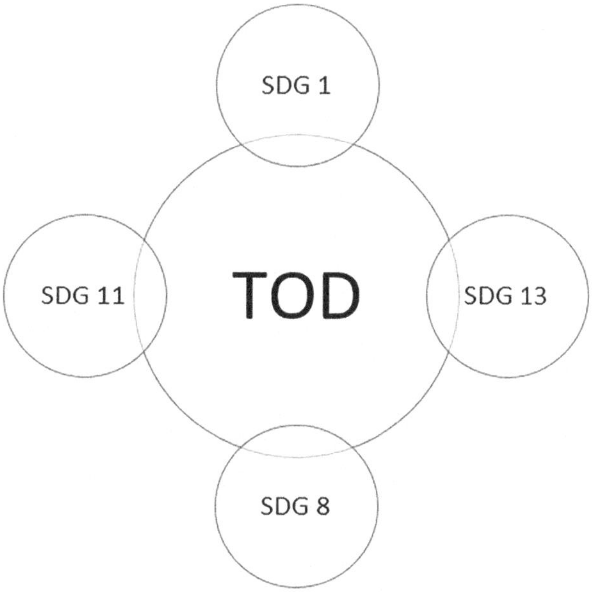

Fig. 2.1 Linking transit-oriented development (TOD) to the UN Sustainable Development Goals

2.4 Case Studies

As global urbanisation and expansion of urban areas keep increasing unprecedentedly, several cities throughout the world are developing transit-oriented development policies to facilitate economic fairness, business development and environmental sustainability. Portland, San Francisco, Vancouver, Toronto, Copenhagen, Denver, Curitiba, Johannesburg, Hong Kong and Ahmedabad, among many other cities, have developed and continue to produce strategic plans and policies, with the aim of increasing the use of public transit and reduce automobile dependency, thereby reducing their carbon footprint while becoming more productive, socially inclusive and more liveable (Salat & Ollivier, 2017).

2.4.1 Copenhagen: Mobility and Connected Urban Development in Copenhagen

Copenhagen, Denmark, is a leader in the concept of transit-oriented development mainly because the city has pursued growth policies that integrated urban development and transportation. In 1947, the city implemented a development plan called Fingerplanen (Finger Plan), which concentrates city growth around five suburban trains that leave downtown (Knowles, 2012). The plan selected development areas and retained green areas where the city was not allowed to expand. This TOD policy has provided wide access to quality public transport: 57% of the population lives within a mile of a transit station. Notably, 61% of jobs are located within this same span. This policy creates disincentives for car use through the implementation of parking metres and parking quotas (for residents) as well as fewer requirements for new construction and on-street parking supply. This planning principle remains in effect to this day, even as the urban plan has evolved to meet changes in the Copenhagen metropolitan area and the construction of new transmission lines. The Fingerplanen was updated in 2007 and more TOD areas have been developed.

2.4.2 Curitiba, BRT and Transit-Oriented Development

For over 50 years, the city of Curitiba has been including TOD in its planning efforts. The Master Plan that was published by the Institute for Research and Urban Planning of Curitiba in 1966 encouraged the creation of activities and jobs in areas away from downtown and inner-city locations (Ramirez & Rosas, 2014). Five structural corridors that leave downtown were created in order to achieve this. A high concentration of housing, shops and services are planned for these corridors. As a result, innovative mass transit lanes were placed along these corridors. This was realised as a way to reduce car use through road space restriction. These corridors also promote mixed commercial use, offices, businesses and housing. In fact, the city bought land and built affordable housing along these corridors and near industries to ensure that low-income people have access to public transport and jobs. These homes were subsidised by the government and designed to integrate low-income housing in areas with residential uses. Curitiba also has an integrated Curitiba bus network that covers 90% of the city's area, ensuring that inhabitants do not walk more than 500 m to access public transport. Approximately 28% of homes in the city are located along these hubs.

2.4.3 TOD Policy for Hong Kong

Hong Kong is internationally recognised for its TOD model that focuses on the joint development of urban rail systems and real estate (Ramirez & Rosas, 2014). The Mass Transit Railway Corporation in Hong Kong has developed 25 stations located in long underground housing, shopping and entertainment complexes that include 70,000 residential units and 15 million cubic metres of retail space (Hong Kong Transportation Department, 2013). This TOD model has resulted in a high population density: 6480 people per square kilometre. Most inhabitants live near public transport: 75% of the population, or about 5 million people, live within a mile of a subway station, and 43% live even closer, within 500 m (or less) of a station. This density around transit stations ensures that the majority of Hong Kong residents regularly use public transport, representing 90% of all motorised travel. Taken together, proximity to transit stations and mixed land use encourage residents to walk and cycle, representing 38% of total trips.

The government plays a critical role in the Hong Kong TOD model. On the one hand, landholdings give the government direct control over how plans are developed. On the other hand, the city has a clear vision of how it wants to grow. The vision and planning strategy for Hong Kong up until 2030 includes strategy development around the subway system and a commitment to prioritise development in outlying areas that are already urbanised. The Hong Kong government has enacted regulations that encourage high-density growth in areas surrounding subway stations, including a required occupancy rate of up to 9. At the same time, it applies restrictive car ownership measures such as a high registration tax and annual ownership fee. This has led to a car ownership rate of 63 cars per 1000 inhabitants, the result of a truly integrated mobility policy that prioritises public transport.

2.4.4 TOD Policy for Denver

The need to limit urban growth and expand the transportation system in the Denver Metropolitan Area prompted the establishment of the Metro Vision 2020 in 2000 (Picard, 2010). Through this plan, 220 new subway and BRT kilometres and 57 new stations were to be constructed. In addition, the Blueprint Denver plan, which adopts the TOD as an organising principle for growth around new subway lines, both within the city and its suburbs was proposed.

Until now, under the efforts of TOD, the Denver government has built more than 24,000 residential units near the stations, 464,000 square metres of retail space and 557,000 square metres of office space as well as areas for cultural, educational, medical and government activities. By 2020, officials expect that approximately 30% of the general population and 70% of jobs will be located within 1.6 km (one mile) of a transit station (representing, in aggregate, an area equivalent to 9% of the city).

The government's main role in Denver's TOD model has been to finance an expansion of the transport system through a 0.4% tax on the sales of daily consumer products. Similarly, the government created the Denver Regional Transportation District as an institutional body in charge of coordinating TOD policy with local governments in order to facilitate its implementation.

2.4.5 Johannesburg: Introducing the "Corridors of Freedom"

The legacy of deep inequality and spatial segregation of the urban poor instigated the introduction of TOD in Johannesburg (Du & Gelman, 2018). In 2013, the flagship TOD initiative, "The Corridors of Freedom" programme, was launched by the former Mayor Parks Tau to extend the city's public transit network to offer more economic and social opportunities to the urban underserved (Pieterse & Owens, 2018; Wood, 2019). In addition, the initiative increased public investments across the city that increased public space, offered social services, increased residential density and integrated retail and commercial space into the new development. It also aimed to reorient private investment towards the new public transport service. So far, the city's Bus Rapid Transit network has been extended, and many areas have seen new or improved infrastructure for non-motorised transport, social facilities and public infrastructure. The project has also attracted a niche group of private developers, and TOD generally has become part of the city's long-term spatial planning (Wood, 2021). However, redevelopment has not been without controversy and criticism. Wealthier neighbourhoods in priority zones have complained about potential overcrowding and "undesirable changes". The plan has also stirred debate over whether redevelopment benefits the most underserved populations. While some believe it is better to concentrate public investment in impoverished areas, advocates of the Corridors of Freedom caution against reinforcing the NIMBY (Not-in-My-Backyard) mindset and further isolating of the urban poor in disconnected areas.

Currently, in Lagos, Nigeria, the National Theatre is a good TOD. At the National Theatre, activities are being developed where shops and various commercial orientations will be located to attract people to the station. Also, transportation connections are progressively being developed around 27 economic activities centres. The city has varieties of policies driving TOD in the transport initiatives and development just like it had been implemented in Maryland, USA, and Aberdeen, Scotland.

2.5 COVID-19: Stakeholder's Reflections

The stakeholders alluded that many things have been affected as a result of the COVID-19 pandemic from health system challenges to the transport sector. It has created challenges on social activities and public transport networks, and many livelihoods dependent on the transport sector were critically affected as a result of government measures of restriction of movement on both intracity and interstate. Many busy streets turned quiet, and social distancing was difficult on crowded buses and bus stops. The public perception is that public transport is not safe despite different measures taken to guide against the spread; this is because people are not adhering to the protocols. The measures taken by the government to make public transport safe include fumigation of luggage, use of face mask, reduction in the carrying capacity of buses to avoid overcrowding and securing adequate ventilation and sanitising of hands before boarding. However, stakeholders affirmed that all the measures are good but the drivers and passengers are not keeping to them. It was suggested that touching of money should be controlled and drivers and passengers' cooperation should be encouraged.

2.6 Thematic-Based Issues and Priorities for Lagos from a Transdisciplinary Lens

Most people migrated to Lagos to have access to economic and livelihood opportunities; to academic, intellectual and personal development and to cultural diversity that is formed through the agglomeration of people. In turn, Lagos' economy thrives on being accessible, businesses gain access to more clients and employers gain access to larger labour pools, enhancing productivity, economic growth and competitiveness. Without accessibility, Lagos, or any city at all, cannot function effectively (Fig. 2.2).

However, Lagos like many rapidly urbanising cities has been experiencing declining accessibility consistently in the past few decades. As the city's urbanisation and motorisation intensify, and the city struggles to plan and provide an effective multimodal transport system, traffic congestion worsens and more residents are left underserved (Land Lagos, 2019). Consequently, many Lagosians spend a significant portion of their incomes and time travelling to their destinations (schools, jobs, restaurants, parks and recreation centres, religious and health centres). This challenge imposes huge economic, social and environmental costs on the city, especially the transport underserved (residents who have lower-than-average access to

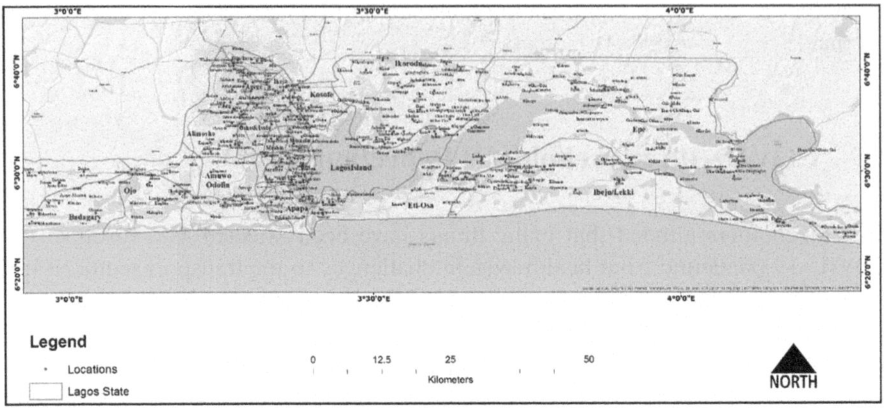

Fig. 2.2 Map of Lagos state. (Source: *Joseph Olusegun)*

destinations in the city, e.g., the low-income groups, spatially marginalised communities and higher-income private vehicle users who face severely congested roads). To be specific, stress, loss of job and lives, decreased productivity, higher vehicle emissions, increased vehicle operating costs, added fuel expenses and decreased annual revenue are some of the impacts on Lagosians (Proshare, 2020), thus contributing to the city's worsening inequality.

So far, the Lagos State Government has responded via a multitude of approaches including the introduction of more high-capacity buses to ply other major corridors in the state, investment in water transport, the ongoing Blue Line, introduction of first- and last-mile mobility and collaboration with the federal government on the use of the Red Line to take away vehicles off the road, expanding road capacity, strategic transport master plan, light rail transport project (in construction stage) and the Bus Rapid Transport (BRT). The private sector has also contributed to the implementation of some initiatives featuring new technologies and service models. These include Oga Taxi, Opay, Bolt (Taxify), GidCab, Jekalo, Uber, Max, inDriver, etc. Each of these approaches has all had great impacts on the economy and people of Lagos.

Since the set-up of the BRT system in Lagos, mobility and accessibility of suburban residents to major activities and services as well as the central business district (CBD) have increased to a large extent. However, in the last few years, the same has equally declined due to inefficiencies in management, and unprecedented urbanisation has exacerbated the problem. With the introduction of the technology-driven transport initiatives, transport options broadened and accessibility also increased, as well as convenience and security for commuters. The technology-driven initiatives have also bridged the gap between commuters and operators through their promptness, fair charges, transparency and accountability. Unfortunately, these responses (the state and private sector) have contributed to traffic and worsening congestion in Lagos due to increasing motorisation. A critical issue of concern is how Lagos can increase accessibility in the face of rapid urban population growth while simultaneously increasing economic productivity, declining spatial inequality and improving environmental quality.

2.7 Recommendations for Government, Investors and Communities

Moving towards an equitable, accessible and environmentally clean city will require multidimensional and transdisciplinary actions. To achieve a sustainable path to urban growth in Lagos, below are recommendations for the government, investors and communities:

Government

- All-weather paved road access should be provided in all neighbourhoods, especially in new and unplanned informal areas with restricted connectivity to the existing urban transport network. Providing timely urban roads in these neighbourhoods will help the government gain better control over urban development as well as increase safe transportation to all users.
- Planning: it is important to have a master plan that looks at critical areas affecting the city upon which TOD must be developed. Hence, the first thing to do before implementing TOD is planning. The planning should accommodate convenient pedestrian connections. A pedestrian-friendly street is visually interesting and makes walking enjoyable. Trees, landscaping, wide, separate sidewalks and on-street parking protect people from vehicle traffic and create a pleasant pedestrian zone. In addition, good urban design is highly necessary; this will involve well-lit and designed stations to accommodate "around-the-clock" activity. Landscaping features can define special precincts and encourage transit patrons to linger and explore the station area. Convenient and legible signs orient visitors to buildings and activities around the station. Also, accommodating vehicles is still critical to the success of a vibrant TOD district. Convenient parking and drop-off zones need to be planned for in all station area plans.
- There is a need to understand the land characteristics, how to channel the travel pattern and which transport type is best suitable to a particular context without running into losses.
- Understanding the appropriate land use—what use is appropriate for which context, etc. The land use should ensure transit-supportive uses. Transit- supportive uses are high pedestrian generators that directly promote greater transit ridership. They also provide opportunities for multi-purpose trips that can be made as a pedestrian. Medium- to high-density residential, offices, high schools and colleges have significant transit-supportive uses. The land use should discourage non-transit-supportive uses. Non-transit-supportive uses generate little or no ridership. They consume large areas of land or create bleak or unsafe environments for pedestrians. Large format wholesale stores, warehouse storage, car dealerships, auto service centres and regional sports fields are examples of uses that are

not transit supportive. Besides, a mix of uses should be encouraged. A mix of residential, office and support services in station areas can generate transit trips throughout the day. The land use should locate the uses as close to the LRT station as possible. Locating a majority of transit-supportive uses within a 400 to 600 m walking distance of the LRT station makes transit the most convenient and attractive travel mode for the site.

- To carry out a development review—when building TOD, there are a lot of reviews required to ensure such development is worthwhile, and existing development plans are not conflicted. The review should increase pedestrian movement development: Frequent, interconnected streets increase the efficiency of transit circulation and offer more choices for pedestrians. Blocks of 100–150 m in length keep walking distances short and provide alternative route options. Consideration should be given to a grid-based street pattern that offers multiple accesses to the station and forms the overall development framework for long-term transit-supportive uses. Also, guidelines for "interim" developments that will maintain the long-term vision for the station area should be established. Interim development can be used to provide key public infrastructure, identify future development patterns and establish quality pedestrian environments.
- Integrate existing informal transit, rail, bus and non-motorised services into a more connected network. This will improve access to the underserved communities and residents as well as enhance opportunities for linking up services and cooperation among stakeholders.
- Prioritise investment in transport infrastructure that will protect high- efficiency vehicles like trains and buses from congestion. This will restore their inherent advantages in transporting more people safely and efficiently.
- Integrate private shared mobility services and informal operators into Lagos' multimodal transport system. This will help harness the entrepreneurship and experience of the informal sector, as well as create a sustainable path to mitigating their negative impacts on the residents of the city.
- Coordinate physical reallocation of road space from cars to pedestrians, buses and cyclists by creating transit lanes, non-motorised ways and pedestrianisation of streets. This should be supported by strategic improvement in public transport options. Doing this will discourage private vehicle use and reduce traffic congestion and environmental pollution. Ultimately, stronger support and partnership for multimodal policy agenda will be created.
- Ensure new development is transit-oriented, located near economic opportunities and well-served by public transport. This will increase cycling and walking; enhance accessibility to housing, jobs and services; and reduce daily commuting expenditures.
- Need for enhancement of urban mobility beyond road transportation, as the ten principles of urban mobility suggests.
- Promote comprehensive multimodal integration transportation programme approach.
- Ensure regular and continuous transportation planning based on effective data gathering.

- Ensure coordinated efforts using effective communication and opinion analysis for integrated transportation policy implementation.
- Ensure public-private participation options for efficient and optimum utilisation and execution of transportation projects.

Investors

- Expand integrated e-hailing and payment services to motorcycle, car rides and auto-rickshaw. Close attention should be paid to areas with poor Internet connection (use of text messages should be promoted in such areas). This form of technology could boost informal operators' productivity by better matching supply with demand,particularly in areas with poor accessibility and during off-peak hours.

Communities

- Local advocacy groups should be set up to advocate for short-term incremental improvements (adequate sidewalks, safe routes to jobs, schools, etc.) while larger-scale interventions in transit networks are underway.
- Demand for more infrastructure for improved transit and non-motorised transport options including cycling and walking.

2.8 Conclusion

Urbanisation enhances investment in transport infrastructure which if properly designed and implemented in combination with transit-oriented development can help to address the challenges of traffic congestion, housing deficit, sprawling, inner-city decay and disinvestment. The benefits of transit-oriented development can make cities and human settlements inclusive, safe, resilient and sustainable (Goal 11), promote a healthy and climate-friendly environment in relation to Goal 13 (Climate Action), enhance decent work and economic growth (Goal 8) and ensure no poverty (Goal 1).

References

Alberti, M. (2008). *Advances in urban ecology. Integrating humans and ecological processes in urban ecosystems*. Research Gate. https://doi.org/10.1007/978-0-387-75510-6.

Bentley, T. A., Teo, S. T. T., McLeod, L., Tan, F., Bosua, R., & Gloet, M. (2016). The role of organisational support in teleworker well-being: A socio-technical systems approach. *Applied Ergonomics, 52*, 207–215.

Berg, C. N., Deichmann, U., Liu, Y., & Selod, H. (2017). Transport policies and development. *The Journal of Development Studies, 53*(4), 465–480.

Bernick, M., & Cervero, R. (1997). *Transit villages for the 21st century*. McGraw-Hill.

Blimpo, M., Harding, R., & Wantchekon, L. (2013). Public investment in rural infrastructure: Some political economy considerations. *Journal of African Economies, 22*(2), 57–83.

Boarnet, M. G., & Crane, R. (1998). *The influence of land use on travel behavior: Specification and estimation issues*. Working paper, University of California Transportation Center, Berkeley (pp. 823–845). http://www.sciencedirect.com/science/journal/09658564

Calthorpe, P. (1993). *The next American metropolis—Ecology. Community and the American dream*. Princeton Architectural Press.

Dorina, P., & Dominic, S. (2015). *Sustainable urban transport in the developing world: Beyond megacities sustainability*. https://www.mdpi.com/2071-1050/7/6/7784

Du, J., & Gelman, V. L. (2018, December 18). *Johannesburg fights inequality with transit-oriented development*. World Resources Institute. https://www.wri.org/blog/2018/12/johannesburg-fights-inequality-transit-oriented-development

Gobillon, L., & Selod, H. (2014). Spatial mismatch, poverty, and vulnerable populations. In M. Fischer & P. Nijkamp (Eds.), *Handbook of regional science* (pp. 93–107). Springer.

Hong Kong Transportation Department. (2013). *Monthly traffic and transport digest*. https://www.td.gov.hk/filemanager/en/content_4610/1302.pdf

Institute for Transportation and Development Policy (ITPD). https://tod.itdp.org/what-is-tod/eight-principles-of-tod.html

Japan International Cooperation Agency (JICA). (2016). *Annual report on Japan's ODA evaluation*. Ministry of Foreign Affairs, Japan.

Jing, Z., & Zhentao, Z. (2007, August). Hierarchy analysis and strategies on the imbalance between supply and demand of urban traffic. *Journal of Transportation Systems Engineering and Information Technology, 7*(4), 24–29.

Khandker, S. R., Bakht, Z., & Koolwal, G. B. (2009). The poverty impact of rural roads: Evidence from Bangladesh. *Economic Development and Cultural Change, 57*(4), 685–722.

Knowles, R. D. (2012). Transit-oriented development in Copenhagen, Denmark: From the finger plan to Ørestad. *Journal of Transport Geography*, 251–261.

Land Lagos. (2019, February 15). *Lagos traffic—A never ending tale*. Land Lagos. https://landlagos.com/blog/lagos-traffic-a-never-ending-tale/

Meyer, M. D., & Miller, E. (2001). *Urban transportation planning: A decision-oriented approach* (2nd ed.). McGraw-Hill.

Ngai, P. (2016). Migrant labour in China. In *Post-socialist transformations* (China Today Series, xi, p. 204 Maps). Polity. cambridge.org/core/journals/interna...

Petersen, R. (2010). *Urban land use and transportation planning*. Wuppertal Institute/GTZ.

Picard, P. (2010). *GIS analysis of population and employment centres in metro Denver served by FTD fast tracks*. University of Colorado.

Pieterse, E., & Owens, K. (2018). Johannesburg: Confronting spatial inequality. In *Towards a more equal city* (pp. 1–28). World Resources Institute.

Proshare. (2020, March 04). *Lagos' traffic congestion—Is there a respite in view?* Proshare. https://www.proshareng.com/news/Transportation/Lagos%2D%2DTraffic-Congestion%2D%2D-Is-There-a-R/49687

Ramirez, S. M., & Rosas, J. V. (2014). *Transit-oriented development: Regenerate Mexico cities to improve mobility*. Institute for Transportation and Development Policy Mexico.

Rico, A. D. (2016). La gestióneducativa: Hacia la optimización de la formacióndocenteen la educación superior en Colombia. *Sophia, 12*(1), 55–70.

Salat, S., & Ollivier, G. (2017). *Transforming the urban space through transit-oriented development: The 3V approach*. International Bank for Reconstruction and Development/The World Bank.

Salvesen, D. (1996). Promoting transit-oriented development. trid.trb.org/view.aspx?id=525811 Supplemental Notes: p. 31–35, 87: Col. Ill. *Urban Land, 55*(7).

Still, T. (2002). Transit-oriented development: Reshaping America's metropolitan landscape. In *On common ground*. Winter, pp. 44–47.

UN DESA. (2014). *World urbanisation prospects: The 2014 revision*. United Nations Department of Economic and Social Affairs.

United Nations. (2011). *Report of the United Nations Secretary-General—UNCTAD*. unctad.org/webflyer/2011-report-uni

United Nations Sustainable Development Summit. (2015). *UNIDO*. unido.org/events/united-nations-sus

Wood, A. (2019). The business of global intermediaries in the promotion of bus rapid transit. In T. Baker & C. Walker (Eds.), *Public policy circulation: Arenas, agents and actions* (pp. 89–102). Edward Elgar.

Wood, A. (2021). The tangible and intangible infrastructure of bus rapid transit in African cities. In N. Appelhans, W. Scholz, & S. Baumgart (Eds.), *Transport planning and mobility in urban East Africa* (pp. 124–135). Routledge.

World Bank. (2011). *The World Bank annual report 2011: Year in review*.

World Bank. https://openknowledge.worldbank.org/handle/10986/2378

World Population by Country. https://worldpopulationreview.com/

Yingqun, Z., Rui, S., Rob, van, N., Shiwei, H., & Weichuan, Y. (2019). Identifying urban structure based on transit-oriented development. *Sustainability, 11*(24), 7241. https://doi.org/10.3390/su11247241

Chapter 3
Transit-Oriented Development and Housing in African Cities

T. Nubi, B. Oyalowo, Y. Ohiro, F. Aiyegbajeje, and A. Egunjobi

3.1 Introduction

Commuting, whether by road, rail or ferry (or a combination), is synonymous with urban lifestyle, just as housing occupies a pivotal location that significantly dictates the mobility pattern of urbanites. Commuting costs and transportation have been identified as factors influencing residential location choices of households (Ng, 2008; Jin & Lee, 2017). Importantly too, housing by virtue of location 'is not adequate if it is cut off from employment opportunities, health- care services, schools, childcare centres and other social facilities, or if located in polluted or dangerous areas' (United Nations, 2014). Hence, planning and development of residential neighbourhoods should be carried out in tandem with the planning of transport infrastructure that provides easy access to relevant areas of needs.

The case for transit-oriented development (TOD) to 'create an urban space able to accommodate new residents in a way that provides high accessibility to jobs at

T. Nubi
Department of Estate Management, University of Lagos, Akoka, Lagos, Nigeria

Centre for Housing and Sustainable Development, University of Lagos, Akoka, Lagos, Nigeria

B. Oyalowo (✉)
School of Built Environment, Oxford Brookes University, Oxford, UK
e-mail: boyalowo@brookes.ac.uk

Y. Ohiro
Department of Estate Management, University of Lagos, Akoka, Lagos, Nigeria

F. Aiyegbajeje
Department of Geography, University of Lagos, Akoka, Lagos, Nigeria

A. Egunjobi
Department of Economics, University of Lagos, Akoka, Lagos, Nigeria

T. Nubi et al. (eds.), *Transit Oriented Development in West African Cities*, Urban Perspectives from the Global South,
https://doi.org/10.1007/978-3-031-58726-9_3

the city level combined with amenities, services, and good-quality environment at the neighbourhood level to give a sense of community' (Salat & Ollivier, 2017) is founded on these concerns. TOD can be conceptualised as a mixed-used area intended to maximise access to public transportation (Boarnet et al., 2017), centred around a public transit station, with high-density real estate development within a walking distance away that tapers off to lower-density developments.

The concept of transit-oriented development emerged from America as a nexus of Transportation-Architecture-Urban Planning-Real Estate Development and environmental design philosophy, although strictly speaking its principles are highly visible in early urban planning ideas such as Ebenezer Howard's Garden City design and Clarence Perry's Neighbourhood Design principles, amongst others. It emerged as a design philosophy at a time in the USA when transit agencies had determined that high-density development near their stations encouraged transit usage, environmental and transportation groups coalesced to fight sprawl and reduce auto-dependence and neo-traditional designers were promoting high-density, pedestrian-friendly neighbourhood design (Carlton, 2009). Ideas and advocacy tools from these diverse stakeholders are to be found in the principles surrounding modern TOD proponents. However, they are also to be found in other planning design concepts such as compact city design, pedestrian pockets and environmentally friendly design.

African cities are in dire need of strategies that are modelled to attain the goals and targets of the Sustainable Development Goals. SDG 11, sustainable cities and communities which has targets for promoting housing adequacy and affordability as well as equitable transportation access, can actually be achieved with the incorporation of TODs. Slum proliferation and transportation congestion are two critical urban problems in Africa, and we contend that both could be addressed by adapting TODs to local realities.

As Carlton (2009) points out, throughout the history of transit development, the housing sub-market of the real estate development sector has been a key component of successful planning. However, the *affordable* housing impetus for successful TOD presents an interesting challenge. The underlying question has been: *Can affordable housing really be achieved in transit-oriented development projects?* Several constraining factors have been identified. In tandem, another question rises, which is: *How can affordable housing be achieved in existing neighbourhoods where transportation investments have been injected but where TOD is yet to take root as an intentional planning-investment regime?* Finally, another question is: *What is to be done in situations where housing development has preceded transportation investment?* These are issues that need to be addressed in the context of the African city for TOD to be a major strategy for solving both housing and transportation challenges. On the other hand, there is a need to define what is practical for each city, especially where a significant proportion travel from one peripheral home location to the city centre and the other end of the city for employment and social activities.

This report presents a preliminary appraisal of these questions with lessons from the literature by establishing the relationship between transportation challenges and housing, especially with respect to haphazard peripheral development. This discussion is furthered by linking the attainment of the SDG 11 to TODs and the pivotal role the housing sector will play in this regard. Thereafter, we draw lessons for successful TOD application from two African cities (Johannesburg and Addis Ababa). With a transdisciplinary lens, we gathered stakeholders to explore the the feasibility of TODs in Lagos, a very important commercial centre for Africa. Putting all these into an African context, we then present recommendations for three core stakeholders for TOD projects: government, investors and communities. A conclusive note on the key considerations for enhancing housing provision through TODs closes the report.

3.2 The Relationship Between Transportation Challenges and Housing: A Literature Review

Transport mode choice is influenced by the characteristics of the decision maker or individual, the number and attributes of alternatives and the decision rule. Under travel choice theories, a decision maker is assumed to rank possible alternatives in order of his or her preference and to choose from the available alternatives the option which he or she considers the most desirable considering his or her preferences and the relevant constraints placed on decision-making.

Transportation and housing are fundamental building blocks of every settlement. Given the services they offer and the ways they are planned and used, they shape and define communities, bringing about physical changes to the spatial patterns and processes which in turn determine the economic progress of cities. Together, they have decisive influence over the way any given region will look, feel, function and serve.

Cervero (2007) opined that housing areas generate the traffic volume that places pressure on the transport network, and mobility plays a particularly prominent role in where commuters choose to live. In the same vein, Yiu and Wong (2005), Gallo (2018), Howell et al. (2018) and Glaeser et al. (2008) held that the capability of transport network affects housing development, and in reverse, urban housing influences the planning of urban transport network systems. They also asserted that households often make trade-offs between how much they spend on housing and how much they spend on transportation. As observed by Tembe et al. (2019), there is also a role for travel choice as it impacts urban transportation. They note that under travel choice theories, a decision maker is assumed to choose from a rank of possible alternatives in order of his or her preference and with consideration to constraints placed on the decision maker such as convenience, safety and cost. There is thus a trade-off between transportation costs and housing location choice, which is determined by the layout of the city, existence of transport infrastructure, access to these facilities and quality/cost of available modes of transportation. On the other hand, these decisions also determine the volume of traffic generation per route, the pressure on traffic and transport facilities and in general, the efficiency of the city. This interdependence and overlapping of transport and housing location decisions has been the subject of many studies and theoretical postulates.

In the past, several studies have documented the relationship between transportation and housing from various perspectives ranging from urban form, and suburbanisation to housing markets (Alonso, 1960, 1964; Muth, 1969; Mills, 1972; Brueckner, 1987). These studies have attempted to explain these trade- offs, albeit from an understanding of the western household. African households are no less bound to make these choices, and due to increasing demand for housing by a rapidly increasing population, some families have gone against the tide and moved to the periphery of metropolitan areas, where housing costs are typically lower but transportation costs into the cities are higher (McMillen, 2003).

In reality, for most African cities, the interaction between housing location and transportation poses a complicated and exponential problem as most cities are growing faster than infrastructure provision. For example, The Urban Age Africa report of the London School of Economics (2018) shows that in 2014 Addis Ababa experienced an hourly urban growth rate (migration plus natural birth) of 23 persons/hour; Dar es Salaam, 48 persons/hour; Lagos, 26 persons/hour; and Nairobi, 28 persons/hour. This is an indication of overwhelming present and future housing and transportation needs. The dependence on public transport is also high, with 73% daily trips made by public transport in Accra, 57% in Dar es Salaam, 58% in Lagos and 40% in Nairobi. The quality of these trips (in terms of cost, time, comfort and safety) influences residents' current and future choice of housing location.

The nexus between housing and transportation converges at the residential density. Residential density is a fundamental measure of urban structure and efficiency because inevitably, commuters' trips start and eventually end at the home. When residential, transportation and employment densities align, city's efficiency and livability improve. The ability of cities to densify improves capacity to provide infrastructure at scale, as sprawling, low-density cities have to expend so much more on transportation infrastructure to ensure that people who live at the fringes and work in the city are able to do so. The framework must also be accessible, safe, efficient and affordable. However, high densities do not necessarily imply efficiency because availability, accessibility and affordability of various modes of intra-city transport could undermine any density gains. As African cities differ in terms of sprawl, from high densities in Kinshasa and Lagos to low densities in Accra and Kampala, their level of investment in transportation infrastructure varies. Also, the factors that led to housing choice and sprawl also varies and the efficiency of these will also differ. Nevertheless, urban sprawl and mobility in many African cities have been difficult to solve as urban dispersal involves increasing travel distances (and costs) and creates problems of unsustainable mobility.

The interaction between housing, transportation and city livability is further compounded by traffic congestion resulting from population influx of new residential areas with inappropriate road networks. The increase in traffic volume along an elongated daily main corridor to the Central Business District, which has arguably led to more serious congestions (Lubell & Salomon, 2010), is also remarkable.

Eisenberg (2001) observes that synergy between housing and transportation in concert with closely related issues of economic development, environmental quality, education and employment can create places of delight, prosperity, justice and opportunity. However, on the flip side when they are misused, or work at cross-purposes, or fail to take advantage of reinforcing linkages, they can create ugly, dysfunctional, unhealthy environments that diminish the quality of life and shut people out of opportunities. On this is founded the concept of transit-oriented development as a form of city design and an investment package capable of providing both social and economic benefits.

Indeed, TOD could be a panacea to mobility poverty in several African cities, where people walk for long distances over an inordinate long distance because they cannot afford the cost of transportation (Godard shows that in Conakry and Douala, the poor walk more and usually longer distances and that women walk much more than men). TOD promotes a clustering of home, work, trade and recreation facilities in a bid to reduce travel over long distances. Like their counterparts in other parts of the world, African cities have sought to link residential areas to employment areas through mass transit systems from the major modes of urban transportation: road and rail. However, executing mass transportation projects in Africa has been generally problematic, with cities like Dakar experimenting unsuccessfully with rail transportation and Abidjan in the 1990s seeking a tramway and urban train which was halted due to the civil war (Godard, 2013). Inefficiencies in public transportation also challenge existing problems of commuting in many African cities. For instance, the failure of government-owned public transportation services in rapidly urbanising Maputo and Nairobi in the 1990s gave rise to the introduction of para-transit operators in both cities. However, the efficiency of these systems in metropolitan areas cannot be relied upon in corridors with higher population density (Tembe et al., 2019). Goddard also points out that there is a need to have adequate institutional and financial framework and preparatory processes before efficient mass transport projects can even be launched. These are generally unavailable in most African cities.

On the other hand, there are few ongoing attempts. From Cairo's metro system, to the light rail under construction in Lagos and its Bus Rapid Transportation (BRT) system which is also utilised in Cape Town and Lagos, to Addis Ababa's LRT system, investment in transportation has been made to ensure that long distances are closed conveniently. TOD seeks to collect from these long trips and agglomerations of people, homes and employment in transit stops, to reduce travel times and to enable urban residents work and play within a manageable nexus. An ideal just as Ebenezer Howard's Garden City and Clarence Perry's neighbourhood concept, TODs are enclaves of communities that seek to improve city livability, and at the centre of this is the housing investment (Fig. 3.1).

Although transit-oriented development as we know it does not have its origin in Africa, it portrays a combined effort at providing density scales smaller than cities. It seeks to provide a convergence of mixed use, densities, employment and transportation. Whether the ambition of compact enclaves in various nuclei reflects the cultural conditioning, governance priority and economic realities of African cities is

Fig. 3.1 Usual traffic gridlock at peak periods in Lagos. (Source: Brand Spur. https://www.google.com/amp/s/brandspurng.com/2020/03/05/lagos-traffic-congestion-is-there-a-respite-in-view/%3famp)

debatable, as the Johannesburg case study reported in Sect. 3.4 shows. Also, more intense questions of the centrality and feasibility of affordable housing in this nexus also arise, and the reality of institutional harmonisation of transportation provision and housing provision- oft managed by competitive government agencies remains to be tackled (see also the Johannesburg case study below). On the one hand, TODs seem to be the answer to the quest for a sustainable community, one that has a balance of social, physical and environmental attributes. Despite this, there is a need to establish some basis for its application both generally and in African cities in particular. In the next section, the interaction of housing and transit-oriented development with development goals encapsulated in the SDGs is presented as a means of establishing the useability of the concept as a sustainable development intervention.

3.3 Housing, the SDGS and Transit-Oriented Development: Conceptual Linkages

SDG 9.1 and SDG 11 are two goals that stand on the need for better transportation and better housing, respectively. SDG 9.1, 'develop quality, reliable, sustainable and resilient infrastructure, including regional and transborder infrastructure, to support economic development and human well- being, with a focus on affordable

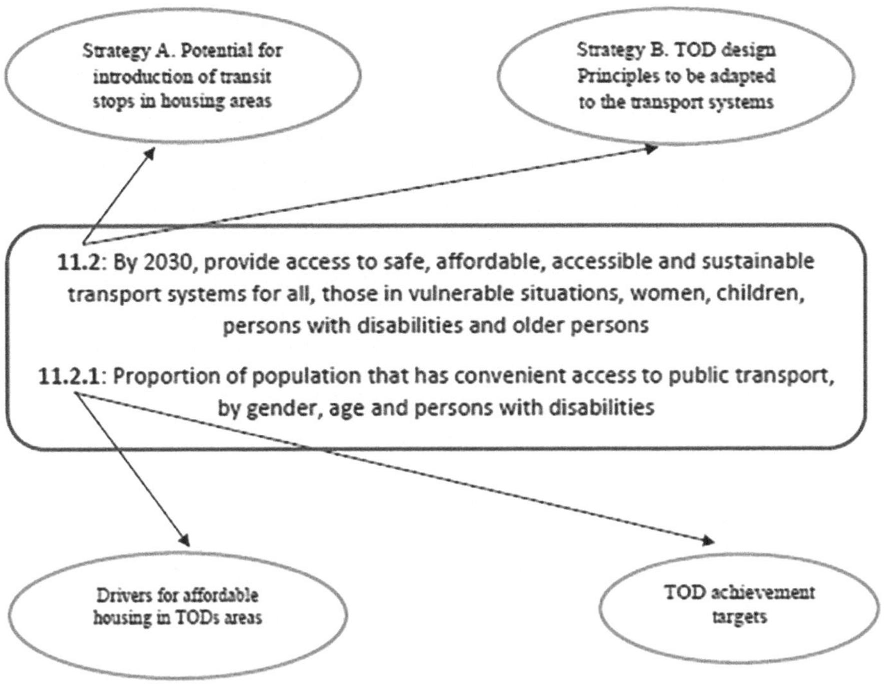

Fig. 3.2 Conceptual linkages between the SDGs, housing and transportation

and equitable access for all', focuses more on inter-city transportation accessibility, while SDG 11 (make cities and human settlements inclusive, safe, resilient and sustainable) is found to have more direct focus on transportation and can be achieved through the use of TOD. Target 11.2 and indicator 11.2.1 provide conceptual linkages between housing, TOD and the SDGs. This is depicted in Fig. 3.2.

According to Carlton (2009), the key components of TOD as espoused in academic literature are built on the works of Calthorpe and are a synthesis of ecological, aesthetic, pedestrian, anti-sprawl, regional and equitable planning principles. These are the following:

1. Organise growth on a regional level to be compact and transit-supportive: compact and accessible and ACCESSIBLE.
2. Place commercial buldings, housing, jobs, parks and civic uses within walking distance of transit stops: ACCESSIBLE MIXED USES.
3. Create pedestrian-friendly street networks that directly connect local destinations: WALKABLE AND ACCESSIBLE.
4. Provide a mix of housing types, densities and costs: AFFORDABLE HOUSING.
5. Preserve sensitive habitat, riparian zones and high-quality open space: RESPECT NATURE.
6. Make public spaces the focus of building orientation and neighbourhood activity: COMMUNITY BUILDING.

Fig. 3.3 Components of TOD

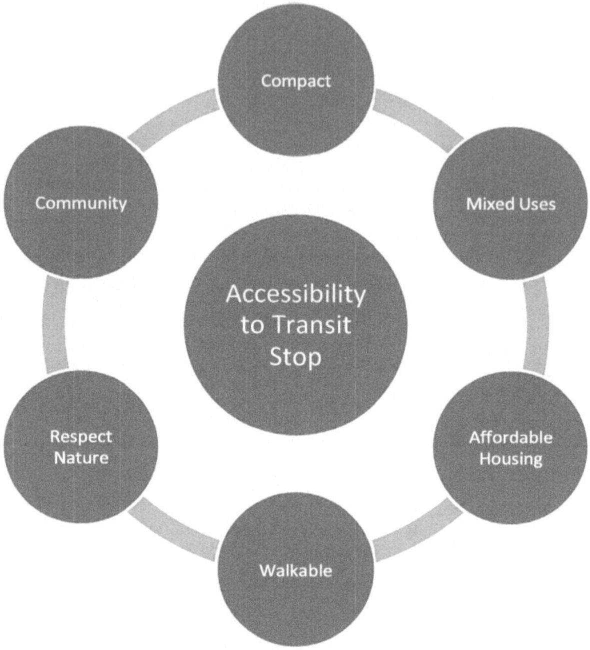

As shown in Fig. 3.3, accessibility therefore features as a very prominent component of TOD and evidently as a geographical condition to accessing the public transit that is the central impetus for such developments.

Of particular importance to this thematic review project is the infusion of affordable housing as a component of TOD. This strongly links the TOD concept back to SDG 11.1 and enables its adoption (or better still adaptation) as a strategy towards actualising the Sustainable Development Goals in African cities. As shown in Fig. 3.3, one of the indicators for attaining SDG 11 is that it measures equitable public transport accessibility, and this is a goal of TODs. In these lie the interactions and connections between housing, transit-oriented development and transportation. Further, by addressing and integrating the issues of transportation accessibility, mixed uses and respect for nature with affordable housing, transit-oriented development represents a scalable, actionable agenda towards actualising SDG 11.2 and 11.2.1.

The conceptual linkages between housing and TOD can be furthered in analysing the probable externalities of the deliberate infusion of housing units in close proximity to transit corridors, as a means of achieving mixed-use development aspirations, affordable housing, reduced automobile-based travel and reduced travel costs of households as they enjoy closer proximity to work and environmental sustainability. It has been experimented within various cities across the world and a few in Africa. The next section provides examples of its use in two African cities.

3.4 Case Study Review of Transit-Oriented Development (TOD)

TOD is an urban planning strategy that has been successfully implemented in cities like Curitiba, Singapore and London, but it is rather new in Sub-Saharan Africa. Reference has also been made to Addis Ababa, Cape Town and Johannesburg as three cities that have attempted some form of TOD in Africa. This section presents a case study of Addis Ababa and Johannesburg.

3.4.1 TOD Case Study of Johannesburg, South Africa

To correct an urban system of spatial segregation and socio-economic mismatch, Johannesburg's Mayor Parks Tau in 2013 presented transit-oriented developments as an integral part of the long-term 'Growth and Development Strategy (GDS) Johannesburg 2040' (Pieterse & Owens, 2018).

Corridors of Freedom (COF), which are strategic locations within the city of Johannesburg, were mapped out for the implementation of TOD. COF are well-planned transport arteries linked to interchanges where the focus will be on mixed-use development – high-density accommodation – supported by office buildings, retail development and opportunities for leisure and recreation (Group Communication and Tourism Department, n.d.). The intent was to make people live closer to jobs and other areas of socio-economic interests, thereby reducing travel time and distance as well as dependence on private motorised transport. Government demonstrated commitment by creating incentives through development levies and planning approval for high-density developments to woo the private sector. Furthermore, 45% of the city's capital budget was devoted to the development of COF (Pieterse & Owens, 2018).

TOD efforts, however, created institutional traction across municipal departments and entities, as well as interdepartmental coordination and alignment of capital investments between 2013 and 2016. As regards development of COF areas, response by private developers was initially slow because the COF did not offer viable real estate market potentials. However, in September 2017, a new crop of private developers found COF viable for the development of 4500 units of affordable rental housing for young professionals and students of tertiary institutions. Today, the provision of bike lanes, sidewalks, clinics, libraries, community centres and mixed-income developments in the COF has connected poorer districts and townships to the central business district (Du & Gelman, 2018; Pieterse & Owens, 2018).

The Johannesburg experience illustrates that creating affordable and efficient mobility options between housing and jobs through TOD works better in overcoming the spatial mismatch than the option of bringing jobs to the people (Du & Gelman, 2018). The Johannesburg case also evidences the centrality of housing as an economic stimulus and the fact that installation of efficient urban mobility systems relies on housing locations.

But there are also challenges. Pieterse and Owens (2018) identified the absence of land market reforms and lack of gap[2] (social) and mixed-income housing as major issues inhibiting the implementation of TOD in Johannesburg. Mixed-income housing failed because the idea was alien to urban lifestyles in the city , and there was no precedent of such housing models (Pieterse & Owens, 2018). Thus, the Johannesburg case implies that in-depth analysis of how urban lifestyles fit foreign or new ideas of housing models have implications for the success of TOD in Africa.

3.4.2 The Case of Light Rail Transit as a TOD Strategy in Addis Ababa, Ethiopia

Addis Ababa is the capital and largest city in Ethiopia with a rapidly growing population of 3.5 million inhabitants, representing 25% of Ethiopia's population (Tarrosy & Vörös, 2019; Pojani, 2020). Mobility pattern in the city is characterised by low rate of private car ownership with public transportation representing 85% (Institute of Transportation and Development Policy [ITDP], 2019; Pojani, 2020). Although existing mobility pattern offers incentive for sustainability, rapid population increase coupled with recent increase in private car ownership has led to daily challenges of traffic congestion, frequent accidents, air pollution and lack of infrastructure to accommodate non-motorised transport (NMT) (Addis Ababa City Administration, 2018; Tarrosy & Vörös, 2019).

In response to these challenges, Addis Ababa's Light Rail Transit (LRT) was launched to set the pace for transit-oriented development (Aklilu & Necha, 2018). The electrified LRT is linked to 41 stations connecting inner-city business and transport hubs with rapidly developing commercial centres and new housing developments at the urban periphery (Alade et al., 2020; Rode, 2018).

Notably, the LRT system has considerably enhanced mobility and reduced travel times. For example, the Addis Ababa-Djibouti journey of 3 days was reduced to less than 1 day (Tarrosy & Vörös, 2019). In Addis Ababa, LRT has also proved to be a pole of development as seen in the gradual development of apartments, malls, small businesses and condos along its lines (Tarrosy & Vörös, 2019). However, there are several flaws.

First, the failure of the LRT project to reutilise existing city centre stations by developing new stations at the outskirts farther away from the city centre defeats the advantage of direct city-centre rail connectivity (Rode, 2018) and contrasts with the project's aspiration of adhering to sustainable transportation principles.

Realising how TOD can induce high property value that could limit housing affordability, the government of Ethiopia has tried to ensure efficient connectivity between existing communities and the LRT stations through non- motorised infrastructure (EMBARQ Network, 2015; Arup, 2016). However, the LRT could not sufficiently provide the service level required to connect peripheral public housing estates to the city centre (Rode, 2018).

Unlike the Johannesburg case where the planning was inclusive and informative, the case of Addis Ababa has been criticised for its top-down institutional arrangement because they were directed from the highest political level (Rode, 2018; Rode et al., 2020). More networked governance structures, elaborate planning processes and integration instruments were required. Also, the LRT was established in isolation with little consideration for connectivity to areas of need, particularly housing. However, the Addis Ababa case is gradually meeting up with expectations. In the long run, if the benefits outweigh the costs, the Addis Ababa LRT case could establish a TOD process worthy of emulation in other African cities.

In these case studies, there are some key lessons. The need for inclusive, culturally sensitive models is strongly indicated, while the generation of negative externalities as a result of TOD can also not be overlooked. In addition, government support and private sector involvement are key to the achievement of tangible goals. In the next section, we examine the feasibility of TODs in Lagos, which is one of Africa's commercial centres and one of its most densely populated cities.

3.5 Housing and Transit-Oriented Development (TOD) Priorities for Lagos from a Transdisciplinary Lens

To ascertain the practicability of TOD in Lagos, an interactive session of academics, representatives of government agencies, private sector and community took place at the University of Lagos, Akoka, on 3 March 2020. Areas of integral concern were:

1. The impact of current transportation challenges on housing access and livelihoods in Lagos
2. Existing plans to integrate current transportation investment with TODs to enhance housing provision in Lagos
3. Workable strategies for the adoption of TODs in Lagos

Although rail and water transportation exist, unanimous conclusion indicated road transportation as the most accessible and widely used means of movement in Lagos. Overdependence on road transportation is the bane of the high cost of transportation, traffic congestion, man-hour loss and several health problems that Lagos residents' experience on a daily basis.

The problem is exacerbated by rapid urbanisation in Lagos as physical development of areas (for residential purposes at the peripheral areas) is always ahead of infrastructural development. In any case, the road system in terms of network, capacity and connectivity is grossly inadequate.

This problem was captured by participants as follows:

> The issue of transport and housing is like supply and demand, where you put in a transport infrastructure, the demand for housing goes up. (Rep Future Cities Nigeria)

And another:

> The current transportation system is not good enough for assessing housing in Lagos due to traffic congestion. A situation where all other means of transportation are being developed will allow people from Badagry work in different parts of the state. It will enable them get to their place of work and go back home quickly. Also, it will reduce traffic congestion. The cost implication is that the further you are away from public transportation centres, the cheaper the housing becomes. (Rep Urban Planning Policy Maker)

It is also the case, however, that the farther people live from the city, the more they spend on transportation. There is thus an underlying connection between housing location choice, accessibility to work and cost of transportation as identified in the literatures. An underlying factor cementing this trade-off is preference for home ownership amongst all income classes (Ebekozien, 2020); this provides some empirical evidence for middle-income class in particular. However, due to high costs of land ownership and construction within Lagos, most aspiring homeowners have recourse to only peripheral areas of Lagos, the boundaries between neighbouring Ogun State. As many work in commercial hubs on the Mainland and the Island, they spend hours travelling in between, with increased transport cost, reduced productivity and increased health costs. Another outcome is that even though people build their homes in the periphery, they would rent or squat in the city during the weekdays and go back home at the weekends. This also represents additional costs of renting and cost of living which impacts disposable income and ability to pay for other services.

To alleviate dependency on road transportation, alternatives such as water and rail transport systems have been explored. Efforts have recently been diverted to the waterways, but much success has not been recorded because from an investment perspective, the breakeven point is high, which translates into higher fares, thus discouraging users with modest incomes. Rail transportation then becomes a substitute because it can accommodate a large number of commuters and is cheap. Multiple train coaches will not only decongest passengers but also reduce the traffic caused by movement of goods. There are ongoing investments on light rail transportation in the city, but this has not been completed. In any case, targeted users of both rail and water transports have to travel through the road to use these services, adding to overall commute costs.

3.5.1 Existing Plans to Integrate Current Transportation Investment with TODs to Enhance Housing Provision in Lagos

There are no current, conscious plans for TOD interventions in Lagos State, even though the state has recently made significant investments in housing development, water transportation, road and rail transportation as well as in updating city master plans. Investment in transportation and investment in housing do not appear to be strategically linked.

As one government representative explained (Fig. 3.4):

In 2009, a Strategic Master Plan for Lagos was developed to guide transportation, investments and so on. That plan was extended in 2014 to include neighbouring states in terms of transportation development up to 2032. The cable car, the rail, and the ferry are all contained in the master plan. That master plan was developed using the Model City Plan for various locations in Lagos. All the plans were fused into a holistic investment plan; which included seven rails and the mono rail…. 14 other BRT routes (with one already in place, and two being developed at the moment: Abule-Egba to Oshodi, Oshodi to Apapa. There are plans to do another two within the next two years, towards Iyana-Ipaja. (Rep LAMATA)

Thus, the current focus of the state is on mass transportation, facilitated by investment in BRT. In addition, it was reported that a transportation demand model which included development, population, employment and so on was developed and tested in different scenarios to know what needs to be done to decongest and provide a multimodal system. The cost of using different kinds of transportation was also included. However, the present transportation master plan does not include

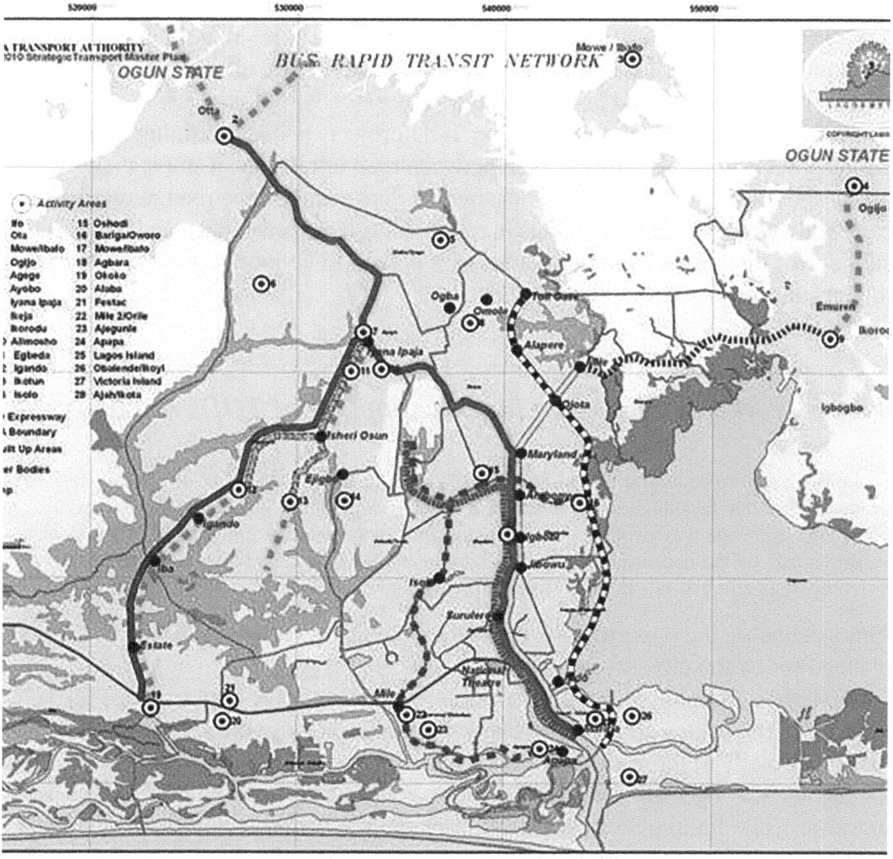

Fig. 3.4 Lagos BRT route. (Source: Udoji & Szpytko, 2020)

constructing residential accommodation along the corridors, but there is an opportunity to achieve this since the plan will be reviewed in 2 years.

The Lagos State Government has however responded to current housing accessibility and transportation cost challenges with projects that seek to provide homes within the city. Examples are the Lagos HOMs, which are designed as a mixed density housing with one-, two-, and three-bedroom apartments in various locations across the city. Other housing projects aimed at improving housing accessibility within the city are the LSDPC (Lagos State Development and Property Corporation). This engaged in a redevelopment project of compaction and densification; some areas were redeveloped, occupants compensated and multiple flats constructed. These gave people the opportunity of living in the urban centre and to have easier access to work and reduced cost of transportation. However, there were no connections between the new developments, BRT transportation routes and stops and also new investment in rail transport systems.

Also, it is noticeable that housing development often follows new industrial development, even though this will usually not follow planning guidelines. For example, the Dangote Refinery at the Lekki Free Trade Zone has led to the construction of residential homes by private individuals, stalls, a regional highway and the dredging of the coastal ways opening up of the waterways, all these in expectation of the influx of people and goods into that axis.

In recognising the lack of conscious TOD projects and ascertaining appropriate basis for such investment as a choice between road or rail transit stops, the problem of transportation in Lagos being inordinately dependent on the road network again came into focus. While investments in rail projects have been made, and projects not completed, it appeared to be preferred as it is seen to be more efficient than roads, though currently heavily underfunded.

3.5.2 Workable Strategies for the Adoption of TODs in Lagos

> Looking at it from a different point of view—looking at the location of our businesses—most corporate headquarters are located in Lekki axis, Victoria Island and Ikoyi, thereby, driving everyone that works for them in (that) one way, towards this axis. If this is done differently by the decentralising of the commercial hubs; it will reduce the concentration going to a particular place. (Rep, Transportation Agency)

This establishes the opportunities for TOD, even though it has not been consciously planned for by the city government.

First, the expansion of the BRT road network offers some opportunities for TOD inclusion. The routes are being expanded, more buses procured and road linkages made especially to those parts of Lagos with high population densities but far from the city centre. This will reduce road traffic congestion and amount of time wasted in traffic while linking areas known to provide low-income housing with effective road network. High urban growth rates estimated at up to 26 persons/hour make decongesting strategies difficult as the space created will be used up especially by migrants from surrounding towns. Thus, a strategy of linking up heavily populated areas to BRT access offers a more practical approach.

Second, the government is also interested in regeneration through its Urban Renewal Agency, and this also offers another avenue for TOD inclusion. Existing enclaves of substandard housing can be upgraded to decent standards, neighbourhood facilities provided, greens included and employment investments made, using TOD principles:

> Part of the strategies would include.... Trying to linkup various communities where there are no roads. For example, government is looking at linking up Ijegun to Festac through construction of a bridge -Isawo to Berger, this will take away concentration from the main Ikorodu road, Badagry Express Way and that will also enhance development along that axis, hooking up these areas would decentralise the traffic in Lagos to an extent. (Rep., Transportation Agency)

Third, in retrospect, in recognition of the massive development that has already taken place in the city along existing highways (e.g., Ikorodu Road, Agege Motor Road, both of which hosts BRT lanes) and the proposed light rail routes (Apapa-Badagry to Lagos Island and onward extension to Lekki in the nearest future), there might be a place for 'development-oriented transit' rather than a fixation on TOD, where Lagos is concerned:

> If we bring in new transport systems – for instance we bring in the red line and blueline – that is an opportunity for businesses to spring up along that transit line. So, there should be housing along that transit corridor too. That is how I understand it. Automatically, commercial activities will spring up. So, the question is, 'How do we encourage housing development along these corridors?. (Rep., LAMATA)

In this wise, current development can be made more transit-friendly through additional infrastructure investment following TOD principles. In this strategy, TOD plans will come after development has occurred, rather than the 'ideal' case of planned TODs in new areas which might be practical for new town development projects.

Finally, whatever 'modern' ideas are conceived, the culture of the people must be respected and protected. The city of Lagos, for instance, is heterogenous in culture, people and economy. Its indigenes are conservative and will not allow the destruction of their historic buildings and places, and its migrants include low-income dwellers. Both groups are highly business-oriented and possess a sense of civic rights. The state must be ready to embrace the 'new without destroying the old':

> Some people cannot do away with their "Ijoko Ale".... So, we are not saying we should not be modern, but whatever we are doing, we should maintain our historic places conserved. (Rep., Urban Planning Policy Maker)

Therefore, any strategy for TOD inclusion must take cognisance of this.

In conclusion, TODs are necessary for connectivity and there are current opportunities to adopt it in Lagos. However, the underlying principles need to be fine-tuned to embrace the peculiarities of the people and their culture. Thus, for existing neighbourhoods, the focus should be on cheap, functional and effective mass transportation alongside construction of affordable high-density housing. For new route transportation investments and new town development projects, however, TODs that align with the realities of Lagos might be feasible.

3.6 Recommendations for Government, Investors and Communities

Government In many African cities, governments have significant investments in transportation infrastructure and significant investments in housing. However, both capital-intensive and people-centred investments may not sufficiently be linked. TOD provides an opportunity to link these investments in a purposeful way to increase the housing stock and through well-packaged housing products possibly at affordable rates. Firstly, for governments, the onus is to declare a TOD intervention area at existing locations and neighbourhoods deemed to be convergence points where transportation investments and high-density housing needs meet. Through the joint working of the Ministry of Transportation, Ministry of Housing and Ministry of Physical Planning, the collaboration of existing community groups and civil society would ensure that the negative impacts of increased housing costs that are often associated with TODs are mitigated.

Secondly, as existing urban plans are reviewed, and new ones developed, governments should incorporate TODs in new development areas. Sprawling is not only unsustainable but is also expensive. Hence, concrete legislations that discourage sprawling will shift development attention towards existing communities and innovate ideas for urban housing regeneration. Incentives for affordable rental housing options within TODs are low hanging fruits that can discourage sprawling and encourage city regeneration. Governments must provide enabling environment and subsidies to entice and provoke private sector investment in TOD.

Investors There is the need for investors to be guided towards the provision of high-density, affordable housing near major transportation routes whether in the inner city or at the peripherals with the inclusion of schools, markets, shops and gardens in such projects.

In the inner city, but in partnership with existing communities, real estate investors can leverage on TOD concept to create compact communities abutting existing major transportation routes. Thus, existing residential areas can be used as a guide to identify viable investments in new transportation provision. Spatial analysis and management are a neglected aspect of real estate that the private sector can invest in to implement TOD. Analysing functional space, required housing needs and neighbourhood businesses will help to reorganise existing residential communities for TODs with minimal altering of the complex social and economic environment that define such areas.

For peripheral areas, and with collaboration with government and local land-owning families to ease land assembly, designing housing affordability options that meet with a variety of socio-economic and cultural status while offering reduced transportation costs would be viable. High-density housing areas provide opportunities for establishing small-scale businesses whose services and products prevent residents from travelling far from home as well as providing employment opportunities within close proximity. This can be explored further.

Communities The communities that are to be affected by TODs must be seen as equal partners in such development, not to be forcefully evicted, dislocated or displaced. The strengthening of social capital is very important to ensure they have a voice. Community development associations (CDAs) need to be encouraged and given more government recognition and support so that they can play a vital role in the partnership.

3.7 Conclusion

Indeed, the targets of SDG 11 for sustainable communities, affordable and accessible transportation options and environmental protection align closely with modern TOD principles, and this is where it holds some relevance for Africa. However, 'sometimes the market and people's choices have already created (seeming) transit-oriented development'. This comment, raised by a participant at the research meeting, underscores the existing relationship between housing and transportation. The 'transit-oriented development' we see in most of Lagos, as an example, is a reflection of unguided housing development and sprawl along existing transportation routes, with minimal government interface (until physical planning wields the big stick of development control). Thus, what might actually be required at this point is development-oriented transit, which reflects TODs that are incorporated into existing development.

Similarly, the expansion of transportation routes along designated areas, with minimal connection to new housing projects in the city, is also another scenario that is currently being played out. TODs also offer an opportunity to formally connect new transportation expansion, BRT provision and new housing projects. Thus, as an intervention to create sustainable cities following SDG 11 aspirations, TODs can be useful strategies for city regeneration and the guided development of peripheral areas.

However, there are concerns that in achieving diverse yet connected aspirations of affordable housing, accessible places, transportation access and green neighbourhoods with private sector involvement, TODs would generate several negative externalities that impact housing affordability. Research evidence shows that TODs typically result in higher development costs as a result of increase in land values which arises when new infrastructure (especially transportation) is invested in a neighbourhood. While this can be offset by the intentional subsidisation of housing development costs by the government to enable housing access by low-income households, there is a risk of gentrification also occurring when these same households are priced out of new developments to the advantage of middle-to-high-income groups who can afford to offset the developers' costs. This is a particular issue where there are no social safety nets to protect lower-income households and there are no direct subsidies to housing suppliers. The alternatives are for the government to provide direct construction of affordable housing units with targeted allocation systems (such as the one attempted in the Lagos-HOMS project) or for the government to support land acquisition in TODs by low-income housing groups such as co-operative societies. In the USA, there are mechanisms such as covenants on rent control in an elongated period of years between developers, landlords and government agencies in the introduction of TODs to prevent these externalities, and these may be adapted in African cities.

In the final analysis, TODs would certainly not solve all the problems associated with cities and urban life and, if not carefully guided, might lead to the creation of new ones. However, it certainly represents opportunities for improving current haphazard development of the peripheral areas, strategies for regeneration of the city and improving the living conditions of the people who reside in them, an intervention that has the capacity for enhancing joint working amongst government agency while also bringing in two capital intensive investment threads of housing and transportation into a structure for achieving sustainable communities. The successful adoption of transit-oriented development in Africa therefore requires further localised studies.

References

Addis Ababa City Administration. (2018, November). *Addis Ababa Non-motorised Transport Strategy 2019-2028*. https://www.itdp.org

Aklilu, A., & Necha, T. (2018). Analysis of the spatial accessibility of Addis Ababa's Light Rail Transit: The case of east–west corridor. *Urban Rail Transit*. https://doi.org/10.1007/s40864-018-0076-6

Alade, T., Edelenbos, J., & Gianoli, A. (2020). A sustainable approach to innovation adoption in light rail transport. *Sustainability, 12*(3), 1262. https://doi.org/10.3390/su12031262

Alonso, W. (1960). *A model of the urban land market: Location and densities of dwellings and businesses*. Dissertations. Pro Quest. AAI6003564. https://repository.upenn.edu/dissertations/AAI6003564

Alonso, W. (1964). *Location and land use. Toward a general theory of land rent*. Harvard University Press.

Arup. (2016). *A Transit-Oriented Development (TOD) in Africa's fastest growing city*. Arup.com. https://www.arup.com

Boarnet, M., Bostic, R., Williams, D., Santiago-Bartolomei, R., Rodnyansky, S., & Eisenlohr, A. (2017). *Can affordable housing in Transit-Oriented Development help solve California's housing crisis while also addressing environmental goals?* National Centre for Sustainable Transportation. https://escholarship.org/uc/item/31t851c6

Brueckner, J. K. (1987). The structure of urban equilibria: A unified treatment of the Muth-Mills Model. In *Handbook of regional and urban economics* (Vol. 2, Issue 20, pp. 821–845). https://doi.org/10.1016/SI574-0080(87)80006-8

Carlton, I. (2009). *Histories of transit-oriented development: Perspectives on the development of the TOD concept* (Working paper series). Institute of Urban and Regional Development (IURD).

Cervero, R. (2007). Transit-Oriented Development's ridership bonus: A product of self-selection and public policies. *Environmental and Planning A, 39*(9), 2068–2085.

Du, J., & Gelman, V. L. (2018, December 18). *Johannesburg fights inequality with Transit-Oriented Development*. World Resources Institute. http://www.wri.org/

Ebekozien, A. (2020). Homeownership in a Sub-Saharan Africa city: Exploring self-help via qualitative insight to achieve sustainable housing. In T. Nubi, I. Anderson, B. Oyalowo, & T. Lawanson (Eds.), *Housing and the SDGs in urban Africa: Springer nature's advances in 21st Century Human Settlements Series*. Forth coming.

Einsberg, A. C. (2001). *The housing transportation connection: Building better communities through better housing and transportation*. govinfo.library.unt.edu>mhc

EMBARQ Network. (2015). *How Addis Ababa is on the frontier of sustainable transport for African cities*. https://www.smartcitiesdive.com/ex/sustainablecitiescollective/fr

Gallo, M. (2018). The impact of urban transit systems on property values: A model and some evidences from the city of Naples. *Journal of Advanced Transportation, 22*, 1767149. https://doi.org/10.1155/2018/1767149

Glaeser, E. L., Gyourko, J., & Saiz, A. (2008). Housing supply and housing bubbles. *Journal of Urban Economics, 64*(2), 198–217. https://doi.org/10.1016/j-jue.2008.07.007

Godard, X. (2013). Comparisons of urban transport sustainability: Lessons from West and North Africa. *Research in Transportation Economics, 40*(1), 96–103. https://doi.org/10.1016/j.retrec.2012.06.033

Group Communication and Tourism Department. (n.d.). *Corridors of freedom: Re-stitching our city to create a new future*. Group Communication and Tourism Department. www.joburg.org.za>corridors

Howell, A., Currans, K. M., Gehrke, S., Norton, G., & Clifton, K. J. (2018). Transportation impacts of affordable housing: Informing development review with travel behaviour analysis. *The Journal of Transport and Land Use, 11*(1), 103–118. https://doi.org/10.5198/jtlu.2018.1129

Institute for Transportation & Development Policy (2019, April 11). *Addis Ababa is boosting economic growth with an ambitious walking and cycling strategy*. https://www.itdp.org/2019/04/11/addis-ababa-walking-and-cycling

Jin, J., & Lee, H. (2017). Understanding residential location choices: An application of the urban sim residential location model on Suwon, Korea. *International Journal of Urban Sciences*. https://doi.org/10.1080/12265934.2017.1336469

London School of Economics and Political Sciences. (2018). *Developing urban futures*. https://www.lse.ac.uk/cities/publications/urban-age/Developing-Urban-Futures

Lubell, J., & Salomon, E. (2010). *How transportation reform could increase the availability of housing affordable to families with a mix of incomes near public transit, job centres, and other essential destinations*. Centre Housing Policy. https://community-wealth.org/sites/clone.community-wealth.org/files/downloads/report-lubell-salomon.pdf

McMillen, D. P. (2003). The return of centralisation to Chicago: Using repeat sales to identify changes in house price distance gradients. *Regional Science and Urban Economics, 33*(3), 287–304.

Mills, E. S. (1972). *Studies in the structure of the urban economy*. The Johns Hopkins Press.

Muth, R. F. (1969). *Cities and housing: The spatial pattern of urban residential land use*.

Ng, C. F. (2008). Commuting distances in a household location choice model with amenities. *Journal of Urban Economics, 63*(1), 116–129. https://doi.org/10.1016/j.jue.2006.12.008

Pieterse, E., & Owens, K. (2018, December). *Towards a more equal city Johannesburg: Confronting spatial inequality*. World resources report. https://wrirosscities.org

Pojani, D. (2020). South-south policy transfer? Transit-Oriented Development from China to Sub-Saharan Africa. *Journal of Handbook on Transport and Urban Transformation in China*. https://www.eelgar.com/shop/gbp/handbook-on-transport-and-urban-transformation-inchina-9781786439239.html

Rode, P. (2018, November). Ethiopia's railway revolution. *Urban Age*. https://urbanage.lsecities.net/essays/ethiopia-s-railway-revolution

Rode, P., da Cruz, N. F., & Terrefe, B. (2020). *Policy brief 2020. How to connect cities and transport infrastructure governing infrastructure interfaces for Ethiopian cities*. LSE Cities, London School of Economics and Political Science.

Salat, S., & Ollivier, G. (2017). *Transforming the urban space through Transit-Oriented Development: The 3V approach*. World Bank. https://doi.org/10.1596/26405.

Tarrosy, I., & Vörös, Z. (2019, January 26). Revisiting Chinese transportation project in Ethiopia. *The Diplomat*. https://theiplomat.com/2019/01/revisiting-chinese-transportation-projects-in-ethiopia/

Tembe, A., Nakamura, F., Tanaka, S., Ariyoshi, R., & Miura, S. (2019). The demand for public buses in Sub-Saharan African cities: Case studies from Maputo and Nairobi. *IATSS Research, 43*(2), 122–130. https://doi.org/10.1016/j.iatssr.2018.10.003

Udoji, C. U., & Szpytko, J. (2020). The concept of a sustainable urban transport system model for developing countries on the example of Lagos. *Journal of KONBiN, 50*(1), 10.2478/jok-2020-0003.

United Nations. (2014). *The right to adequate housing*. https://www.ohchr.org

Yiu, C. Y., & Wong, S. K. (2005). The effects of expected transport improvements on housing prices. *Urban Studies, 42*(1), 113–125.

Chapter 4
Transit-Oriented Development and Urban Health and Liveability in African Cities

Immaculata Nwokoro, Tinuola Odugbemi, Oluwatoyin Fatunsin, Oluyemi Akindeju, and John Agwu

4.1 Introduction

Globally, transportation infrastructure experienced transformation between the 1760s and 1900s. This period marked the ending and beginning of two important transport eras: motorised and non-motorised eras. In Africa, the Industrial Revolution of the late eighteenth and early nineteenth centuries and colonial invasion of the nineteenth and twentieth centuries brought major changes in transportation infrastructure. Prior to these two events, pre-colonial Africa largely depended on non-motorised modes of transportation to move people, goods and services from one place to another. During this era, transportation networks were limited, and few available ones were dominated by road and water networks that are mostly limited to local travels involving intercity and intracity movements for agricultural activities. In this era, track roads, shallow inland waterways and other inefficient trip attributes and network conditions limited the impact and influence of transportation. However, with the advent of Industrial Revolution and colonialism, Africa embraced motorised transportation, and the impact of transport increased as the need for transportation services expanded transportation modes and advanced fuel type.

The advent of motorised transportation in Africa resulted in impacts such as reconfiguration of human settlement, stimulation of urban living, upsurge in interstate interaction and increase in deforestation and wanton resource exploitation.

I. Nwokoro (✉) · O. Akindeju · J. Agwu
Department of Urban and Regional Planning, University of Lagos, Lagos, Nigeria
e-mail: inwokoro@unilag.edu.ng

T. Odugbemi
Department of Community Health and Primary Care, College of Medicine, University of Lagos, Lagos, Nigeria

O. Fatunsin
Department of Industrial Chemistry, University of Lagos, Lagos, Nigeria

T. Nubi et al. (eds.), *Transit Oriented Development in West African Cities*, Urban Perspectives from the Global South,
https://doi.org/10.1007/978-3-031-58726-9_4

Settlement reconfiguration, through the construction of wagon roads, ports and rail tracks, can facilitate the activities of manufacturing and extractive industry, leading to the development of transit, commercial and administrative town/cities at the hinterlands and port cities at the coasts or beaches. For instance, the introduction of port facilities, railway tracks and locomotive engine in South Africa, Ghana, Nigeria, Morocco, Algeria and Tunisia between the 1850s and 1890s saw the development of cities like Cape Town, Stellenbosch, Wellington, Durban, Port Elizabeth, Tarkwa, Kumasi, Accra, Achiasi, Apapa, Otta, Offa, Oshogbo and Ibadan. In the words of Gwilliam (2011), transport facilities in Africa were built primarily for the colonial exploitation of mineral and agricultural resources, and the trend of transport development in East Africa, North Africa, South Africa and West Africa is similar.

Thus, as transportation infrastructure in Africa advanced, economic and political interests of Europe intensified, leading to significant demographic and economic growth. This changes gradually produced traffic congestion, population explosion, air and noise pollution, traffic accidents, advent of sedentary lifestyle and other environmental problems like erosion and flooding. Despite occasional efforts to address it, the trend continued, and in less than a decade after the departure of the colonial masters, passing through difficult periods of civil war, military rule, apartheid policy, corruption and political instability, transportation system in Africa further nosedived. Considering these facts and the importance of regional connectivity and cooperation in achieving the African Urban Agenda and SDGs, this scoping study interrogates the body of knowledge on transport, land use, the environment and health nexus and uses the same to benchmark transport infrastructure in Africa.

The study evaluated the state of transportation infrastructure using cost, technology, traffic density, safety, service satisfaction, growth, distribution, connectivity and governance as key indicators, across road, rail, water and air transports. The transport-health nexus framework adopted green space and aesthetics, physical activity, access, mobility independence, contamination, social exclusion, noise, urban heat islands, vehicle crashes, air pollution, community severance, electromagnetic fields, stress and greenhouse gas emissions to examine priority issues in Lagos.

4.2 Transportation and Health: A Conceptual Model and Literature Review

Transportation and health are related in several ways, and this relationship has been documented in both public health and transport literature. Across these disciplines, the relationship between transport and health can be categorised into two: detrimental and beneficial relationships. The detrimental relationship is seen as challenges and treated as a negative health impact of transportation, while the beneficial relationship is considered as important and treated as health benefits of transportation. Each of these categories has several pathways through which they manifest. These pathways are framed as linkages between transportation and health. This section reviews the linkages and highlights parameters often employed in literature to conceptualise and quantify health impacts associated with transportation. According to Khreis et al. (2019), the four factors that shape the impact of transport on health are land use and the built environment, transportation infrastructure, transportation mode choice and transportation technologies and disruptors. Environmental exposures and lifestyles produced by these four factors and referred to as pathways by Khreis et al. (2019) are determined by transportation. In all, there are 14 pathways through which transportation impacts health, namely, green space and aesthetics, physical activity, access, mobility independence, contamination, social exclusion, noise, urban heat islands, vehicle crashes, air pollution, community severance, electromagnetic fields, stress and greenhouse gas emissions (Khreis et al., 2019). The first four pathways are associated with beneficial health impacts, while the others are associated with detrimental health impacts. These pathways in conjunction with their underlying factors form the basis of this review. Although this review is designed to identify a conceptual model through which health impacts of transportation are quantified and tracked in literature, the main objective is to promote holistic solutions that integrate human health into transport planning and policy particularly in the urban areas.

4.3 Health Benefits of Transportation

A large number of studies have established that transportation systems provide health benefits to people, and most of these studies recognise land use and the built environment as conduit through which these benefits are transmitted. Drawing from these studies, several models have been used to demonstrate transport-health interaction. Some of the models are specific dealing with only one single channel (e.g. physical activity and active travel model, green space and mental health model, ecosystem function model), while some are broad and comprehensive incorporating more than one channel (mode). Examples of multimodal models include Integrated Transport and Health Impact Modelling Tool (ITHIM), Spatial Network Analysis for Multi-Modal Urban Transport Systems (SNAMUTS), Capability-Opportunity-Motivation-Behaviour model (COM-B), Theoretical Domains Framework (TDF), etc. Focusing on green space (land use and built environment), studies by Ying et al. (2015), Gascon et al. (2016) and Zijlema et al. (2018) have established that green spaces integrated with transport infrastructures such as roads encourage physical activities (primarily walking, cycling and use of public transport) and reduce the likelihood of mental health outcomes, diseases and premature mortality.

Also, studies by Hartig et al. (2014) and Nieuwenhuijsen (2016) still focusing on green space revealed that green spaces reduce the adverse effects of harmful transportation-induced environmental exposures such as urban heat islands, air pollution and noise. This also elicits positive health effects depending on the scale of visual integration (Khreis, et al., 2019). Using environmental restructuring and behavioural change framework, studies conducted by Panter et al. (2016) and Rafiemanzelat et al. (2017) established that land-use policies that promote high density, connectivity and active transportation infrastructure can boost physical activity. Specific health outcomes associated with a green transport environment include decreased risk of Alzheimer's disease, cancer, cardiovascular disease, cognitive decline, dementia, diabetes, hypertension, depression and anxiety, obesity, stroke, stress, premature mortality, mental health and well-being, improved pregnancy outcomes, improved self-reported health, improved sleep patterns, etc.

Access as another pathway through which transportation provides health benefits relates to land-use decisions that increases individual's ability to reach destinations and to protect and improve their health, including health facilities and services, healthy food, green space, physical activity facilities, jobs and education (Litman, 2015b). Examples of strategies to increase access include development practices like complete streets (Litman, 2015a), densification and transit-oriented development (Renne et al., 2016). These strategies can decrease distance to public transportation and increase active transportation and reduce morbidity and mortality (Nieuwenhuijsen, 2016). Mobility independence being another pathway through which transportation provides health benefits to people refers to the ability to use various transportation modes to access commodities and neighbourhood facilities and to participate in meaningful social, cultural and physical activities without assistance or supervision. In 2013, Rantanen reported that mobility independence might promote healthy aging and mental well-being through physical activity and engagement in community activities which sustain cognitive function.

4.4 Detrimental Health Impacts of Transportation

Documented evidence in literature shows that most motorised forms of transportation have detrimental health effects. These health effects may be directly linked to transportation challenges suggesting that the frequency, intensity and severity of transportation challenges increase its negative impact on health. All the four factors that shape the impact of transport on health earlier identified are sources of transportation challenges that produce negative health effects. For instance, land use and the built environment can increase vehicle trip and travel time, leading to traffic congestion and air pollution. The number, the spread and the condition of transportation infrastructure, such as stations, bridges, roads, road surface material, ports, airports and waterways, create challenges such as traffic accidents, oil spillage, community severance, noise and urban heat island. On the other hand, transportation mode choice, transportation technologies and disruptors create transportation issues such as social exclusion, air pollution, stress, greenhouse gas emission and electromagnetic fields. Each of these challenges produces negative health impacts.

According to Burant et al. (1987) and Khan and Strand (2018), oil, gasoline, heavy metals, particulate matter and polycyclic aromatic hydrocarbons found on surfaces of congested roadway contaminate water sources, soils and air. These pollutants potentially end up in what humans consume (Adamiec et al., 2016) resulting in different health issues. Similarly, Julien et al. (2015) reported that culmination of transportation-related inhibitions and/or deprivations limit the opportunity to participate in community activities and as such promote social isolation, loneliness and living alone. In a study conducted by Holt-Lunstad, Smith, Baker, Harris and Stephenson in 2015, social isolation, loneliness and living alone resulted in 29%, 26% and 32% increase in mortality, respectively. According to Khreis et al. (2019), other health conditions associated with limited transport opportunities include cardiovascular diseases, mental health issues, premature mortality, etc.

In addition, studies by Zuo et al. (2014), Ostro et al. (2009), the World Health Organization (2018a, b) and Lelieveld et al. (2015) have shown that noise, urban heat island, vehicle crashes and air pollution, etc., associated with transportation, can lead to insomnia, dementia, premature mortality, premature birth, etc. Studies by Ding et al. (2016), Mindell et al. (2017), Watts et al. (2018) and Anciaes et al. (2019) also have documented evidence showing that health effects associated with transportation include hindered cognitive development in children, nerve stimulation, reproductive complications, retinal phosphene occurrence, anxiety, depression, fatigue, heart disease, hypertension, obesity, unhealthy diet, stroke, substance abuse, malnutrition, physical injury, etc. The transport-health nexus as a two-way phenomenon amplifies the need to maximise the health benefits through deliberate transit-oriented development projects.

4.5 Conceptual Linkage Between Transit-Oriented Development, Sustainable Development Goals and Urban Health

The essence of exploring the transport-health relationship is to develop a comprehensive frame through which sustainable transport, improved urban health and livability can be achieved in Africa. Through this lens of improved urban health and livability, TOD provides a meeting point between SDGs and transportation.

The elements that characterise TOD include mixed-use development, compact development and high-density development, open spaces, near excellent public transport (transit) and walking and cycling mode of transportation. These elements suggest that TOD is about the size and form of land use, transport development and human behaviour.

Focusing on land use, TOD is measured in terms of use types, diversity of use patterns and density types. In terms of transportation development, TOD is measured by transportation mode (public or private, active or inactive), accessibility (distance, time and cost), distribution and availability (number of people served at a time). From a behavioural lens, TOD is evaluated using population census (distribution of residents), ridership (traffic demand around the transit nodes), population mobility and social diversity (types of human activities). According to Zhang et al.

(2019), this view implies that TOD is a key concept in analysing urban spatial structure (Fig. 4.1).

Seen from an outcome perspective, TOD can be understood as an urban planning strategy designed to achieve social, economic and environmental sustainability in cities by restructuring land use to minimise urban sprawl and to create more compact cities (Bajracharya & Khan, 2006). The compact city outcome, when created, is expected to resolve social and environmental problems by reducing car dependence, conserve energy, promote social interaction and subsequently improve air quality by causing reduction in greenhouse gas emissions (Newman & Kenworthy, 1999; Newman, Urban Design and Transport in In Search of Sustainability, 2005). Reduced car dependence outcome on the other hand is expected to promote active living, improve health conditions and prevent deaths by traffic accidents and pollutant emissions. Beyond public life and social interaction, public spaces and transit developments such as BRT and rapid transit rail lines are expected to increase property value and promote economic prosperity. These outcomes correspond with

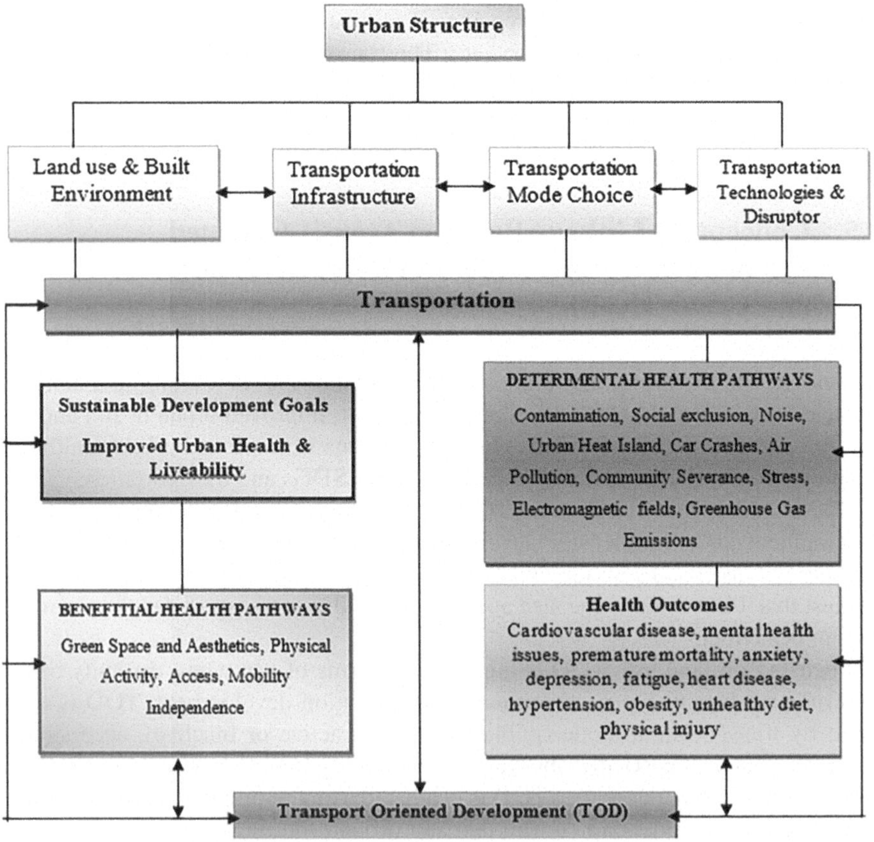

Fig. 4.1 TOD, SDG and urban health conceptual model adapted from Khreis et al. (2019)

Goals 3, 11 and 13 of SDGs specifically Target 3.4, Target 3.6, Target 3.9, Target 11.6 and Target 13.2. TODs can also increase the use of public transport by creating higher residential densities integrated with bicycle roots, walkways and greenspaces.

In a broader perspective, these TOD outcomes also enhance the health and livability of cities through green spaces, physical activity, increased access, functional mobility independence, reduced contamination, social inclusion, noise elimination, decreased urban heat island, reduced vehicle crashes, minimal air pollution, increased community severance and decreased stress (Bernick & Cervero, 1996; Katz, 1994). These broad outcomes correspond to SDG targets mentioned earlier including Targets 5.1 and 11.7. SDG Goal 3 is a global commitment to ensure healthy lives and promote well-being for all at all ages; Goal 11 commits to make cities and human settlements inclusive, safe, resilient and sustainable, while Goal 13 pledges to take urgent action to combat climate change and its impacts. Climate change has significant health impacts including stress, premature mortality, anxiety, depression and fatigue. Integrating climate change measures into TOD reduces these health impacts and promotes liveability.

4.6 State of African's Transport Infrastructure: Case Study of Johannesburg, South Africa

According to Gwilliam (2011), the state of transport infrastructure in Africa is consequence of history. This section analyses the condition of transport infrastructure in Johannesburg from the TOD perspective. This is to determine the trend of growth in transport infrastructure and its vibrancy and sustainability.

Johannesburg is a South African city with a history of racial discrimination that was physically and socially imprinted. As observed by Griffith (2017), apartheid spatial and racial segregation in Johannesburg resulted in the exclusion of black and coloured populations from business centres and traditional white areas. For example, Soweto, which is one of the most affected town in Johannesburg, lies 15 km southwest of Johannesburg's central business district. The apartheid policy was also accompanied by disinvestment in critical infrastructure such as public transport to deny black residents' access to workplaces and other basic needs including right to the city. As noted by Griffith (2017), the apartheid policy also involved forceful removal of non-white people from inner cities and the relocation to urban fringes, away from the economic opportunities present in the central city (Figs. 4.2 and 4.3).

Post-apartheid spatial policy in South Africa has compelled the national and provincial government to embark on strategic intervention to reverse the situation by constructing transit arterial connections to link apartheid-isolated areas to other parts of the city since 1990 (City of Johannesburg Metropolitan Municipality Council, 2018). Despite these efforts, the city of Johannesburg still has transportation challenges. TOD implementation in Johannesburg raises critical questions on land-use pattern, publicness of public transport and degree of participation.

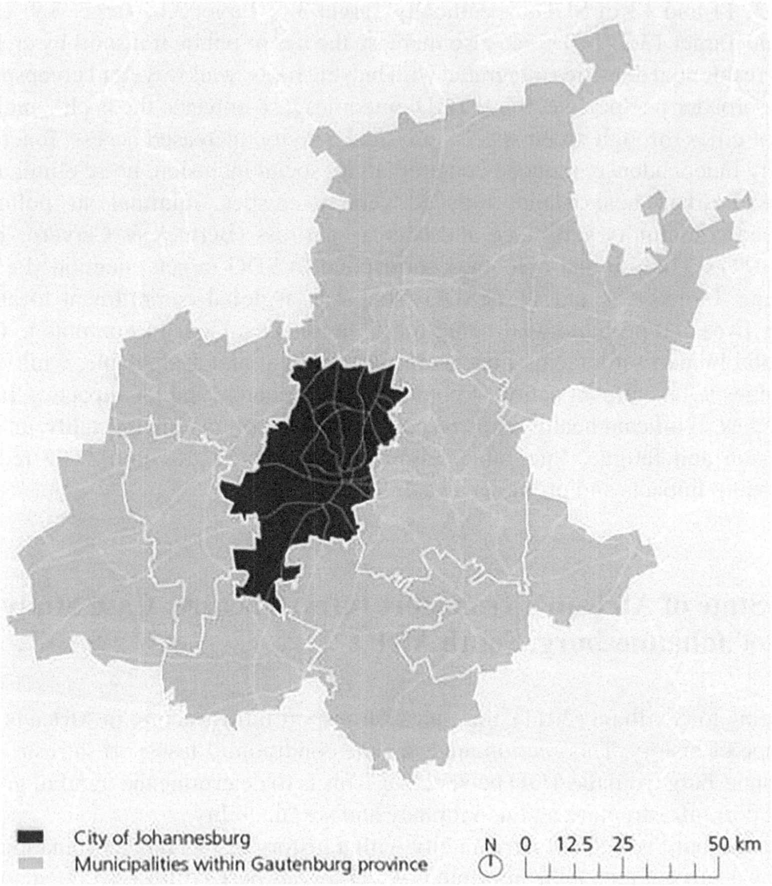

Fig. 4.2 City of Johannesburg. (Adapted from Global Future Cities Programme Johannesburg—City Context Report, 2018)

4.6.1 Diversity of Land-Use Activities in Johannesburg

A study on spatial distribution of land use in Johannesburg revealed that apartheid and modernist planning historically influenced land-use activities in the city. As observed by (SACN—South African Cities Network, 2017) and Maina (2015), apartheid policy enforced segregated and decentralised land-use system which ensured that black population are located at a significant distance from the predominantly white urban centre. With this spatial order, black townships like Soweto were located at urban periphery, while economic activities were concentrated in industrial areas, CBDs and from the 1970s, in decentralised office and shopping complexes, close to 'white' middle- and upper-income areas. This meant that transportation links between townships were provided entirely for daily commute (World Bank, 2017). Although a land-use diversity study conducted by Du Plessis

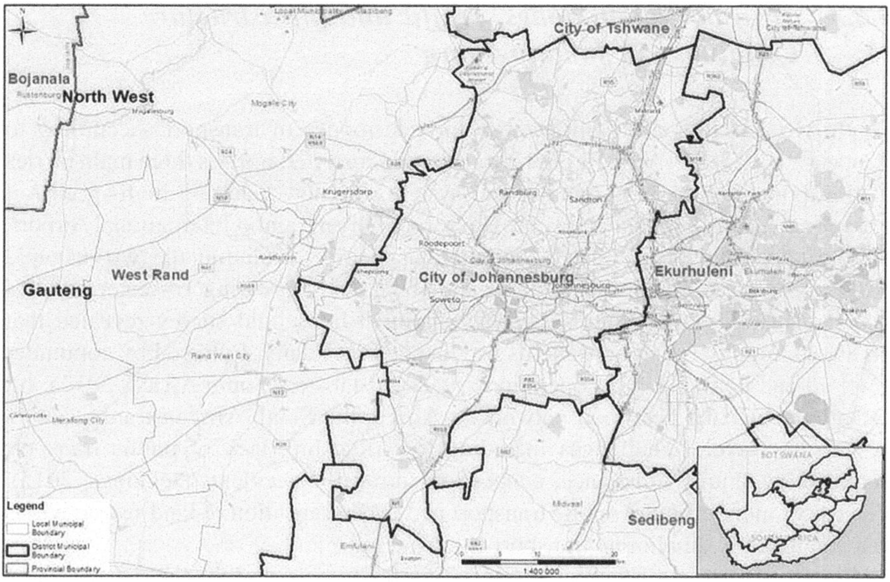

Fig. 4.3 The city of Johannesburg metropolitan municipality. (Adapted from profile: City of Johannesburg Metro, 2019)

and Boonzaier (2015) confirmed high levels of land-use mix within focus areas identified by the 1994 spatial development framework, the overall city-level intensity of land-use mix has only experienced marginal changes. Despite deliberate spatial policy to reverse the spatial order in Johannesburg, the transport inequity implication of land-use patterns persists, and urban form in Johannesburg still creates long commuting distances for workers and overreliance on motor car (Todes et al., 2017).

4.6.2 Land Use and Population Density of Johannesburg

In a submission given by SACN (2017), low built densities at the urban centre and high built densities at the peripheral townships defined Johannesburg's urban form. According to Todes et al. (2017), the most significant impact of this density shift was occupancy density and is mostly pronounced in the inner city. Other forms of density increase such as growth of informal backyard housing especially in and around the former townships and new Reconstruction Development Programme (RDP) areas, growth of large town house complexes and RDP housing took place at the city's edge. In clear terms, the World Bank (1991) stated that this pattern contrasts with standard density gradients and imposes considerable social and economic costs in terms of infrastructure requirements, transport costs, travel times and access to urban facilities and economic opportunities.

4.6.3 *Transportation Modes, Traffic and Infrastructure Conditions in Johannesburg*

South African cities are dominated by private modes of transport. According to Aropet (2017), South Africa's public transport industry comprises three main modes of transport: traditional commuter rail system, the new Gautrain high-speed rail between Johannesburg and Tshwane (Pretoria), Oliver Tambo International Airport, the subsidised and unsubsidised commuter bus industry, including the two bus rapid (BRT) systems in Johannesburg and Cape Town, and a growing 16-seater minibus taxi industry. Findings obtained from a national household survey revealed that 68.8% of South African households use taxi services daily, followed by commuter (21.1%) and commuter rail operations (9.9%) (Statistics South Africa, 2014a, b). Despite having the best road network in Africa, the South African transportation system has several challenges including low ridership, lack of public transport accessibility, equity imbalance, congestion and traffic accidents (Jennings, 2015). The use of more efficient public transport and decentralisation of land-use activities can significantly ameliorate transport challenges.

Report on users' satisfaction of transport infrastructure in Johannesburg revealed that of the 42% of South Africans who are train users, only 37% are more than satisfied with train services (Statistics South Africa, 2014a, b:8). Again, the survey reported that before making use of available transport modes, commuters consider service punctuality, level of crowding, distance from the station and security of the trains. This report implies that trains are often overcrowded, underpoliced and distanced from homes. In Johannesburg, a study by Mtizi (2017) reported that operators of low-capacity vehicles (6-seater minibus taxis) operate without licenses, and the driving behaviour of these taxi drivers is usually reckless. The taxi drivers break traffic rules and change taxi fares in response to peak time and weather. Mtizi (2017) also argues that bus services do not cover certain routes, leaving commuters with the option to walk long distances or use other forms of transportation to reach their destination. However, to an average commuter, bus services are regarded as a safer option when compared to other modes of transport despite the fact that it translates to long-distance travel and high cost.

4.7 Transport and Urban Health Issues in Lagos

Several urban health challenges are associated with transportation, evidence of this association has been reported at city level and Lagos is not left out. In terms of urbanisation and urban form, Lagos shares similar fashion with other African countries. However, the constraint imposed by the intercontinental water body at the South and the physical limit offered by boundary states at the east and west means that resolving transportation problems in Lagos requires a unique approach. Lagos has a total area of 3577.28 square kilometres of which 779.56 square kilometres representing about 22% is wetland and lagoon. Based on a 2013 report, Lagos has a population of over 18 million with a population density of 6515 persons per square kilometre (Lagos State Government, 2013). Physically, Lagos is compelled to demographic and spatial expansion towards the north where it dynamically interfaces with Ogun State. The physical structure of Lagos also simply implies that movement in Lagos is naturally constrained to flow in two directions: from north to south and from south to north. With a huge urban population (highest in Africa), infrastructural and governance inefficiency, Lagos is confronted with huge transportation challenges.

In terms of transport infrastructure, the city exhibits the shortest public transport system of many megacities (Salau, 2018). Its rail system runs from the south to the north, and two water routes run from south to east and south to west. The city also has one airport with three wings: international airport, cargo airport and domestic airport. With heavy reliance on roads, the Lagos-Ibadan Expressway and Lagos-Abeokuta Expressway are the major corridor highways that connect the southern and northern part of the city and also serve as the interstate highways to adjourning states of Oyo and Ogun states, respectively (Salau, 2015). The Lagos-Badagry Expressway serves outlying towns to the west of Lagos, while the Lagos-Abeokuta Expressway serves outlying towns to the east of Lagos, in all Lagos has 2600 km of roads (Lagos Metropolitan Area Transport Authority [LAMATA], 2014). Estimates of transport demand in metropolitan Lagos in 1990 ranged from seven to ten million passenger trips daily, but this figure increased to 22 million in 2014. In both years, over 95% of these trips were undertaken by road, primarily by private cars, buses and taxis and commercial motorcycles. Walking accounted for 40%; BRT, 0.41%; LAGBUS, 1%; private cars, 11%; minibuses (Danfo), 45%; train, 1%; ferry, 0.34%, and non-data mode, 1% (LAMATA, 2014).

Report shows that Lagos has one of the most extensive road networks in the West African subregion but with the worst vehicle statistics. For instance, Lagos has approximately two million vehicles with over 100,000-vehicle annual increase, 264-vehicle density and over 100,000 commercially operated motorcycles. With these statistics, the transportation system in Lagos has operational, technical and design challenges (LAMATA, 2020). Health impacts of these challenges are enormous and range from physical health and mental health to environmental health problems. According to Oshodi (2016), poor driver behaviour, indiscipline, unsafe vehicle condition, poor road conditions, poor street lighting, lack of pedestrian facilities and poor traffic enforcement all combine to produce an accident rate that is probably among the highest in the world. In addition, low-density urban form, poor urban design and segregated land-use pattern produce long-distance travel, unbearable car dependency and use of low occupancy vehicles (LOV) which results in serious traffic congestions, air pollution, stress, accident, insecurity and anxiety (Morimoto, 2009; Adbelhamid et al., 2018).

In a study conducted by Atubi in 2006, Lagos State recorded 148,485 road accidents resulting in 10,132 deaths and 18,972 injuries between 1970 and 2001 (31 years). In 2016 alone, Lagos recorded 441 road accidents involving 3244 persons, 90 fatal cases, 165 deaths, 971 injured and 1136 casualties (World Health

Organization, 2020). Between January and September 2019, the Federal Road Safety Corps (FRSC) of Nigeria reported that Lagos had 419 road accidents involving 209,221 persons and 84 deaths (Federal Road Safety Corps, 2019). Associated with the carnage on Lagos roads is environmental pollution from vehicular emissions. Evidence of households' vulnerability to vehicular emission abounds in literature. According to the United States Environmental Protection Agency (USEPA, 2008), road emissions contribute a larger share of 53% in urban areas where most people live. From Cohen et al. (2017) and the Global Initiative for Chronic Obstructive Lung Disease (2018), long-term exposure to PM2.5 is a major contributor to human mortality; morbidity; ischemic heart disease; tracheal, bronchus and lung cancer; and diabetes mellitus.

Air pollution monitoring around residential and transportation sector in Lagos Mainland by Obanya et al. (2018) revealed that most air quality parameters including PM10 and PM2.5 were significantly higher than the World Health Organization (WHO) guideline around bus stops ($p < 0.05$). Ezeh et al. (2018) estimated that the average population weighted PM2.5 concentration in Lagos is 68 ug/m^3. A study on residential population exposure to vehicle emission in Ikeja, Lagos, by Agwuin (2014) revealed that 133 (33.3%) of the respondents live in a distance less than 50 m off congested arterial road. This analysis is important in deducing the percentage of the sampled population that are vulnerable to high levels of vehicular emission using proximity to congested roads as exposure metrics. Using this approach, Croitoru et al. (2019) estimated that exposure to ambient PM2.5 was responsible for about 11,200 premature deaths in Lagos in 2018. In monetary terms, this is about US $1.9 billion for PM2.5-related mortality and US $0.2 billion for related morbidity.

Again, the use of roadside mechanics and second-hand vehicles and boats for road and water transports in Lagos pose a serious threat to environmental stability through soil and water contamination from oil, diesel and premium motor spirit (PMS) leakages. These leakages are also associated with fire incidents and car crashes due to slippery surfaces. Soil and water contamination are also of serious concern to aquatic life, human health, ecosystem function and livelihood. Proliferation of abandoned vehicles and vehicle carcasses is also of serious concern to urban aesthetics, pedestrian safety and intramodal connectivity, which presents a great gap due to lack of walkways and street canopies. Lagos commuters are constantly exposed to harsh weather conditions, accidents and robbery, which affects their health and well-being. The lack of intermodal connectivity between pedestrian and vehicle modes therefore creates physical difficulties, social marginalisation and financial burden to the vulnerable in the society especially the children, physically challenged, aged and pregnant women.

4.8 Recommendations

Dissecting the state of transportation infrastructure in Africa from the TOD perspective establishes beyond reasonable doubt the need for urgent attention, particularly in the cities and thematic area studied.

Focusing on transport-oriented development (TOD) as mandated by study package I, a transdisciplinary meeting was specifically convened to foster structured and documented transdisciplinary views on transportation challenges in Africa and its health impacts, solutions and priorities for action. The purpose of the meeting was achieved by identifying transport challenges through individual contributions and constructive evaluation; harvesting expert opinions on health impacts of transport infrastructure; sharing views on existing and emerging solutions from academics, practitioners and policymakers; cross- examining the challenges, solutions and practice; and establishing priorities from stakeholders' perspectives.

The interactive session focused on the questions generated from literature search and included: What are the factors driving transport infrastructure development in Lagos, impacts transport infrastructure has on health problems and by what means are the constraints and opportunities for TOD to resolve health issues in Lagos? How best can TOD be planned, designed and implemented in Lagos and other Africa cities? Between operational and technological efficiency, which one should be prioritised? How do we resolve the constraints imposed by communicable disease on shared public spaces and facilities promoted by TOD?

At the end of the interactive session, various issues that formed the basis of the recommendations were highlighted. These include population increase, which was unanimously accepted as a factor driving socio-economic activities and co-founder of high travel demand, high vehicular density, spatial expansion, over-utilisation of available transport facilities/personnel and monocentric migration behaviour. Other factors identified as transport infrastructure development incentives are climate change imperatives, Sustainable Development Goals (SDGs) and socio-political agenda. Most participants also shared similar views on factors responsible for transport challenges in Lagos. The outstanding factors were lack of modern and quality transport facilities; insufficient funding; lack of discipline on the part of road users, law enforcement agents and residents around transportation corridors; illiteracy; lack of coordinated regulatory framework; design and planning inefficiency; lack of maintenance; corruption; and policy inconsistency. It was generally observed that the transportation system has become a pathway through which communicable and non-communicable diseases are transmitted.

Considering these challenges, participants suggested innovative, early and continuous transport education, stiffer regulation, coordinated regulatory framework, culturally and sustainable adaptive planning and design measures, robust feedback mechanism, intermodalism, appropriate funding, policy review, public-private partnership and proactive demand management as solutions. In addition, participants agreed that every aspect of the transport system should be prioritised, but some might be considered as short-term, intermediate or long-term goals. An advice from a member of the Lagos State Task force on Coronavirus reminded participants to observe personal hygiene and self-isolate themselves pending when the pandemic is finally curtailed or a reliable vaccine is confirmed.

Considering the drivers, constraints and opportunities in African cities, this paper makes the following recommendations:

1. The socio-spatial structure of African cities has a significant impact on the current state of their transport infrastructure. Adequate steps must be taken to identify areas and locations that need social and spatial restructuring using TOD. Hence, each city should conduct a TOD feasibility and viability study focusing on land -use diversity, density types, active transport, transport accessibility, ridership, population mobility and social diversity. Outcome of such studies should measure urban vibrancy or urban growth poles, sustainability or city balance to show that TOD will occur.
2. City managers in collaboration with relevant stakeholders should rely on evidence provided by TOD feasibility and viability studies to develop sustainable TOD strategy that will moderate or control urban development and reverse health challenges associated with transportation, urban sprawl and low- density urban forms in African cities. These TOD strategies should focus on improving transportation justice, encourage non-motorised transport modes, develop high-capacity public transport and promote urban vibrancy using complete streets and public spaces.
3. African cities, as a matter of urgency, should embark on comprehensive socio-technical innovations to improve operational and technological efficiency of existing and new transportation infrastructure. To improve operational efficiency, TOD strategies should be integrated with demand management, parking management, transport regulation and technologically efficient transport facilities such as autonomous vehicles and standalone photovoltaic bus stops.
4. Transport infrastructures such as super rail lines, technologically efficient airports and seaports should be seen as a tool for regional integration and economic development. Apart from increasing African interregional trade, facilitating easy movement between countries and reducing time spent on transit, such transport facilities will greatly reduce air pollution and traffic accidents on interstate and transnational highways.
5. Cities that do not have transport authorities should create one to improve enforcement and compliance of road users to road transport and road safety regulations.

4.9 Conclusion

With the commitment to provide a deeper and a more holistic view of the nature and depth of transportation challenges identified as an intractable problem impeding the functionality of African cities, the environmental health and livability cluster explored transdisciplinary views on transportation challenges in Africa and their health impacts, solutions and priorities for action. This was achieved by identifying transport challenges through individual contributions and constructive evaluation; harvesting expert opinions on health impacts of transport infrastructure; sharing views on existing and emerging solutions from academics, practitioners and policy-makers; cross-examining the challenges, solutions and practice; and establishing priorities from stakeholders' perspectives.

Based on findings from this scoping study, the main factors driving the demand for sustainable transport infrastructure development in African cities include urban structure, population density, regional interaction, economic development, sustainable development agenda and New Urban Agenda. A strong nexus between transport and health in African cities was firmly established, and the socio-spatial structure of African cities has significantly impacted on the current state of their transport infrastructure. The transit- oriented development (TOD), therefore, stands as an urban planning strategy designed to achieve social, economic and environmental sustainability in cities with broad outcomes corresponding to targets that actualise the Sustainable Development Goals (SDGs).

References

Adamiec, E., Jarosz-Krzemińska, E., & Wieszala, R. (2016). Heavy metals from non-exhaust vehicle emissions in urban and motorway road dusts. *Environmental Monitoring and Assessment, 188*, 369. https://doi.org/10.1007/s10661-016-5377-1

Adbelhamid, M. M., Elfakharany, M. M., & Elfakharany, A. M. (2018). *Reducing private cars dependency to achieve sustainable urban environment in congested cities*. In 7th International conference on modern research in civil engineering, architectural & urban development (pp. 1–13).

Anciaes, P., Stockton, J., Ortegon, A., & Scholes, S. (2019). Perceptions of road traffic conditions along with their reported impacts on walking are associated with well-being travel behaviour and society. *Travel Behaviour and Society, 15*, 88–101.

Aropet, R. (2017). *Southern African solutions to public transport challenges*. In 36th Southern African Transport Conference (SATC, 2017). Safer Spaces.

Atubi, A. O. (2006). *Road traffic accident patterns in Lagos State from 1970 to 2001*. Ph.D. thesis. Department of Geography, University of Nigeria, Nsukka.

Bajracharya, B., & Khan, S. (2006, September). Sustainability framework for Transit-Oriented Development in South-East sustainability framework for Transit-Oriented Development in South-East Queensland Bhishna Bajracharya and Shahed Khan.

Bernick, M., & Cervero, R. (1996). *Transit villages in the 21st century*. McGraw-Hill.

Burant, A., Selbig, W., Furlong, E. T., & Higgins, C. P. (1987) *Trace organic contaminants in urban runoff: Associations with urban land-use*. Environmental Pollution (Barking, Essex). https://doi.org/10.1016/J.Envpol.2018.06.066.

City of Johannesburg Metropolitan Municipality Council. (2018). *City of Johannesburg Land Use Scheme, 2018. South Africa*. www.joburg.org.za

Cohen, A. J., Brauer, M., Burnett, R., Anderso, R. H., Frostad, J., Estep, K., & Forouzanfar, M. H. (2017, May 13). *Estimates and 25-year trends of the global burden of disease attributable to ambient air pollution: An analysis of data from the global burden of diseases study 2015*. https://doi.org/10.1016/S0140-6736(17)30505-6

Cooperative and Governance Affairs. (2019). *Profile: City of Johannesburg metro.*

Croitoru, L., Chang, J. C., & Kelly, A. (2019). *The cost of air pollution in Lagos.* http://documents.worldbank.org/curated/en/980031575616020127/pdf/The-Cost-of-Air-Pollution-in-Lagos.pdf

Ding, D., Lawson, K. D., Kolbe-Alexander, T. L., Finkelstein, E. A., Katzmarzyk, P. T., van Mechelen, W., & Pratt, M. (2016). The economic burden of physical inactivity: A global analysis of major non-communicable diseases. *The Lancet, 388*, 1311–1324.

Du Plessis, D., & Boonzaier, I. (2015). The evolving spatial structure of South African cities: A reflection on the influence of spatial planning policies. *International Planning Studies, 20*, 87–111.

Ezeh, G. C., Obioh, I. B., Asubiojo, O. I., Onwudiegwu, C. A., Nuviadenu, C. K., & Ayinla, B. S. (2018). Airborne fine particulate matter (PM2.5) at industrial, high- and low-density residential sites in a Nigerian megacity. *Toxicological & Environmental Chemistry*, 326–333. https://doi.org/10.1080/02772248.2018.1485924

Federal Road Safety Corps. (2019, November 3). Lagos FRSC records 419 road crashes, 84 deaths in nine months. *Premium Times*, p. Front Page.

Gascon, M., Triguero-Mas, M., Martinez, D., Dadvand, P., Rojas-Rueda, D., Plasencia, A., & Nieuwenhuijsen, M. J. (2016). Residential green spaces and mortality: A systematic review. *Environment International, 86*, 60–67.

Global Initiative for Chronic Obstructive Lung Disease. (2018). *Global strategy for diagnosis, management and prevention of chronic obstructive pulmonary disease*. Global initiative for Chronic Obstructive Lung Disease.

Griffith, J. (2017). Equitable access to public transport: Corridor plans for Transit-Oriented Development in Soweto, South Africa and Boston, Massachusetts compared. *Journal of Comparative Urban Law and Policy, 1*, 22–63.

Gwilliam, K. (2011). Africa's transport infrastructure mainstreaming maintenance. In V. Foster & C. Briceño-Garmendia (Eds.), (First). The International Bank for Reconstruction and Development/The World Bank. https://doi.org/10.1596/978-0-8213-8456-5.

Hartig, T., Mitchell, R., Vries, S., & Frumkin, H. (2014). Nature and health. *Annual Review of Public Health, 35*, 207–228. https://doi.org/10.1146/annurev-publhealth-032013-182443

Jennings, G. (2015). *Public transport interventions and transport justice in South Africa: A literature and policy-775 review.* In Southern African transport conference (SATC) (pp. 764–775). Southern African transport conference.

Julien, D., Richard, L., Gauvin, L., Fournier, M., Kestens, Y., Shatenstein, B., Daniel, M., Mercille, G., & Payette, H. (2015). Transit use and walking as potential mediators of the association between accessibility to services and amenities and social participation among urban-dwelling older adults: Insights from the VoisiNuAge study. *Journal of Transport & Health, 2*, 35–43. http://www.sciencedirect.com/science/article/pii/S2214140514000759

Katz, P. (1994). *The new urbanism.* McGraw-Hill.

Khan, R. K., & Strand, M. A. (2018, April). Road dust and its effect on human health: A literature review. *Epidemiology and Health, 40*, e2018013. https://doi.org/10.4178/epih.e2018013

Khreis, H., Glazener, A., Ramani, T., Zietsman, J., Nieuwenhuijsen, M., & Mindell, J. (2019). *Transportation and health: A conceptual model and literature review.* Centre for Advancing Research in Transportation Emissions, Energy, and Health.

Lagos Metropolitan Area Transport Authority (LAMATA). (2014). https://www.lamata-ng.com

Lagos Metropolitan Area Transport Authority (LAMATA). (2020). Sustainable safe mobility in a mega city in Africa. Presentation by LAMATA managing director, Lagos, Nigeria. https://www.lamata-ng.com/presentation/

Lagos State Government. (2013). *Lagos state development plan 2012–2025.* Ministry of Economic Planning and Budget.

Lelieveld, J., Evans, J. S., Fnais, M., Giannadaki, D., & Pozzer, A. (2015). The contribution of outdoor air pollution sources to premature mortality on a global scale. *Nature, 525*, 367–371. https://doi.org/10.1038/nature15371

Litman, T. (2015a). *Evaluating complete streets. The value of designing roads for diverse modes, users and activities*. Victoria Transport Policy Institute.

Litman, T., (2015b). *Transportation demand—Evaluating the amount and type of travel people would choose under specific conditions*. Victoria Transportation Policy Institute. https://www.vtpi.org/tdm/tdm132.htm.

Maina, M. (2015). Densification trends and typologies in the city of Johannesburg. In I. A. Todes, D. Weakley, & P. Harrison (Eds.), *Resilient densification: Four studies from Johannesburg* (pp. 14–37). University of the Witwatersrand, Gauteng City-Region Observatory.

Mindell, J. S., Anciaes, P. R., Dhanani, A., Stockton, J., Jones, P., Haklay, M., Groce, N., Scholes, S., & Vaughan, L. (2017) Using triangulation to assess a suite of tools to measure community severance. *Journal of Transport Geography, 60*, 119–129. ISSN 0966-6923. http://www.sciencedirect.com/science/article/pii/S0966692316305026. https://doi.org/10.1016/j.jtrangeo.2017.02.013.

Morimoto, A. (2009). Transportation and land use. *IATSS Review*, pp. 50–62.

Mtizi, C. (2017). *Southern African solutions to public transport challenges*. In 36th Southern African Transport Conference (SATC 2017) (pp. 73–76). Southern African transport conference.

Newman, P. (2005). *Urban design and transport in search of sustainability* (Eds. J. Goldie, B. Douglas, & B. Furnass, pp. 234–267). CSIRO Publishing.

Newman, P., & Kenworthy, J. (1999). *Sustainability and cities: Overcoming automobile dependence*. Island Press.

Nieuwenhuijsen, M. J. (2016). Urban and transport planning, environmental exposures and health—New concepts, methods and tools to improve health in cities. *Environmental Health Journal, 15*, S38.

Obanya, H. E., Amaeze, N. H., Togunde, O., & Otitoloju, A. A. (2018). Air pollution monitoring around residential and transportation sector locations in Lagos mainland. *Journal of Health & Pollution, 8*(19).

Oshodi, L. (2016, August 12). *International development, urban infrastructure and governance*. https://oshlookman.wordpress.com/2016/08/12/transportation-and-mobility-system-in-lagos/

Ostro, B. D., Roth, L. A., Green, R. S., & Basu, R. (2009). Estimating the mortality effect of the July 2006 California heat wave. *Environmental Research, 109*, 614–619.

Panter, J., Heinen, E., Mackett, R., & Ogilvie, D. (2016). Impact of new transport infrastructure on walking, cycling, and physical activity. *American Journal of Preventive Medicine, 50*, e45–e53.

Rafiemanzelat, R., Emadi, M. I., & Kamali, A. J. (2017). City sustainability: The influence of walkability on built environments. In *Transportation research procedia* (Vol. 24, pp. 97–104). Elsevier. https://doi.org/10.1016/j.trpro.2017.05.074

Renne, J., Voorhees, A., Bloustein, E., & Jenks, C. (2005). *Transit-Oriented Development: Developing a strategy to measure success*.

Renne, J. L., Hamidi, S., & Ewing, R. (2016). Transit commuting, the network accessibility effect, and the built environment in station areas across the United States. *Research in Transportation in Economics, 60*, 35–43. http://www.sciencedirect.com/science/article/pii/S0739885917300240

SACN (South African Cities Network). (2017, December 2). *Towards an alternative financing model for metropolitan cities in South Africa*. Johannessburg, Gauteng, South Africa. http://www.sacities.net/wp-content/uploads/2017/02/201702_Towards_an_alternative_Metropolitan_Municiple_Finance_Model_Report_FINAL.pdf

Salau, T. (2015). Public transportation in metropolitan Lagos, Nigeria: Analysis of public transport users' socioeconomic characteristics. *Urban, Planning and Transport Research, 3*(1), 1–8. https://doi.org/10.1080/21650020.2015.1124247

Salau, G. (2018, July 1). A decade after, Lagos Rail Project still crawling. *The Guardian*, p. Cover Page.

Statistics South Africa. (2014a). *National Household Travel Survey 2014, 2016*. Statistics South Africa.

Statistics South Africa. (2014b). *National Household Travel Survey, February to March, 2013, Statistical Release P0320*. Statistics South Africa.

Todes, A., Weakley, D., & Harrison, P. (2017). *Journal of Housing and the Built Environment, 33*(2), 281–299. https://doi.org/10.1007/s10901-017-9561-6

United Nations Human Settlements Programme. (2018). *Johannesburg city context report*. Global Future Cities Programme.

United State Environmental Protection Agency-USEPA. (2008). *EPA's report on the environment*. United State Environmental Protection Agency—USEPA.

Watts, N., Amann, M., Ayeb-Karlsson, S., Belesova, K., Bouley, T., Boykoff, M., Byass, P., Cai, W., Campbell-Lendrum, D., Chambers, J., Cox, P. M., Daly, M., Dasandi, N., Davies, M., Depledge, M., Depoux, A., Dominguez-Salas, P., Drummond, P., Ekins, P., Flahault, A., Frumkin, H., Georgeson, L., Ghanei, M., Grace, D., Graham, H., Grojsman, R., Haines, A., Hamilton, I., Hartinger, S., Johnson, A., Kelman, I., Kiesewetter, G., Kniveton, D., Liang, L., Lott, M., Lowe, R., Mace, G., Odhiambo Sewe, M., Maslin, M., Mikhaylov, S., Milner, J., Latifi, A. M., Moradi-Lakeh, M., Morrissey, K., Murray, K., Neville, T., Nilsson, M., Oreszczyn, T., Owfi, F., Pencheon, D., Pye, S., Rabbaniha, M., Robinson, E., Rocklöv, J., Schütte, S., Shumake-Guillemot, J., Steinbach, R., Tabatabaei, M., Wheeler, N., Wilkinson, P., Gong, P., Montgomery, H., & Costello, A. (2018). Countdown on health and climate change: From 25 years of inaction to a global transformation for public health. *The Lancet, 391*, 581–630. https://doi.org/10.1016/S0140-6736(17)32464-9

World Bank. (1991). *South Africa urban sector reconnaissance*. World Bank Mission Aide Memoire. World Bank.

World Bank. (2017). *Tookkit on fare collection system for urban passenger transport*. World Bank. https://www.ssatp.org/sites/ssatp/files/publications/Toolkits/fares%20 Toolkits%20content/case-studies/johannesburg-south-africa.html

World Health Organisation. (2018a). *Environmental noise guidelines for the European region*. http://www.euro.who.int/en/publications/abstracts/environmental-noise-guidelines-for-the-european-region-2018

World Health Organisation. (2018b). *Global status report on road safety 2018*. https://www.who.int/violence_injury_prevention/road_safety_status/2018/en/

World Health Organisation. (2020). *World Health Organisation. Road traffic injuries*. https://www.who.int/news-room/fact-sheets/detail/road-traffic-injuries

Ying, Z., Ning, L. D., & Xin, L. (2015). Relationship between built environment, physical activity, adiposity, and health in adults aged 46–80 in Shanghai, China. *Journal of Physical Activity and Health, 12*, 569–578.

Zhang, Y., Song, R., van Nes, R., He, S., & Yin, W. (2019). Identifying urban structure based on transit-oriented development. *Sustainability, 11*, 7241. https://doi.org/10.3390/su11247241

Zijlema, W. L., Avila-Palencia, I., Triguero-Mas, M., Gidlow, C., Maas, J., Kruize, H., Andrusaityte, S., Grazuleviciene, R., & Nieuwenhuijsen, M. J. (2018). Active commuting through natural environments is associated with better mental health: Results from the PHENOTYPE project. *Environment International*, 721–727.

Zuo, F., Li, Y., Johnson, S., Johnson, J., Varughese, S., Copes, R., Lui, F., Jiang Wu, H., Hou, R., & Chen, H. (2014). Temporal and spatial variability of traffic-related noise in the city of Toronto, Canada. *Science of the Total Environment, 472*, 1100–1107.

Chapter 5
Transit-Oriented Development and Urban Infrastructure in African Cities

Iniobong Beauty John, Olayinka Oluyemi Agunloye, Michael Adeloye Adebamowo, and Adedoyin Babalola

5.1 Introduction

Urban infrastructure is critical to the functionality of African cities and requires deep theoretical and cultural insights for sustainable developments. Calthorpe (1993) used the term transit-oriented development (TOD) for the first time to address the urban sprawl caused by the usage of too many cars and the disappearance of community centers in America. TOD principles present the theoretical foundation for new urbanism and focuses on effectively integrating mixed-use communities and encouraging public transit use. Globally, TOD has been deployed to redefine urban infrastructure development. The dearth of urban infrastructure in African cites requires deploying structured models like TOD for planning and development.

The provision of access to safe, affordable, accessible, and sustainable transport systems for all as contained in SDG 11 target 2 will ensure inclusive, safe, resilient, and sustainable human settlements. However, sustainable transportation is mainstreamed across several SDGs and targets (SDGs 2, 3,6,12, and 13 to cover issues on food, security, energy, health, infrastructure, and economic growth), thus indicating its significance to the attainment of other SDGs. In the light of sustainable transportations' cross-cutting relevance, it is inevitable to understudy how African cities can effectively harness its transportation potentials for better integration of the economy UN (2015).

I. B. John (✉)
Department of Quantity Surveying, University of Lagos, Lagos, Nigeria
e-mail: ijohn@unilag.edu.ng

O. O. Agunloye
Department of Urban and Regional Planning, University of Lagos, Lagos, Nigeria

M. A. Adebamowo · A. Babalola
Department of Architecture, University of Lagos, Lagos, Nigeria

T. Nubi et al. (eds.), *Transit Oriented Development in West African Cities*, Urban Perspectives from the Global South,
https://doi.org/10.1007/978-3-031-58726-9_5

Lagos, the most populous metropolitan city in Sub-Saharan Africa, has over 50% of Nigeria's industrial and commercial establishments, and 70% of the country's manufacturing activities is domiciled in Lagos. Abuja became the Federal Capital Territory in 1991, but Lagos was the capital of Nigeria in 1914 and would remain the commercial capital of Nigeria. Although Lagos is Nigeria's smallest state with 0.4% of the nation's total land mass (approximately 357,700 hectares), 22% of its total area are lagoons and creeks (180-km-long coastline). Lagos is bounded on the south by the Atlantic Ocean, in the west by Republic of Benin, and on the north and east by the Ogun State of Nigeria (Oteri & Ayeni, 2016).

Lagos Metropolitan Area constitutes 37% of Lagos State's total land area but houses about 85% of the state's total population as an urban agglomeration (Lagos State Information, 2015; Olukoju, 2015). The two ports operating in Lagos collectively handle about 75% of Nigeria's import by weight and about 90% of non-oil exports by weight. Lagos international airport handles about 80% of airborne exports and imports and about 80% of the passenger movements in and out of Nigeria. The ability to deploy various transport systems presents important possibilities for exploring intermodal transportation in Lagos.

Intermodal transportation is key to providing sustainable urban mobility, improving the quality of life for residents and providing adequate enabling environment for businesses in Lagos. Lagos transportation system deploys intermodal transport facilities, but unfortunately, the intermodal operation cannot be felt when compared to best global transport practice and standards. This could be as a result of "majoring-in-minor" in some salient transport issues, which can be consulted for. It could also be a result of the lack of state transport policy, which could have shown a construct direction for the future of transportation in Lagos, Nigeria (Agunloye, 2020). One of the ways to resolve the Lagos transport issues is intermodalism, with benefits such as low costs, low carbon footprints, and standardized transit schedules particularly for freighters who are able to streamline their reverse logistics.

5.2 Correlates of Transportation Challenges and Urban Infrastructure

The Ministry of Transportation is the primary organ for managing the transportation function sector in Lagos. Public transportation functions were initially performed by the Transportation Unit under the planning division of the old Ministry of Works and Planning. The history of trying to solve the transport problems in Lagos was not good. In the 1980s, the Lagos State Ministry of Works and Transport (LSMWT) was responsible for managing the local transport sector. The LSMWT established a publicly owned passenger transport operator, the Lagos State Transport Corporation, the Lagos State Ferry Services Corporation, the Lagos State Parking Authority, and the Central Licensing Authority. These transport establishments were expected to focus on different modes of transportation and different transport challenges. In

1994, the LSMWT was split into the Lagos State Ministry of Works mainly responsible for the state road network and the Lagos State Ministry of Public Transportation responsible for overall policy development and planning and later renamed the Lagos State Ministry of Transport.

The fragmented Lagos bureaucracy proved wholly incapable of responding to the escalating pressures on the transport network. A string of one-off initiatives that were carried out unsystematically and ignoring critical interfaces failed to implement any long-term sustainable solutions. Poor institutional arrangements, political pressure, and the lack of financial resources also meant that many transport plans never got off the ground.

In 2002, the World Bank and Lagos State Government jointly established the Lagos Metropolitan Area Transport Authority (LAMATA) to provide effective implementation of policies and provision of sustainable infrastructure in the transportation sector. The proliferation of slums, increasing urban population, and the degradation of urban facilities have intensified several problems affecting all transportation modes in Lagos. Further, a poorly structured public transportation system characterized by a highly atomized sector, a high number of nonregulated routes (which tend to concentrate along the main corridors) and an equally high number of public transport services using small vehicles (Danfo) lead to inefficient public transport service. This situation has led to a state of growing traffic congestion, which represents a major challenge of transportation in Lagos. The chaotic transportation system supports the use of private vehicles as the best transport option in Lagos. The use of private cars further impacts on congestion, thereby making it a recurrent problem.

5.2.1 Transportation Challenges (Fig. 5.1)

The relationship between transportation challenges and urban infrastructure can be grouped into direct urban service, effects of intermodalism, and challenges affecting intermodal transport. The direct urban intermodal service challenges are a consequence of the service that is being provided. These include traffic gridlock, inadequate capacity, higher user cost, high facility cost, and lack of safety for users, among others.

Challenges of effects of intermodalism entails air pollution, noise pollution, visual intrusion, excessive right of way, and inordinate change in land value, among others. Challenges affecting intermodal transport include increased automobile ownership, peakness in traffic, and concentration of industry in an area, among others (Agunloye, 2020; Agunloye et. al, 2018). In conclusion, the relationship between transportation challenges and urban infrastructure in Lagos is encased in the absence of an urban policy, inadequate urban planning, weak intermodal coordination, insufficient urban public transport to cope with the ever-increasing demand for movement, urban traffic congestion, security, and environmental friendliness (Sumaila, 2013).

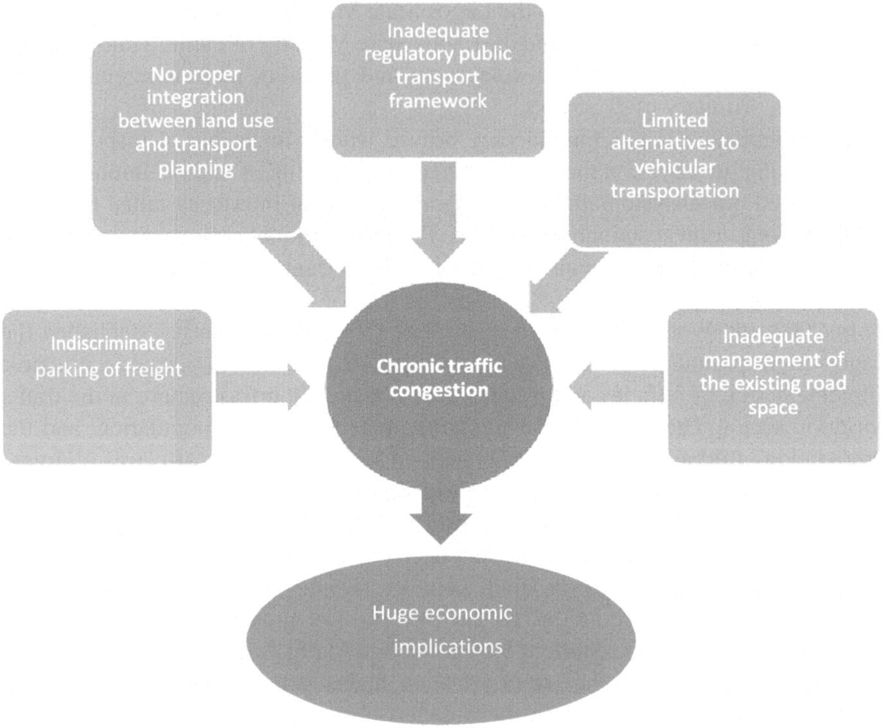

Fig. 5.1 Infrastructural challenges of transportation and economic implications

5.2.2 *Sustainable Road Infrastructure*

Sustainable urban roads are inevitable for effective road usage. Key processes needed to develop sustainable road infrastructure include (Table 5.1):

- Having sustainable road design and construction manuals that outline sustainable practices while ensuring all design team members are on board and informed about the sustainable goals and objectives of the project.
- Considering whole life cycle costs of roads will enhance sustainability.
- Nurturing innovation in the design team by doing something better than how it was done before.
- Improved landscaping and rehabilitation of natural habitats to enhance the natural value of the project area.
- Enhancing storm water management for effective management of runoff.

Transportation infrastructure directly and indirectly affects the overall competitiveness of the economy. Nigeria's transportation infrastructure was ranked 130 out of 141 countries in the Global Competitiveness Report of 2019. The report presents Nigeria's transportation infrastructure scorecard and indicates that there is a need to

Table 5.1 Nigeria's transport infrastructure index component for economic profiles of Global Competitiveness Index 4.02019 ranking

Index Component		Value	Score	Rank/141
Second pillar infrastructure 0–100		–	39.7	130
Transport infrastructure 0–100		–	31.6	130
2.01	Road connectivity	77.5	77.5	63
2.02	Quality of road infrastructure	2.5	22.5	130
2.03	Railroad density	4.2	10.4	73
2.04	Efficiency of train services	1.8	13.2	98
2.05	Airport connectivity	26,395.1	43.2	78
2.06	Efficiency of air transport services	3.4	39.6	125
2.07	Liner shipping connectivity	19.0	19.0	68
2.08	Efficiency of seaport services	2.5	24.7	122

Source: Schwab (2019) The global competitiveness report, p. 431

invest in transportation infrastructure to enhance the overall competitiveness of the nation.

5.3 Conceptual Linkage of Transit-Oriented Development (TOD), Sustainable Development Goals, and Urban Infrastructure

Transport drives economic progress while linking places and people for prosperity. Innovative support of transit-oriented development and sustainable urban infrastructure is key to achieving the Sustainable Development Goals. Lagos has constantly attempted to build innovative and sustainable infrastructure that will meet the transport needs of the teeming population. Intermodal transportation is key to providing sustainable urban mobility, improving the quality of life for residents, and providing enabling environment for businesses in Lagos. Tackling these challenges requires a clearcut understanding of how each infrastructure is categorized: soft or hard infrastructure.

Soft infrastructures are institutional capabilities and policies that can address emerging challenges. These are human capital that are capable of providing solutions through programs and policies/plans that will yield the desired results. Soft infrastructure challenges are addressed through the initiatives of development of the strategic transport master plan (STMP), development of integrated intermodal transport guidelines, nonmotorized transport (NMT) policy, freight transport master plan, and STMP marketing document. Hard infrastructure refers to "brick and mortar" infrastructure such as bridges, terminals, roads, railways, tunnels, and a host of others. Hard infrastructure challenges in Lagos State are addressed through construction of BRT, construction of LRT, construction of bus terminals, e-ticketing/ ITS, and water transport initiative, among others. Sustainable transport

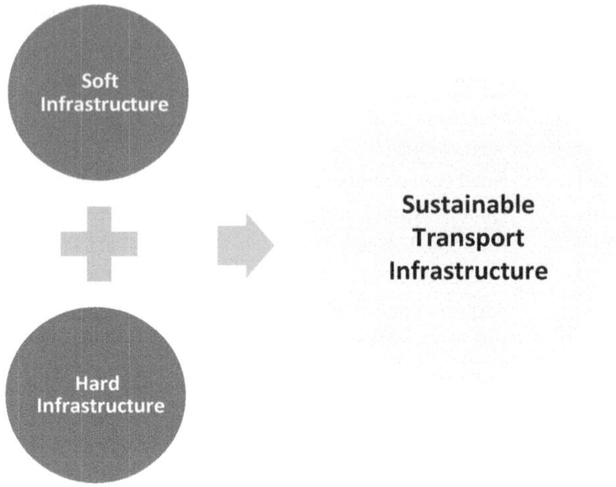

Fig. 5.2 Sustainable transport infrastructure model

infrastructure is dependent on a combination of soft and hard infrastructure. Sustainable policy framework alongside the construction of roads, bridges, and rail tracks, among others, will support sustainable transport infrastructure as indicated in the conceptual model in Fig. 5.2.

The Ministry of Transportation (MoT) is strategic to providing soft infrastructure to the transport sector in Lagos. MoT encourages private sector participation in the provision of transport infrastructure. It further promotes transparent access to subsidiaries for the provision of transport services for social and economic benefit.

MoT improves coordination in intermodal transport while also increasing access to transport services in rural areas. In synchronizing soft infrastructure in the transport sector, MoT is saddled with the integration of transport policies with planning appraisals and implementation of transport policies across various transport modes such as development of the strategic transport master plan (STMP), development of integrated intermodal transport guidelines, development of nonmotorized transport (NMT) policy, provision of freight transport master plan, and provision of STMP marketing documents.

Infrastructure provision falls within the purview of the Lagos State Ministry of Works and Infrastructure (MOWI). MOWI is strategic to the development of first-class infrastructure to support the state's growing population.

This is in addition to serving as the critical development driver of Lagos State's vision of being Africa's model smart city and a global economic/financial hub. Some of the hard infrastructure currently provided for in the Lagos transport sector includes the construction of BRT lane, construction of LRT, construction of bus terminals, e-ticketing, and provision of water transport initiative. These initiatives are intentional and are driven in coordinated manner by resource personnel within Lagos State agencies of transportation. These agencies provide the required human capital needed for the continuous development of public transportation in Lagos.

Sustainability plans are built into all projects as this is necessary for continuous maintenance, functionality, and upgrade of projects.

Major projects funded from budgetary provisions in Lagos between May 2019 and May 2020 include the following:

- Rehabilitation of 116 inner roads (completed).
- Completion of Agboju-Trade Fair section, Lagos-Badagry Expressway (ongoing).
- Rehabilitation and upgrading of Ijede Road (ongoing).
- Reconstruction and upgrading of Igbogbo-Bola Ahmed Tinubu-Igbe Road, Ikorodu (ongoing).
- Construction of Agric-Access Road, Phase II, Igando (ongoing).
- Construction of Agric-Isawo Road, Ikorodu (ongoing).
- Rehabilitation of Oba Sekumade Road, Ipakodo, Ikorodu (ongoing).
- Ijegun-Egbe Road and Isuti Road, Igando (ongoing).
- Construction of 31 network of roads in Ojokoro LCDA (completed).
- Construction of Pen Cinema Bridge nearing completion (completed).

Aside from several rehabilitation works undertaken by the Public Works Corporation, the provision of hard infrastructure in Lagos which is majorly sourced from budgetary provisions and bonds has huge financial implications for the state. Other options used in funding the provision of hard infrastructure in Lagos include the private sector partnership (PSP)/private infrastructure improvement partnership (PIIP), and CBN intervention using dedicated deposits at single-digit financing rate with cash reserve ratios (CRR).

5.4 Case Studies from Foreign Settlement Case Studies of North America, China, and Australia

North America

After experiencing car-oriented development, the developed cities in North America are facing serious suburbanization, functional decline of old city centers, massive private cars, low land utilization, and dispersed urban spatial layout. The spread of private cars has made the residents in North America adapt to car travel and single detached homes, and to a large extent, they would not prefer public transportation and high-density accommodations. Based on this situation, the implementation of TOD is affected by the adjustment of market mechanisms and may confront enormous risk and uncertainty.

Beijing, China (Fig. 5.3)

Beijing is densely populated with widely distributed job opportunities. The TOD index for Beijing is centralized in the core areas of buildings, business centers, and the main modes of transportation (Zhang et al., 2019). However, rail nodes and

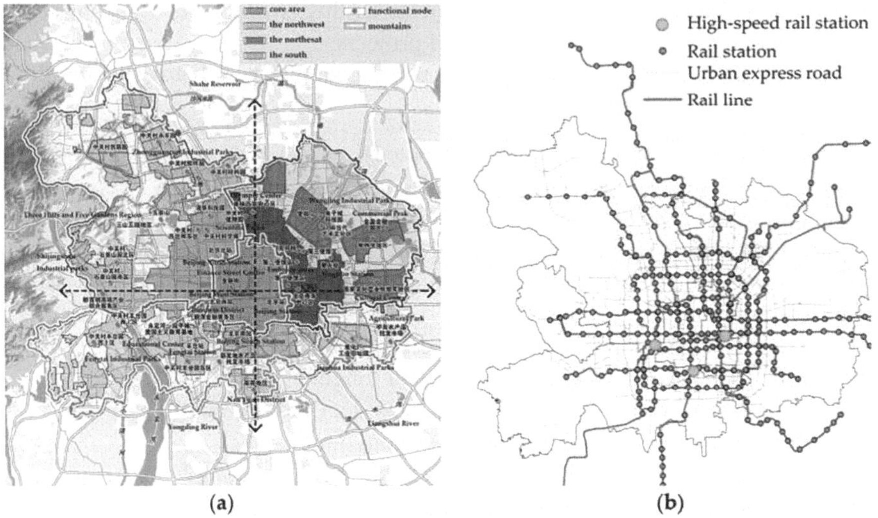

Fig. 5.3 (**a**) The master plan and functional zone for Beijing central districts (2016–2035); (**b**) the Beijing rail network (2019). (Source: Zhang et al. (2019), p. 10)

business centers have the best performance for TOD index. Nonetheless, there is a deficiency in public transportation network due to the large number of commuters resulting in the inadequate transportation facilities around residential and working areas. In order to reduce the tension in the Beijing city center, there is a need for a diversified TOD community.

Stellenbosch, South Africa (Fig. 5.4)

South African urban space houses more than 60% of its populace despite the challenge of poverty, lack of basic services, inadequate supply of suitable land, and poor access to economic opportunities (MoCoGTA, 2014). Stellenbosch is located in South Africa's Western Cape Province and is surrounded by affluent towns, a central business district, informal settlements sited north of the town, and the Du Toit Station. The station presents a TOD site capable of enhancing public transport nodes, although it was built to serve poor African communities (Christopher, 1992). However, the main Stellenbosch train station was built to serve the Dutch and Victorian Architecture of Stellenbosch (Rock, 2011). TOD was deployed to enhance urban development using a multistakeholder approach.

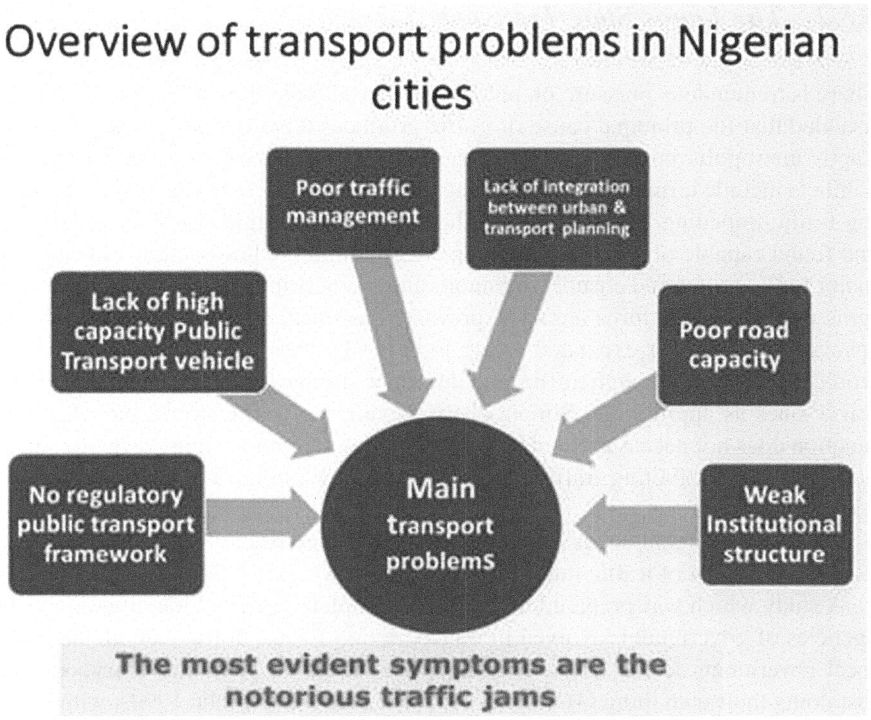

Fig. 5.4 Map of Stellenbosch showing suburbs and location of train stations. (Source: Mafame (2017) p. 40)

5.5 Thematic-Based Issues and Priorities for Lagos

5.5.1 Overview of Transport Problems in Nigerian Cities

The lack of planning has led to the proliferation of slums, degradation of urban areas and urban infrastructure, and several transportation problems affecting all modes. This situation has led to a state of growing congestion, which represents a major challenge for the megacity. A poorly structured public transportation system does not help in easing the current congestion problem: a highly atomized sector, a high number of nonregulated routes (which tend to concentrate along the main corridors), and an equally high number of public transport services using small vehicles (Danfo) leading to inefficient public transport service and modal change from public to private transportation. Logically, this situation generates additional congestion and a degradation of public transportation in Lagos. The result is a chaotic transportation system where the use of private vehicles becomes the best option and congestion becomes a recurrent problem. Several transportation problems exist in Nigerian cities.

5.5.2 *The Lagos State Transport Sector*

There is tremendous pressure on public transportation in the city of Lagos. Studies revealed that the principal cause of traffic gridlocks at various locations across the Lagos metropolis is the presence of conflicts at major road intersections. These conflicts include turning traffic impeding each other, particularly left turns and turning traffic impeding through traffic. The following strategies have been deployed and found capable of reducing the intensity of conflict at intersections: identify the major traffic routes and ensure continuous and uninterrupted traffic flow, allow right turns only while left turns are to be prevented as much as practicable, introduce a specialized U-turn or expanded return loop (ERL) with separate travel lanes that would not obstruct through traffic, and introduce storage (deceleration and merging) travel lanes as appropriate. Simple overpass carrying traffic over a busy highway junction does not necessarily solve the problem of the intersection. Very often there is unresolved conflicting traffic under the bridge resulting in gridlock (Olusanya, 2020) (Fig. 5.5).

An analysis of challenges posed by urban traffic intersections is a necessary prelude to the design of traffic interventions.

A study which was prepared in 1989 and completed in 1992 identified over 100 agencies of government involved in transportation issues, at the federal, state, and local government levels. There was no coordination, no planning; everybody was just doing their own thing (*Managing Director LAMATA*). In the 1990s, with Lagos

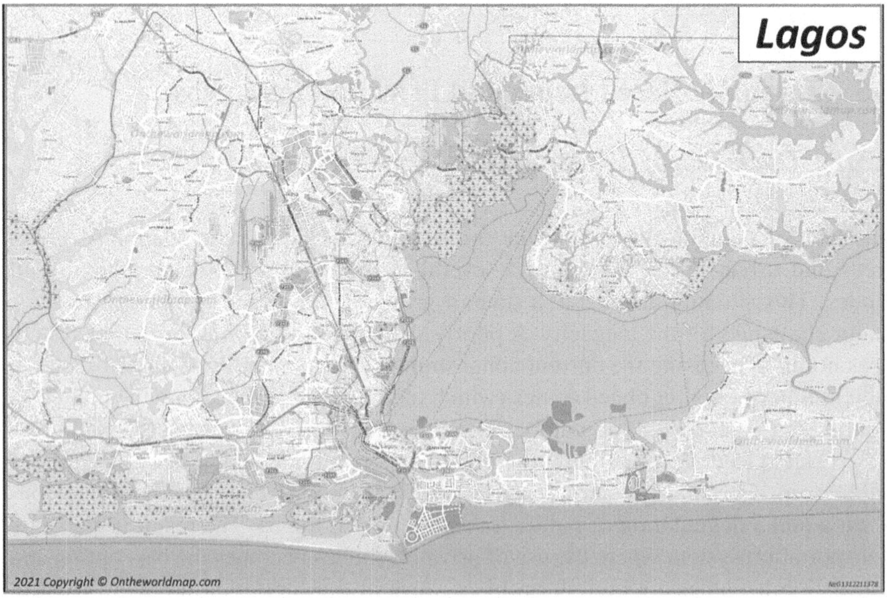

Fig. 5.5 Lagos map. (Source: lagos-area-map.jpg (2514 × 1693) (ontheworldmap.com) accessed on February 2, 2023)

facing a real threat of the transport system deteriorating to intolerable levels, the governor of the time, Bola Tinubu, appointed a special adviser on transportation issues and sought help from the World Bank. The Lagos Mass Transit and Transport Systems Programme Study was then developed, and the central recommendation was the creation of the Lagos Metropolitan Area Transport Authority (LAMATA).

5.5.3 LAMATA

The 1992 study conducted by LAMATA showed that metropolitan Lagos saw about 7.6 million trips daily with 80% which took place on road public transport despite the fact that Lagos metropolitan has one of the lowest road provisions in the West African subregion. The total passenger traffic per day is over seven million. It later became necessary to define what the ideal transport system should be and the strategic transport master plan (STMP) was developed to serve as a guide for all existing and proposed development. A presentation by Olusanya (2020) indicated that the principal cause of congestion on Lagos roads is the presence of conflicting traffic at major highway intersections throughout the metropolis.

Traffic lights are inadequate for management at intersections beyond certain traffic volumes. Simple grade separation at intersections requires a level of infrastructure that is generally unaffordable for widespread use in developing economies and does not always resolve on-grade traffic below. Olusanya (2020) posits that cloverleaf interchange is highly efficient but requires expensive and extensive infrastructure and land use which is not always available in built-up urban locations.

5.5.4 Lagos State Government Actions Toward Actualizing the SDGs in the Transportation Sector

Lagos, with a population of over 18 million people and 7 million passengers per day has over 50% of Nigeria's transport problems. Road congestion, air pollution from transport activities, limited transportation facilities, and greenhouse gas emissions constitute the major transport problems in Lagos. Sustainable transportation enhances economic growth, promotes trade opportunities, and improves access to basic facilities. The United Nations defines sustainable transport as a transport means that better integrates the economy while respecting the environment (ThistlePraxis, 2016). Nonmotorized transport (NMT) occupies an important position in Lagos transport system (UN Environment Programme, 2019). However, it is not properly planned to accommodate pedestrians' walkways, bicycle lanes, zebra crossings, footbridges, underpasses, and signs.

The UN Environment Programme (2019) identified five targets of the SDG framework that are directly related to the transport sector:

- Target 3.6. Halve the number of global deaths and injuries from road traffic accidents.
- Target 7.3. Double the global rate of improvement in energy efficiency.
- Target 9.1. Develop quality, reliable, sustainable, and resilient infrastructure, including regional and trans-border infrastructure, to support economic development and human well-being, with a focus on affordable and equitable access for all.
- Target 11.2. Provide access to safe, affordable, accessible, and sustainable transport systems for all, improving road safety, notably by expanding public transport, with special attention on the needs of those in vulnerable situations, women, children, persons with disabilities, and older persons.
- Target 12.c. Rationalize inefficient fossil fuel subsidies that encourage wasteful consumption by removing market distortions, in accordance with national circumstances, including by restructuring taxation and phasing out those harmful subsidies, where they exist, to reflect their environmental impacts, taking fully into account the specific needs and conditions of developing countries and minimizing the possible adverse impacts on their development in a manner that protects the poor and the affected communities.

Although transport is also a critical enabler of achievement in other SDG targets besides the aforementioned, it is critical to the economic, social, and environmental transformation of Lagos into a megacity. LAMATA developed a strategic transport master plan for metropolitan Lagos in 2009. The plan aimed to move Lagos beyond its current challenges of traffic congestion and a chaotic public transport system. It identified possible transport facilities and services to manage the growing travel demand in Lagos by 2020. However, the 2009 STMP did not consider the real mobility dynamics of metropolitan Lagos, with a significant number of trips coming from the furthest areas in Lagos State and some parts of Ogun State. Lagos should be understood as a megacity region covering the entirety of Lagos State and most immediate areas of Ogun State and should be analyzed in a unitary way, aiming at full accessibility and interconnectivity within its whole territory.

The list of proposals for road (private and public) transportation includes the following projects:

- Thirteen public transportation projects already envisaged in the current STMP (BRT and LRT lines).
- Two new public transportation projects currently envisaged by Lagos State in Lagos Island and Victoria Island.
- Twenty-two new projects for private and public transportation, which include extensions of the original STMP projects and new corridors.
- A whole new system of inland waterway routes.
- A system of 14 interchanges, to be established in order to guarantee the correct operation and efficiency of the transport system. The mainland central area will play a key role in this regard: four of the proposed interchanges will be located

within this area, and they will play a fundamental role in the organization of the MRT routes of the megacity.

The development of Lagos transportation system has been phased in different stages, according to the time horizons agreed with the State: 2017, 2022, and 2032. The distribution of the different projects integrating the proposal has been guided by two main criteria:

(a) The urban development pace established in the harmonized vision for the megacity.
(b) The transportation projects will have to serve the urban developments to which they are strongly associated.

The results of the model assignments are as follows: the passenger load results have determined the final choice of the time horizon for each project.

Specific Plans Associated with the Extension of the STMP
In order to guarantee the success of the megacity's road network and public transport proposal, complementary plans must be defined in the domains of freight, nonmotorized transport (NMT), road safety, and climate change. The following highlights the main outputs of these plans.

Freight Plan
New key elements in the megacity's freight system present new challenges for Lagos' freight system, as they will bring additional freight flows and more capacity and connectivity:

- New seaports and free trade zones at Badagry and Lekki.
- New airport (to specialize in agro-cargo) at Shagamu.
- New international airport at Lekki.
- Future inland container terminal at Ibadan.

These new elements also present new opportunities for the organization of freight flows in the megacity, based on a system of freight corridors and distribution centers.

Nonmotorized Transport Plan
Urban areas of African cities and Lagos, particularly, can improve working and living conditions through NMT plans, not only because it presents an environmentally friendly and socially inclusive mode of transportation but also because it presents a fundamental part of the urban transportation chain, particularly for access to public transportation modes and modal interchange. The NMT plan for Lagos should respond to these challenges through several measures and actions, which can be grouped in the following categories:

First level Facilitation of pedestrian movements and community integration within neighborhoods
Second level Nonmotorized access to public transportation corridors
Third level Specific nonmotorized access plans to transportation hubs/modal interchanges

Road Safety Plan

The vision of a whole new transport system for the megacity requires the establishment of an adequate traffic and road safety plan considering all users of transport modes, especially nonmotorized transportation, through the creation of a pedestrian- and bike-friendly urban environment, as well as by adequate law enforcement and efficient health care systems. Additionally, the road safety plan for Lagos will have to be supported by the knowledge of road traffic accidents in the city through an adequate data collection and analysis system.

In order to achieve these objectives, three main domains of activity are identified.

Definition of Road Safety Baseline for Lagos

This includes the definition (and calculation) of road safety indicators and procedures for data collection and analysis. Geographical information system (GIS) represents a key tool in this regard (identifying spatial distribution of road accidents, managing related information such as type of incidence, type of vehicle and casualties, etc.).

Establishment of Road Safety Policies and an Implementation Strategy

This refers to the establishment of specific targets regarding the road safety indicators, for the mid and long term, a road safety policy direction, and subsequent implementation strategy.

Establishment of a Monitoring System

Through realistic policy indicators, a monitoring system will allow data gathering on the level of achievement of the aforementioned policies.

Climate Change Plan

With a significant part of the city built in the surroundings of a large water body like Lagos Lagoon, Lagos is particularly vulnerable to the impacts of climate change, particularly flooding caused by the rise of the sea level, as it might imply a migration of population from the coast to inland areas. Saltwater intrusion into freshwater sources is also a possible consequence from climate change that Lagos might have to tackle. Therefore, concerted and urgent action is needed in order to reduce future effects of climate change. Vehicular emissions represent a major source of pollutants and greenhouse emissions in Lagos. This is a result of constant growth in transport activity (direct consequence of economic growth) not accompanied by appropriate planning. Continuous population growth and urban sprawl contribute to bigger commuting distances. This circumstance, together with the high dependence on fossil fuels and private transport, as well as a high percentage of old vehicles, has led to a remarkable growth of pollution levels, serious health effects, and contribution to climate change. The introduction of multimodal transport represents the contribution of the STMP Extension project to climate change mitigation and its change plan which is based on the establishment of mass transport systems comprising a combination of bus rapid transit (BRT) and rail and water transport services, with larger passenger capacity and lower emission rate than those of private cars.

The objectives of the STMP Extension Climate Change Plan can be summarized in three different stages.

Reduction of GHG emissions by 45% in the 2030 horizon, in accordance with the 2009 STMP.

The Climate Change Plan includes an analysis of the GHG emissions reduction in each one of the implementation stages of the road and public transportation proposal, comparing the base scenario with the outcomes obtained in the different horizon years of the STMP Extension (2017, 2022, 2032).

Further reductions for the period between 2022 and 2052 (equivalent to 30%) due to the progress in technologies, development of massive public transport systems, happening in urban contexts that favor concentration of the population as opposed to the urban sprawl of the second half of the twentieth century.

5.6 Recommendations for the Government, Investors, and Communities

Based on the findings of this research, systematic planning and adoption of TOD will enhance the functionality of and upgrade the infrastructural projects in Lagos. The provision of mass, large-capacity buses, bus rapid transit, and light rail will enhance intermodal transport in Lagos. Policy frameworks that can ensure the deployment of sustainable transport systems across various modes to cater for the needs of Lagos should be adopted.

5.7 Conclusion

This scoping study established that sustainable intermodal transport infrastructure is inevitable for Lagos. TOD within the best proximity to transport conveyance points encourages development along heavily used bus and bus rapid transit corridors as well as development of a mixed use including housing development. TOD further encourages development, walkability, and pedestrian-friendly connections to the surrounding communities among others. TOD enhances quality of life, economic development, community character, environmental quality, and sustainable urban livelihood.

References

Agunloye, O. O. (2020). The future of road and rail transportation in Lagos. *Ekotecture Quarterly Magazine*. EQ Publications.

Agunloye, O. O., Ajakaiye, O. O., Akinola, A. O., Okagbue, H. I., & Afolabi, A. O. (2018). Datasets on factors influencing the urban environmental quality of intra- urban motor parks across density areas of Lagos metropolis. *Data in Brief, 19*, 2109–2118.

Calthorpe, P. (1993). *The next American metropolis: Ecology, community, and the American dream*. Princeton Architectural Press.
Christopher, A. J. (1992). Segregation levels in south African cities, 1911-1985. *The International Journal of African Historical Studies, 25*(3), 561–582.
Lagos State Information. (2015). National Bureau of Statistics.
Mafame, T. (2017). *Transit–Oriented Development (TOD) as a facilitator for urban development integration: Case study: Du Toit Train Station Precinct, Stellenbosch* [Thesis presented for Masters of Philosophy]. The Faculty of Economic and Management Sciences, Stellenbosch University.
MoCoGTA. (2014). *Towards an integrated urban development framework: A discussion document*. MoCGTA.
Olukoju, A. (2015). *The travails of migrant and wage labour in the Lagos Metropolitan Area in the inter-war years*. Liverpool University Press.
Olusanya, O. (2020). *Alternative design solutions for traffic gridlock in the Lagos megacity*. A presentation at the ARUA Webinar on Intermodal Transport Infrastructure for Inclusive, Safe, Resilient and Sustainable Human Settlement.
Oteri, A. O., & Ayeni, R. A. (2016). *Water megacities & global change*. The Lagos Megacity. http://eaumega.org/wp-content/uploads/2016/05/EN-Lagos-Monograph.pdf
Rock, D. (2011). *The Location shall be called Kaya Mandi: A history of Khayamandi. Stellenbosch* [Masters thesis]. University of Stellenbosch.
Schwab, K. (2019). *The global competitiveness report 2019*. World Economic Forum. pp. 1–666. http://www3.weforum.org/docs/WEF_TheGlobalCompetitivenessReport2019.pdf
Sumaila, A. F. (2013). Road crashes trends and safety management in Nigeria. *Journal of Geography and Regional Planning, 6*(3), 53–62. https://doi.org/10.5897/JGRP2013.0318
ThistlePraxis. (2016, September 1). *Sustainable transportation: Addressing transport challenges in Nigeria*. https://medium.com/@ThistlePraxis/sustainable-transportation-addressing-transport-challenges-in-nigeria-b8603191726d
UN Environment Programme. (2019). *Nigeria*. https://www.unenvironment.org/explore-topics/transport/what-we-do/share-road/nigeria
United Nations. (2015). *A/RES/69/213—Role of transport and transit corridors in ensuring international cooperation for sustainable development*. United Nations General Assembly.
Zhang, Y., Song, R., Nes, R., He, S., & Yin, W. (2019). Identifying urban structure based on transit-oriented development. *Sustainability, 11*, 7241. https://doi.org/10.3390/su11247241

Chapter 6
Transit-Oriented Development and Land in African Cities

M. M. Omirin, E. O. Thontteh, J. B. Olaleye, H. A. Koleoso, A. S. Afolayan, J. B. Babade, and O. B. A. Idowu

6.1 Introduction

This study was undertaken in line with the objective of the ARUA Centre of Excellence in Urbanisation and Habitable Cities to investigate, promote, and provide local knowledge and workable solutions to seemingly intractable problems impeding functionality of African cities in the area of urban infrastructure. Specifically, from the thematic area perspective, the scoping study sought to:

- Identify land related challenges of transit-oriented development (TOD) particularly as it relates to land acquisition for transportation infrastructure provision and expansion in Lagos and land use optimization around transportation hubs.
- Evaluate possibilities of adopting international best practices that can result into better and more proactive channeling of land resources into sustainable transportation infrastructure provision and maintenance within the context of TOD.
- Investigate the role of land-based taxation in revenue generation for TOD with a view to highlighting key challenges surrounding the issue.

Transit-oriented development (TOD) seeks to achieve functional integration of land uses and transits by creating compact, livable, mixed-use urban communities that cluster jobs, housing, services, and amenities within half a kilometer distance around public transport hubs (Carlton, 2009; Thuo, 2010; Bevilacqua et al., 2013). Apart from increasing choices in mobility and interconnectivity for citizens of all socioeconomic classes, TOD facilitates land use efficiency, optimization, and

M. M. Omirin · E. O. Thontteh (✉) · H. A. Koleoso · A. S. Afolayan · J. B. Babade
O. B. A. Idowu
Department of Estate Management, University of Lagos, Lagos, Nigeria
e-mail: ethontteh@unilag.edu.ng

J. B. Olaleye
Department of Surveying & Geo-informatics, University of Lagos, Lagos, Nigeria

T. Nubi et al. (eds.), *Transit Oriented Development in West African Cities*, Urban Perspectives from the Global South,
https://doi.org/10.1007/978-3-031-58726-9_6

compactness, thereby reducing reliance on personal car usage and the attendant carbon dioxide emission problems. Socially, it promotes integration and cohesion among citizens and enhances quality of life. Economically, it stimulates productive activities around the transport hub, thereby increasing property values and boosting revenue generation potentials of the real estate. Further, it increases housing choices while reducing living expenses for residents. It is a holistic approach that boosts public health, promotes livability, and optimizes land values.

While the benefits of TOD are already being realized in cities like Hong Kong, Singapore, and Cape Town, South Africa, the introduction and use of TOD is relatively new in Lagos where transportation problems are endemic. The inadequacy and poor quality of road transportation in particular is well documented (see, e.g., Atubi, 2013; Lagos Ministry of Transport et al., 2018; Edema, 2019). Many neighborhoods have developed without planning and lack proper roads and drainages. Most existing transport hubs developed informally and are not properly integrated with surrounding land uses in a structured manner that can promote livability. In fact, they are prone to causing serious time-wasting traffic congestion.

Further complicating the problem is the fact that as a result of exorbitant rents in core areas many citizens have moved to the urban fringe from where they commute long distances to their employment bases at great inconvenience and high risk to their health. Such commuters spend virtually 8 h daily stuck in traffic while moving from home to work and back, in uncomfortable privately operated minibuses with debilitating effect on their productivity.

The existing rail service has had little impact as a means of intra-city transportation, and mass transit projects like the light rail and monorail projects embarked upon over the years have proceeded in a slow and haphazard manner. Although the bus rapid transit (BRT) system is now fairly established and well patronized, it is yet to have full coverage in the city. These projects have focused principally on moving people, goods, and services and are not part of an overarching TOD policy that aims at integrating them with housing and commercial developments to form compact, livable, mixed-use urban communities clustered around the transport hubs. Perhaps the closest example is the Oshodi regeneration scheme which is about 70% completed now even though it was flagged off in 2016.

In realization of the need to address this anomalous situation and prevent further deterioration, the Lagos State government through its Ministry of Transport and the Lagos Metropolitan Transportation Agency (LAMATA) in conjunction with the United Nations Environment Programme drew up the nonmotorized transport (NMT) policy in 2018. The policy aims principally to discourage the use of personal vehicles in order to reduce congestion, loss of man-hours wasted through traffic jams, environmental pollution, loss of lives, and associated psychosocial stress. Implied TOD focus is discernible in the policy's emphasis of walking, cycling, and the use of public transport which will require compact communities with mixed land uses.

Although TOD has a multidimensional relevance and use in ensuring sustainable cities, the critical role of land policy, planning, and market-led articulations of uses must be clearly understood along with the peculiar challenges they pose if the benefits are to be fully realized and preserved. In the face of unprecedented urbanization rates and densification of existing settlements, the potentials of TOD and the land related issues become even more critical.

This report presents the results of enquiries into the existing state of affairs, the potential for successful land resource infusion into TOD, the attendant potential problems, and the key areas identified for further enquiry to generate useful research agenda after comparison with some best practices in other countries.

6.2 Transportation Infrastructure and the Urban Environment

In facilitating the movement of goods and people, transportation infrastructure not only promotes activities that generate economic productivity, employment, social integration, and cultural interchange; they also act as the gel that glues local and regional settlements with national and international markets. They link residential areas with work and trade bases and facilitate distribution of goods and services. Zubairu (2019) noted that transportation infrastructure is a major cost center which often forces governments to foster bilateral relations to be able to exploit international trade, foreign exchange earnings, and financial interdependencies. As a contributor to national GDP, it is also a major sustainer of livelihoods and critical element in national defense. On a microscale, transportation means and costs are important determinant of households' choices of residential location, rent, and improvements in quality of life. It is therefore a major factor in dynamic growth of informal settlements in marginal areas. For commercial and industrial land uses, transportation is a vital consideration in locational choices of operational bases, land acquisition, and pricing. Its importance for sustainability cannot therefore be overemphasized. Further, transportation infrastructure is a major land use competing with other uses for valuable land in core and fringe areas. It is an important tool for directing city expansion, preventing sprawl, ensuring orderly development, and opening up new areas for development.

The deficit in transportation provision has increased due to the lack of strategic vision under investment, neglect of maintenance, ad hoc approach to infrastructure development planning, failure to properly integrate socioeconomic requirements with land use and transportation planning, political instability, and to a large extent the lack of political will.

To put the issues in proper context, it is vital first to gain some insights into the peculiar nature of urban areas and their growth. As stated by Bobylev and Jefferson (2014) and Schrenk et al. (2011), urban infrastructure is facing increasing challenges globally in sustaining its services under pressures of urbanization and climate change factors. Africa is urbanizing more rapidly than any other part of the planet, and its 1.21 billion population is expected to double by 2050. More than

80% of this increase is expected to occur in cities, especially in the slums (Gutman & Patel, 2018; Fleming & Makalima-Ngewana, 2013).

Already, about 40% of the population in Sub-Saharan Africa live in urban areas. This proportion is predicted to double within the next 25 years, raising unprecedented challenges for the region particularly because the bulk of the increase is in low-income areas lacking infrastructure. As Muggah (2018) points out, the region already confronts critical deficits in infrastructure and related funding. The African Finance Corporation (2019) conjectures that it would require no less than USD 47 billion annually to cover the deficit in transportation infrastructure alone. This is against an overall estimate of USD 170 billion for meeting the deficit in all forms of infrastructure generally. After water, sanitation, and electricity, transportation infrastructure has been flagged up as a most critical area where government intervention is urgently needed to promote social welfare and economic development. In fact, it is asserted that poor road, rail, and harbor infrastructure add 30–40% to the costs of goods traded among African countries (Oyesola, 2021; Effiom & Ubi, 2016).

The situation in Lagos is made peculiar because the city accommodates over 65% of national commercial and industrial activities. Even though Lagos is only 0.4% of Nigeria in land area, the population is currently estimated at over 20 million which is more than 10% of the national population growing at 3.2% p.a. due more to rural/urban migration. In fact, the rate of growth in the metropolis has been put at 6% per annum.

LAMATA (2015) estimated that road infrastructure is 75% of available transport infrastructure moving 90% of commuters on daily basis. Private cars move 2.5 million (11%) commuters, while the bus rapid transit system moves only 90,000 commuters (0.42%). Private informal transporters move 9,982,000 (45%) albeit inefficiently.

6.3 The Role of Land in Transportation Infrastructure Development

Land is the most crucial prerequisite for ensuring sustainable transportation infrastructure development. While finance is needed to embark on each project, acquire the land, and develop the facility, land itself is usable as a source of revenue through which existing infrastructure can be maintained and new projects initiated. Apart from its role as a major input and the receptacle for huge capital investments involved, land defines the location, accessibility, and spatial linkages as well as the economic desirability of the settlements. On the other hand, a TOD situated in a neighborhood will exert substantial influence on the real estate market by increasing rental and sales values of surrounding property due to increased attractiveness, enhanced business potentials, convenience, and livability. However, access to land for TOD and infrastructure provision is fraught with problems of varying proportions. Except in countries where land is fully nationalized, acquisition of land in adequate quantity and at the right place and cost is a challenge due to:

- The lack or inadequacy of land records to confirm entitlement of compensation claims by displaced persons as a result of prevalent informality of land holdings and preponderance of unregistered land with no titles.
- Exorbitant land prices and high cost of compensation for acquired land.
- Difficulty of resettlement of affected persons.
- Inadequacy of compensation offered to displaced persons.
- Resistance by holders of customary interests in the land.

Other challenges include:

- Raising the requisite finance and ensuring cost recovery in order to facilitate replication of projects in other areas.
- Determining the optimal level of compactness and preventing overconcentration of activities in a manner that can degenerate into new levels of congestion.
- Resettling displaced persons or integrating existing users into new schemes using such paradigms as land readjustment.
- Strategically addressing the rapid pace of urbanization by embarking on balanced development of rural settlements.

Successful prosecution of TOD projects requires that these challenges be fully understood and dealt with before they constitute permanent impediments to project conceptualization, planning, and execution. Another key issue is the need to align TOD project with the United Nations Sustainable Development Goals—particularly Goal 11 as shown below.

6.4 TOD, Land Matters, and the SDGs: Some Conceptual Linkages

TOD is particularly relevant to the second target of the 11th United Nations Sustainable Development Goal, which requires that by the year 2030, all households should be provided with access to safe, affordable, sustainable transport systems and improved road safety by expanding public transport with equal attention to the needs of those in vulnerable positions particularly women, children, persons with disabilities, and older persons. This goal is clarified into targets with indicators that measure (Fig. 6.1):

- Road traffic accident deaths per 100,000 people.
- Extent of access to all weather roads.
- Number of people within 0.5 km of public transit running every 20 minutes.

The focus of TOD on provision of housing and commercial land uses in clusters within walking and cycling distance of transport hubs also has bearing on Target 1

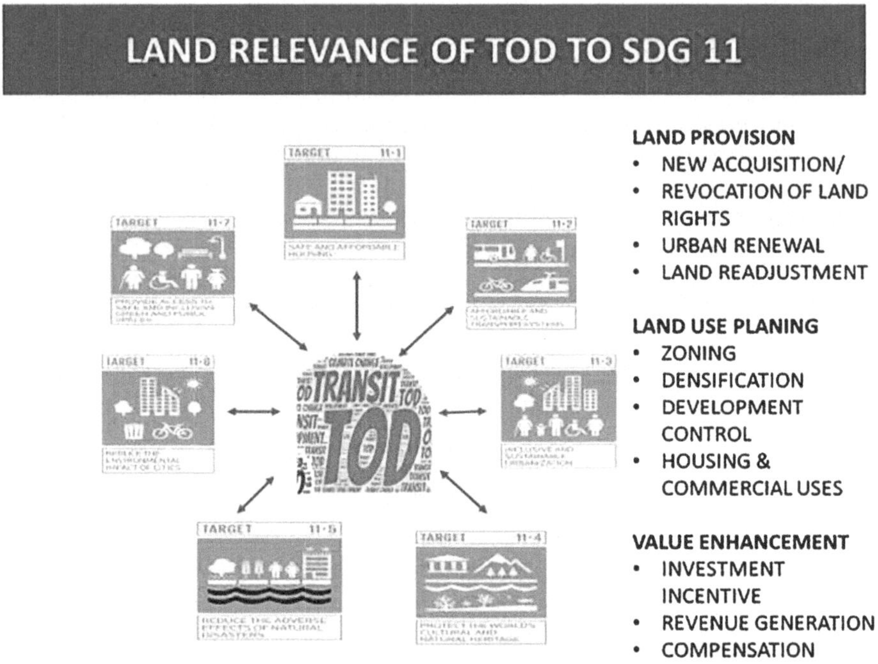

Fig. 6.1 Conceptual linkages between Land, TOD, and sustainable development goal no. 11. (Source: Authors' conceptualization)

(safe and affordable housing) and Target 3 (inclusive and sustainable urbanization). Residing within walking distance of work places, markets, and other areas reduces cost of living and offers health benefits for citizens. The implied land use efficiency in the compact neighborhoods will also ensure conservation of land resources by greatly reducing sprawl and the tendency to expand haphazardly into viable agricultural land in the periphery. This is the focus of Target 4. The potential of TOD to reduce carbon footprint of cities by reducing dependence on vehicular traffic has a bearing on Target 5 and is in accordance with Target 6. Ensuring that parks and other public spaces are an integral part of the environment around the transport hubs, needless to say, greatly promotes achievement of Target 7.

To achieve these targets, Lagos, along with other African cities, needs to be able to meet up with existing huge deficits and strategize to ensure resource availability for dynamic and sustainable provision. Further pressure arises from the drive for global competitiveness and attractiveness for foreign direct investment and employment generation. TOD enhances efficient spatial structures, adequate availability of infrastructure and urban services, affordable housing, and healthy environments. As Dowall (1999) and Teklemariam and Shen (2020) pointed out, effective urban land management is necessary to ensure these and to promote urban regeneration and development of new industrial and commercial districts, investments to upgrade and expand critical infrastructure systems, programs to enhance and protect the environment, and initiatives to upgrade social overhead capital (housing, education, healthcare).

To assess TOD projects and the contextual role and challenges of land management, a few case studies from the United States, Asia, and Africa are examined.

6.5 Brief Insights into TOD Projects in Selected Locations

United States

While many successful TOD projects have been undertaken in the United States since the 1990s, the Miami-Dade Transit's South project bears mentioning because of peculiar land-related challenges involved. As reported by the Transit Cooperative Research Programme (TCRP) (2004), land assembly was made difficult by the multiplicity of owners having only small parcels of land and the large number of small local authorities governing zoning and land use matters. Many of the private landowners refused to sell or release their land, thereby increasing the attendant risks and costs of the project. Eventually, the project was undertaken by private developers based on a 60-year lease of the land renewable for a maximum term of 99 years. Other constraints revolved around lack of political support, poor location, and inadequate demand for the commercial spaces provided. Moreover, economic provision of parking spaces for the benefit of users of the transport facilities, on the one hand, and the determination of optimal retail mix to promote viability, on the other, proved problematic. Another constraint arose from poor public investment in infrastructure around the transit stations. Among key lessons learned through the project is that mixed land uses in TOD achieve benefits of complementarity, while commercial uses are market driven and not transit driven as is often presupposed.

China

Based on Doulet et al. (2017) review of relevant literature on TOD in China, a number of hinderances to TOD were discovered to include, among others:

- The absence of real land management measures. Land speculation was found to be high due to increasing demand for home ownership and rising prices, independent of mass transit services in the TODs. In addition, local government objectives of short-term value maximization often worked against the overarching aims of the TOD.
- The functionalist approach to planning which results into transit infrastructure not being designed to drive and shape urbanization. Lot sizes and spatial placements were found not to be in synch with the ideals of compactness and accessibility aspirations of TOD.
- Design misapplications are common as the transit hubs do not reflect the culture of mixed. What obtains is a segmented approach to land use which places the stations far apart and far from the neighborhoods so that the developments are more transit adjacent than transit oriented.

Hong Kong

Hong Kong Mass Transit Railway Corporation operates as a commercial entity and was noted to be controlling 84 stations and 68 light rail services moving 4.5 million passengers about 40% of total public patronage daily (Zhou et al., 2022; Hung, 2014). There is a plan to expand these over 17 years to offer greater facilities by 2031. TOD in Hong Kong has often been referred to as best practice particularly in the area of vigorous pursuit of land value capture to finance TOD and recover the costs of property development.

The World Bank (2014, 2015) eulogized TOD as one of the most effective means of addressing the problems of sprawl, congestion, and emission of greenhouse gases but alluded to the fact that it is very expensive to implement. This necessitates the use of value capture methods to ensure that projects can always pay their way while generating more resources to facilitate maintenance and replication in other locations. Cervero and Murakami (2009) and Banister (2012) reported a rise in house prices to between 5% and 30% due to TOD most of which are developed around rail stations in Hong Kong. Apart from the use of taxes to recover some of this, increased land value capture methods often used include:

- Sales and leases of land developments in the TOD precinct.
- Partnerships between transit agencies and developers.
- Sale of air rights to enable developers build more vertically.
- Partnerships with land owners to promote land consolidation and development of mixed-use projects.
- Sale of portions of pooled land in land readjustment schemes.

Hung (2014) specifically indicated that the MTRC obtains land for its projects by receiving land development rights from the government in return for a premium on green field basis. It designs the stations to encourage ridership and then grants development rights to developers who undertake the investment, renting and selling the properties to cover the capital costs. Income from fares is used for operational costs. The agency uses a Rail + Property (R + P) paradigm to integrate its projects with housing development usually in the form of high-rise apartment blocks within walking/cycling distance of the rail station. The stations themselves are conceptualized as multilevel mixed-use precincts to achieve seamless connection between the stations and the developments.

Singapore

Singapore showed preference for high density development and began to practice TOD long before western countries. It is a small city state occupying just 728 sq. km. with a population of only 5.639 million (World Bank, 2018). Singapore exemplifies optimized urban spatial structure through the integration of transport and land use, successful reshaping of urban landscape through TOD, and successful integration of residential property development with rail transit-oriented development (Blanton, 2000). These characteristics are demonstrated in Bukit Timah which has one of the highest densities of private housing than other areas in Singapore. Its Marina Bay Sands occupies an area of 360 hectares of reclaimed land developed in the 1980s. As of 2017, four mass transit lines were planned along with nine metro stations. The node was designed to be within a 5-min walk from the homes of residents. There is an abundance of cycling paths and broad covered walkways where residents jog, stroll, or simply enjoy the communal space. There are few allusions in the literature to land acquisition problems for TOD projects in Singapore. This is because land there is under government ownership.

6.6 Thematic Based Issues and Priorities for Lagos

The City in Context

Lagos State, with more than 20 million inhabitants out of the nation's estimated 190 million (World Bank, 2018), is Nigeria's biggest urban agglomeration even though it is only 0.4% of the 923.7 sq. km land area. One of the major impediments to expansion lies in the fact that 22% of the 3577 sq. km land area is made up of water bodies including the lagoon and a system of creeks and wetlands. The population growth rate of 3.2% p.a. (6% per annum in the metropolitan core area) is due more to rural-urban migration as a result of the fact that Lagos is the nation's commercial hub accommodating over 65% of national commercial and industrial activities. This therefore calls for proper planning to ease congestion and intense pressure on road networks (Fig. 6.2).

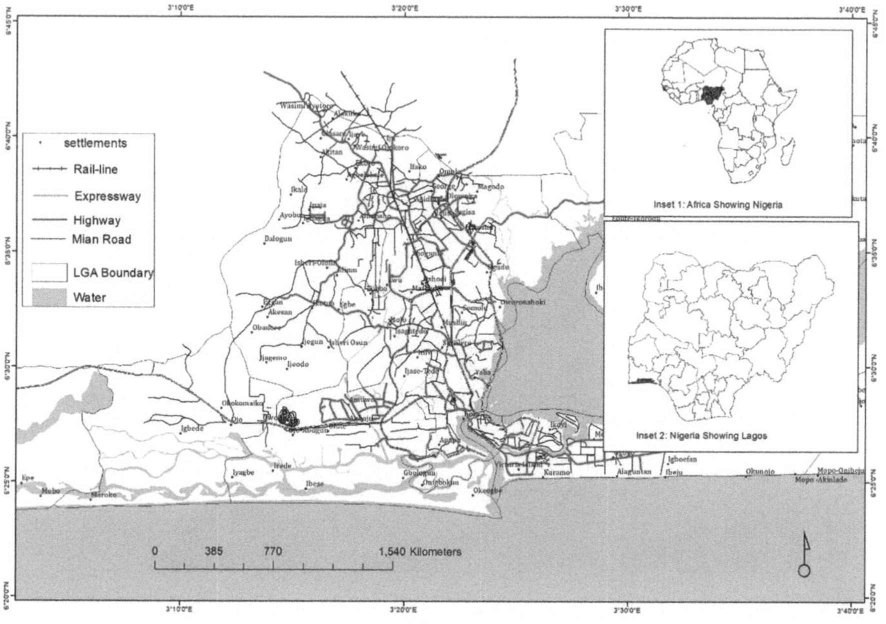

Fig. 6.2 Map showing road networks in some parts of Lagos. (Source: GIS Laboratory, Geography Department (2022))

Factors Necessitating TOD in Lagos

All the factors which necessitate TOD are evident in Lagos—traffic congestion, extreme sprawl, poor or nonexistent roads, and the need to optimize land uses by integrating them with major transportation hubs. The problems are exacerbated by overdependence on private vehicles instead of efficient public mass transit systems, poor prioritization leading to deferred maintenance, focus on regional and intercity links at the expense of intra city roads, the lack of development of other transportation modes, e.g., water and rail, frequent change of government, policies, and land administration leadership leading to delayed completion and abandonment of projects. There is a general lack of foresight, pervasive corruption, and nonadherence to principles of good governance, and overconcentration of activities in core areas so that traffic is unidirectional at peak periods. Apart from the difficulty of raising finance for transportation infrastructure provision, there are serious constraints arising from land acquisition and compensation problems as well as deliberate sabotage by some private interests—the land. In summary, the existing transport networks and hubs evolved spontaneously and are poorly planned and unregulated, quantitatively and qualitatively defective, poorly maintained inadequately integrated with other land uses, reactionary rather than proactive, socially alienating, uncomfortable, unsafe and accident prone, and archaic and outmoded (Figs. 6.3 and 6.4).

LAMATA (2015) asserted that road infrastructure is 75% of available transport infrastructure in Lagos—moving 90% of commuters on a daily basis. Of these, private cars move 2.5 m (11%), BRT moves only 90,000 (0.42%), trains move only 1%, while private informal transporters move 9,982,000 (45%). While private transportation provides employment for their operators, they are often poorly maintained, inconvenient, and uncomfortable for their users.

Fig. 6.3 Characteristic congestion on the roads. (Source: LAMATA, 2015)

Fig. 6.4 Extremely poor condition of roads. (Source: LAMATA, 2015)

Fig. 6.5 Bus rapid transit facilities. (Source: LAMATA, 2015)

6.7 TOD-Related Projects in Lagos

The bus rapid transit (BRT) system was introduced in 2002 under the Lagos Urban Transport Project funded by the World Bank as part of the effort to promote effective movement of people within the state. Fleets of air-conditioned buses were acquired, and exclusive bus lanes were added along the designated routes to free the buses from the effects of traffic jams. This has had limited effects as the LAMATA report shows that they move less than 0.5% of the commuters daily. Moreover, there is no effort to integrate the project with land uses in compact neighborhoods within walking distance for residents as TOD does (Figs. 6.5 and 6.6).

Fig. 6.6 Ongoing light rail project. (Source: LAMATA, 2015)

Fig. 6.7 The Lagos urban rail project. (Source: LAMATA, 2015)

The state government embarked on the urban light rail projects in 2012. As laudable as the objectives are, the pace of work is very slow. A few nodal clusters of shopping facilities are envisaged, but these are not at the level of full-scale TODs (Figs. 6.7 and 6.8).

The Lagos Policy on nonmotorized transport (NMT) was developed in 2018 by the Lagos State Ministry of Transport in conjunction with LAMATA and with the support of the United Nations Environment Programme. This is the closest to TOD. The policy aims to:

- Discourage the use of personal vehicles.
- Encourage walking and cycling.
- Encourage ridership of public transport.
- Ensure equitable distribution of rights of access and use.
- Encourage dense, compact, and mixed-use development.
- Boost economic growth through efficient mobility of goods and persons.

Based on a time scale of 15 years, the project will require street designs that accommodate cycle paths and pedestrian lanes. Further, its potential for use for urban renewal and regeneration of degraded areas is very high.

Fig. 6.8 The ongoing Oshodi TOD. (Source: LAMATA, 2015)

6.8 Land-Related Challenges of Transport-Oriented Development in Lagos

Land matters are governed by the provisions of the Land Use Act (CAP L5) of 2004 which vests all land in every state in Nigeria in the State Governor. While it places far-reaching powers in the hands of the governor to designate urban and rural areas, grant consent for property transfers by holders of rights of occupancy, and revoke rights of occupancy, etc., for overriding public interests, the law is often criticized for its failure to provide for adequate compensation when land rights are revoked.

To gain practical insights into land-based impediments to TOD in the state, a stakeholders' meeting was held in August 2020 followed by a webinar on land-based taxation as a means of generating funds for TODs and other projects under the challenges of the post pandemic situation.

Present at the stakeholder's meeting were panelists comprising representatives of various relevant fields in the academia, directors from the State Lands Bureau, director of the state transport operations, the surveyor general of the state, veteran professors of law, real estate management, as well as surveying and geo-informatics. There was also a member of a traditional land-owning family as well as the director of the Nigerian Railway Corporation and professionals in land-related fields.

Similarly, the webinar was anchored by expert speakers and discussants from within and outside Nigeria.

The Challenges Identified Are as Follows

- In the past, land was often acquired in advance of need as a strategy for ensuring availability for infrastructure and other projects. The global land acquisition scheme carried out under the former western region left acquired areas unutilized for a very lengthy period until they were occupied by squatters who subdivided and sold them informally to others, thus denying their use by the original owners who were often not adequately compensated. The slums and shanties emanating are often later demolished as was the case in Maroko and other areas. Although land banking is a valid strategy, its usefulness in Lagos is hampered by frequent changes in government programs. Moreover, previously acquired land is often not in locations preferred for current projects.
- Land acquisition/revocation of statutory rights of occupancy is often done without due diligence and compensation to customary title holders in the belief that the land is now under government control, customary rights having been extinguished by the Land Use Act. In fact, a policy of "No Formal Title, No Compensation" is pursued. This is inequitable. It impoverishes the affected persons and breeds bitterness and antagonism among them leaving them feeling cheated and disenfranchised.
- Even where the government adopted the people-oriented advocacy approach and requested the customary rights holders and communities to provide land as was done recently in the case of the Oshodi/Abule Egba transport corridor project, they met with resistance. This is because the communities are not convinced as to how the project would benefit them by reducing the congestion in the area. The government eventually had to resort to paying compensation for the land used for the transport corridor and set back. A similar example is the CMS-Ikorodu road expansion project to accommodate BRT lanes.
- The lack of finance is a major impediment to payment of adequate compensation. The use of depreciated replacement cost as basis for compensating affected persons for unexhausted improvement is inimical to their welfare as the amount paid is never enough to erect houses to replace what they are forced to give up. Several proposals for an amendment of this aspect of the Land Use Act are yet to yield any fruit.
- Existing land use control measures are weak. As a result, the current transport corridors suffer encroachment by road side informal markets, street traders, etc. TODs are needed to decongest the transport corridors, create commercial and residential hubs around transportation hubs, and free the roads for seamless commuting.
- The fringe areas are already experiencing spiraling sprawl and unsustainable development particularly in areas overlapping with Ogun State. High and increasing rents in the metropolis have driven many residents to these areas making it necessary to commute long distances to work and back everyday spending long hours in traffic. TODs are needed in these areas to ensure compact self-sustaining

settlements that facilitate short distance movements between homes and work bases.
- Nascent improvements and applications of TOD concepts are not in total conformity with the ideals and best practices.
- Value capture is not being employed to ensure that adequate revenue is gathered for project replication and expansion.
- There is a lack of inclusivity in the activities of government with respect to TODs.

The outcome of the stakeholders' meeting shows that government did not deny their current inefficiencies and shortcomings in executing TOD.

Recommendations

- Concerted efforts should be directed at adopting a multipronged approach which will ensure balanced regional development on the long run to stem the tide of excessive urbanization arising from rural-urban migration. Land policies that incentivize rural land entrepreneurship are necessary in this regard.
- TODs should be better planned to emphasize compactness and mixed-use development at all levels to reduce the need for long distance commuting, increase housing options, and create affordable opportunities for commercial activities for the citizenry. To this end adoption of land use efficiency and smart growth parameters is necessary.
- Private-public partnership approaches to TOD development should be exploited without corrupt practices but rather with inbuilt incentives to dissuade excessive profit expectations which usually stall developments and eventually detract from the social, economic, and environmental benefits.
- Land readjustment options should be used to encourage community and private land owners' buy-in so as to reduce the burden of compensation payments, harness stakeholders' ingenuity, and ensure perpetuation of the benefits.
- Due diligence is required to ensure that the rights of customary title holders are not ignored. The policy of "No Formal Title, No Compensation" should be jettisoned because the lack of documentary evidence of a right may not necessarily indicate nonexistence. Moreover, the process of obtaining certificates of occupancy is still cumbersome, complicated, expensive, and time-consuming.
- Guidelines and framework for proactive development of urban fringe areas should be provided to prevent sprawl.
- Land-sensitive due diligence and alignment with SDG 11 and targets should be assured before embarking on projects.
- To ensure proactive planning, a thorough environmental impact analysis (EIA) and transport impact analysis (TIA) should be conducted for each project to optimize expected benefits.
- The use of land value capture to generate revenue for TOD should be institutionalized following existing best practices identified in Hong Kong and other countries.
- Resolution of morality and legality issues in revocation of land rights is necessary to ensure improved approach to compensation matters using Pareto optimal-

ity concepts that enhance the welfare of all citizens without leaving any group worse off.

- Land rights should be better documented through the application of approaches that are fast and cost effective. The importance of producing sustainable and reliable land records using sophisticated land information systems cannot be over emphasized.

6.9 Thematic-Based Research Priorities for Lagos State

From the study and discussions, the following research areas have been identified:

(a) How to improve land governance in the state to harness resources for positive outcome of TOD projects.
(b) How to ensure improved synergy between government land agencies as well as communities and professionals for land use, land value, and land market-related data generation for use in TOD planning and project execution.
(c) How best to ensure conformity with global best practice in achieving the aim of TOD.

These are subdivided into focused research issues as listed:

- Analyze the relationship between existing transit corridor planning methods and land use behavior and control in both core and peripheral areas in Lagos state.
- Critically assess equity issues in land acquisition for transportation infrastructure development and compensation practice based on extant laws, administrative guidelines, and international best practice.
- Derive effective means of preventing abuse and misuse of land abutting transport routes to control congestion through fiscal and regulatory approaches.
- Identify constraints militating against data collection, analysis, and application through LIS and GIS machinery in transit-oriented development in the state.
- Investigate acceptability of land readjustment approaches to community participation in transport infrastructure provision, expansion, and maintenance within the framework of urban renewal in the city.
- Investigate the impact of new transport corridors and transport infrastructure improvement schemes on land values with a view to determining land-based taxes recoverable from residents.
- Investigate technological and human capacity inadequacies in management of land resources for transit-oriented development.
- Investigate the level of acceptability of current residential land use dispersal through the adoption of mass transit systems linking core area employment bases to peripheral areas (e.g., through light rail, BRT, and other systems).
- Identify best means of boosting competitive housing supply in TOD areas to reduce rents and improve quality and affordability while boosting public health.

6.10 Conclusion

TOD requires proper alignment of policy, planning, and implementation strategies with proactive use of land management approaches that will optimize benefits while minimizing costs and loss of welfare benefits. Successful TOD projects in advanced countries integrate residential and commercial land uses with transportation hubs in a manner that optimizes real estate values, reduce carbon footprints, reduce commuting distances, and promote housing choice. Such projects are undertaken through partnership arrangements that give profitable incentives to private sector investors. Proper land management approaches will encourage seamless release of land for projects through effective land acquisition/land assemblage processes, equitable assessment and payment of compensation, advocacy, and use of land readjustment approaches to ensure community participation. It will also promote land value capture to ensure revenue generation for prosecution of new projects and maintenance of existing ones. This study has identified various ways in which existing practices in Lagos detract from these ideals and recommended methods whereby observed transportation anomalies in Lagos can be corrected using this paradigm. A note of caution must however be sounded on the possibility of exacerbating existing congestion through improper compaction of uses in few hubs. Undue emphasis on the revenue generation alone can also erode envisaged long-term benefits. To achieve the right balance, in-depth research is required into the various areas indicated.

References

Africa Finance Corporation. (2019). GOGLA. https://www.gogla.org/africa-finance-corporation

Atubi, A. O. (2013). An evaluation of transport infrastructure in Lagos state, Nigeria. *Journal of Geographical Earth Sciences, 1*, 9–18.

Banister, D. (2012). Assessing the reality-transport and land use planning to achieve sustainability. *Journal of Transport and Land Use, 5*(3), 1–14.

Bevilacqua, C., Cappellano, F., & Zingali, L. (2013, October). *TOD—Transit-Oriented Development: A sustainable tool towards smart living*. In Proceedings of the 7th conference of international forum on urbanism (pp. 275–284).

Blanton, W. (2000). Integrating land use and transportation. *Planning Commissioners Journal, 40*, 9–14.

Bobylev, N. I. K. O. L. A. I., & Jefferson, I. (2014). *Integration of urban physical infrastructure into land use governance*. In World Bank conference on land and poverty.

Carlton, I. (2009). *Histories of transit-oriented development: Perspectives on the development of the TOD concept* (No. 2009, 02). Working paper.

Cervero, A., & Murakami, J. (2009). Rail and property development in Hong Kong: Experience and extensions. *Urban Studies, 46*(10).

Doulet, J., Delaunay, T., & Delpiron, A. (2017). Taking advantage of a historic opportunity? A critical review of the literature on TOD in China. *Journal of Transport and Land Use, 10*(1), 77–92.

Dowall, D. E. (1999). *Globalisation, structural change and urban land management*. https://www.lincolninst.edu/publications/articles/globalization-structural-change-urban-land-management

Edema, J. (2019). Poor public transport infrastructure in Lagos Nigeria, how sustainable improvement could enhance well-being of the people and provide environmental benefits.

Effiom, L., & Ubi, P. (2016). Deficit, decay and deprioritisation of transport infrastructure in Nigeria: Policy options for sustainability. *International Journal of Economics and Finance, 8*(3), 55. https://doi.org/10.5539/ijef.v8n3p55

Fleming, A., & Makalima-Ngewana, B. (2013). Cape Town's central city development: A strategy of partnership and inclusion. *Town and Regional Planning, 62*, 29–35.

Gutman, J., & Patel, N. (2018). *Foresight Africa viewpoint–Urban Africa: Avoiding the perfect storm*. https://www.brookings.edu/blog/africa-in-focus/2018/01/26/foresight-africa-viewpoint-urban-africa-avoiding-the-perfect-storm/

Hung, W. T. (2014). *Transit-Oriented Development and value capture in Hong Kong*. A paper presented at the UNESCO Regional Experts Group Meeting on Sustainable and Inclusive Transport Development held at Ahmedabad 29th September–1st October.

Lagos State Ministry of Transportation, Lagos Metropolitan Area Transport Authority, United Nations Environment Programme, Institute for Transportation and Development Policy. (2018). *Lagos Non-motorised Transport* Policy. https://www.wedocs.unep.org/bitstream/handle....

LAMATA. (2015). https://lamata-ng.com/. https://lamata.lagosstate.gov.ng/

Muggah, R. (2018). *African cities will double in population by 2050*. https://www.weforum.org/agenda/2018/06/Africa-urbanization-cities-double-population-2050-4%20ways-thrive/

Oyesola, B. (2021). *Urban infrastructure: Urban headaches*. https://www.dandc.eu/en/article/lagos-city-dreams-and-nightmares

Schrenk, M., Popovich, V., Zeile, P., Oduwaye, L., Alade, W., & Adekunle, S. (2011). *Land use and traffic pattern along Lagos– Badagry Corridor, Lagos, Nigeria*.

Teklemariam, E. A., & Shen, Z. (2020). Determining transit nodes for potential transit-oriented development: Along the LRT corridor in Addis Ababa, Ethiopia. *Frontiers of Architectural Research, 9*(3), 606–622.

Thuo, A. D. M. (2010). *Community and social responses to land use transformations in the Nairobi rural-urban fringe, Kenya, Field Actions Science Reports*. Special Issue 1, 2010, Online: http://journals.openedition.org/factsreports/435

Transit Cooperative Research Programme. (2004) *Transit-Oriented Development in USA: Experiences, challenges and prospects*. tcrp report 102. National Academies Press. https://www.nap.edu

World Bank. (2014). *Addis Ababa urban and metropolitan transport and land use linkages strategy review*. https://elibrary.worldbank.org/doi/abs/10.1596/26080

World Bank. (2015). Financing TOD with land values: Adapting land value capture in developing countries. *Transport and Urban Planning Research Reports*. https://www.C40knowledgehub.org/article/Financing-TOD-with-Land-Values

World Bank. (2018). *Global economic prospects, January 2018: Broad-based upturn, but for how long?* The World Bank.

Zhou, H., Liu, Y., & He, M. (2022). The spatial interaction effect of green spaces on urban economic growth: Empirical evidence from China. *International Journal of Environmental Research and Public Health, 19*(16), 10360. https://doi.org/10.3390/ijerph191610360

Zubairu, S. (2019). *Africa: The long road to fixing Africa's infrastructure deficit—Here's how*. https://allafrica.com/stories/201907080646.html

Chapter 7
Transit-Oriented Development and Construction in African Cities

Olumide Afolarin Adenuga, Olajide Julius Faremi, Kudirat Ibilola Zakariyyah, Adegboyega Sunday Sotunbo, Oluranti Olupolola Ajayi, and Simeon Dele Rogers

7.1 Introduction

Besides the challenges of the existing vibrant population, the projections on global urban population increment of over 2 billion by the year 2050 with the anticipated rapid migration to cities and the associated rapid development of urbanization will heighten economic, environmental, and social issues (OECD, 2013; World Economic Forum, 2016; Gomez et al., 2019). In the bid to mitigate the existing challenges, emerging technologies, business models, and concept planning are being reviewed. As transportation is crucial to development, a model of public transport that maximizes spatial/residential space, business, and leisure and harmonizes such a symbiotic effect becomes a sine qua non to cities' urban planning. A typical example of business models/emerging technologies is the inculcation of smart growth into urban development, out of which transit-oriented development (TOD) and infrastructure feature as concept planning. In contributing to social and economic development, therefore, transport infrastructure is considered as a holistic subset of the actual physical network, the vehicles and other means of conveyance, the soft system such as policy and governance, as well as the hard structure that comprises buildings and utilities (Renne & Wells, 2005; Nawaz et al., 2017). Due to its encompassing features, TOD and the building utilities become veritable tools in the concept of sustainable development as they can be used to develop, monitor, redefine, and revitalize settlements by integrating housing and transportation sectors (Jenks, 2005).

O. A. Adenuga · O. J. Faremi (✉) · K. I. Zakariyyah · A. S. Sotunbo · O. O. Ajayi
S. D. Rogers
Department of Building, University of Lagos, Lagos, Nigeria
e-mail: ofaremi@unilag.edu.ng

T. Nubi et al. (eds.), *Transit Oriented Development in West African Cities*, Urban Perspectives from the Global South,
https://doi.org/10.1007/978-3-031-58726-9_7

Depending on peoples' needs and spatial plans, a TOD may range from a central transit stop surrounded by a high-density mixed area, spreading out from the center to a plan of urban and residential spaces integrated with the public transportation system (Jenks, 2005). Transit-oriented development encourages the use of public transport by reducing the dependability on private vehicles while creating a pedestrian-friendly urban environment to achieve a qualitatively livable and sustainable environment (Gomez et al., 2019). TOD is, therefore, being adopted as part of urban development which has resulted in more studies on what makes TOD successful and overall development sustainable. Though interrelationship among economic development, TOD, and spatial development is emphasized, the plethora of studies on TOD converges on the subject that the desired end product of social/economic outcomes are factors of governance which in turn is an offshoot of policy.

Policy, on the other hand, however, stems from the degree of effort that is reflected in land use pattern, zoning, development control, building developments, regulations, standards, guidelines, etc. This indicates that there is a gap between what is expected, what is on ground, and what polices are either in place or implemented in the country. Therefore, considering the rising population of Nigeria as a country with the desire to make the built environment sustainable, this study assesses the opinions of stakeholders on transit-oriented development buildings and utilities. This is intending to arrive at a model that will enhance Lagos State as a metropolitan city, as a major setback to urbanization is uncontrolled development, which often originates when demand and supply are not in synchronism.

7.2 Literature Review

7.2.1 Conceptualization of Transit-Oriented Development Buildings and Utilities

Transportation infrastructure is a network infrastructure that constitutes the channel between nodes, regions, or node region. It covers roads, highways, railways, airports, bridges, waterways, canals, and terminals, which are key players in the transmission of materials and the flow of population during urban agglomeration and diffusion (Jenks, 2005; Gomez et al., 2019). They are critical to economic development at all income levels, supporting personal well-being and economic growth (Gomez et al., 2019; Ibraeva et al., 2020). A condensed and preplanned structure of transport infrastructure for any form of settlement constitutes the transit-oriented development.

Apart from the operating vehicles or media, the buildings and the utilities within the transit-oriented development equally impact the result and define the extent of the social composition; science, arts, and technology of planning; environmental protection; public health; and national security (Gomez et al., 2019; Ibraeva et al., 2020). The buildings and the utilities thus need careful examination in the context of TOD. The types of buildings and utilities coupled with the plan and construction are critical elements of the successful implementation of a TOD project, as these relate to the culture and regional accessibility conditions. Therefore, TOD performance becomes a factor of the complex and multifaceted nature of contemporary urban agglomerations, which a nation needs to understand for the success of the investment. This calls for the inclusion of all stakeholders.

Transit-oriented development is a high-value complement to mass transit development. A TOD typically includes a central transit stop (such as a train station or bus stop) surrounded by a high- or low-density area spreading from the settlement center. The two typical typologies of TOD are the rail transport-oriented development (RTOD) and the bus transport-oriented development (BTOD) such as the bus rapid transit (BRT) system (Currie, 2006). Unlike BTOD or BRT that is provided at most levels of development, RTOD is closely related to high density or large-scale development (Gifford et al., 2001; Rodriguez & Targa, 2004). Both typologies however have a number of challenges and strengths. Some of the challenges for these two typologies include permanence. Both RBOT and BRT can be made permanent based on the density of population, the structure in place, the buildings and the utilities therein, and the policy. For BRT, however, the form of leadership required should be more pragmatic and proactive than for RTOD; this is a major factor to successful implementation of BTOD (Dittmar & Ohland, 2004; Currie, 2006). Other notable challenges or weakness associated with BRT include absence of dedicated and cultured BRT staff, the pollution (noise), and emission (lack of use of clean energy sources) associated with BRT. In addition, they are mostly suited to lower-density environments. On the other hand, though with a higher capital outlay, RTOD's successful implementation outweighs that of BRT/BTOD (Currie, 2006; Xueming, 2010). Despite these challenges both BRTD/BTOD and RTOD are essential as both complement one another. Both concepts should encourage low-density living, in which the policies will not result in a substantial loss in utility or excessive social welfare costs. Besides, the concepts and further remodeling should neither raise the costs of living nor push low- and moderate-income residents farther away from jobs and transit. Rather, an advantage that links workers to employment centers, creates job opportunities in the construction and maintenance sector, and encourages investment in areas that have suffered neglect and economic depression should be the target (Jenks, 2005; Currie, 2006). When this is achieved, coupled with the expenses that would be saved on private transportation, there will be equity and improvement in the housing sector. An example of such is Johannesburg, South Africa. Through good political leadership and governance, dedicated investments were made with the private sector having an interest in the development, as well as government policies that are in place; the construction of the "corridors of freedom" initiative was facilitated (OECD, 2013; Nawaz et al., 2017).

7.3 Transit-Oriented Development: A Case Study of Singapore and Vancouver

Considering Singapore as a case in point, effective urban mobility has often revolved around high-intensity transport technologies. One example is the ever- growing mass rapid transit (MRT), which is consolidated with an extensive public bus network. A fundamental component of Singapore's plan for urban mobility focuses on policy instruments that aim to regulate/influence car ownership and public usage of scarce transport resources or infrastructure (Ibraeva et al., 2020). For over 42 years, Singapore adopts various mechanisms to ration the private ownership and use of cars on its roads. Traffic management policies such as area licensing scheme (ALS) and vehicle quota system (VQS), among others, lay constraints on private transport users (Barter, 2011). The city even recently initiated a "zero-growth" car policy which aims to standardize the annual growth rate of privately owned vehicles to zero.

Vancouver, British Columbia, is not left out on the TODs. Vancouver has built an unconventional history around its SkyTrain lines and building regional town centers at major stations and transit corridors. SkyTrain is a light rapid system in the Metro Vancouver Regional District. With an annual ridership of over 160 million (2018), a 79.6 km of track, and using fully automated trains on grade-separated tracks running on underground and elevated guideways, SkyTrain holds consistently high on-time reliability, connecting metro cities such as Vancouver and surrounding municipalities such as Burnaby, British Columbia. Vancouver going into action designed a dense, coherent, and walkable network of streets, sidewalks, and off-street paths. A pattern of connected streets makes it possible to travel along direct routes, sometimes through walking or cycling, closing the gaps between destinations while encouraging conveniences. These strategies and actions aim to build up the basic social and political harmony needed for long-term, realistic, and appropriate urban transport development for city improvement.

7.4 Conceptual Linkages Between Transit-Oriented Development, the Relevant SDGs, and the Thematic Area of Interest

According to researchers, transit-oriented development (TOD) has five key characteristics, namely, density (compact), design (pedestrian-friendly and sustainable), diversity (mixed-use or multifunctional development), destination accessibility or mobility (efficient public transport service), and distance to transit

(transit stations within the 500 m radius from the center) (Abdullah & Mazlan, 2016; Berawi et al., 2020; Centre for Neighbourhood Technology, 2013; De Vos et al., 2014; Gomez et al., 2019; Jacobson & Forsyth, 2008; Kamruzzaman et al., 2014; Li et al., 2016; Renne, 2016; Widyahari & Indradjati, 2015; Xu et al., 2017). These characteristics are the focal elements of TOD in any communal setting. Transit-oriented building and utilities (TOB&U) in this context refer to the station buildings and utilities that accommodate the ticketing offices, conveniences, loading bays, Internet infrastructure, unmotorized access, and various energy-efficient technologies provided for ease of operations and usage of a transit system. On the other hand, Sustainable Development Goals (Agenda 2030) adopted in 2015 by UN member states consist of 17 goals and 169 targets that are interdependent and focus on people, planet, peace, prosperity, and partnership. These SDGs include themes such as no poverty, zero hunger, good health and well-being, climate action, and partnerships for the goals (Economic and Social Commission for Asia and the Pacific (ESCAP), 2017; Food and Agriculture Organisation, 2017; International Association of Public Transport, 2019; Oliveira et al., 2019; Paoli & Addeo, 2019; Scharlemann et al., 2020; Thilakshan & Bandara, 2019; UN-Habitat et al., 2015; United Nation Development Programme (UNDP), 2015; United Nations, 2018a, b).

The concept of TOD (density, design, diversity, destination accessibility or mobility, and distance to transit) and TOB&U has direct relationship with SDGs. For instance, density (compactness), diversity (mixed-use), and distance to transit of a TOD enhances more pedestrian activities, use of public transport, and reduction in private vehicular activities which reduce road accidents and CO_2 emission from vehicles. These are associated with healthy living and well-being (goal 3), climate action (goal 13), and industrialization and innovative infrastructure (goal 9) (Abdullah & Mazlan, 2016; ESCAP, 2016). Similarly, reduction in the use of fossil fuel and increased usage of alternative energy in TOD promote cleaner energy (goal 7) and responsible consumption (goal 12).

In the same vein, implementing TOD at any level requires transdisciplinary and transnational engagement and collaboration which is in line with SDG goal 17 (partnerships to achieve the goals) (Berawi et al., 2020; International Association of Public Transport, 2019; Thilakshan & Bandara, 2019; United Nations, 2018a, b; UN-Habitat et al., 2015; United Nation Development Programme [UNDP], 2015). As TOD contributes to the realization of SDGs through meeting these goals, so do transit-oriented buildings and utilities (TOB&U) (a subset of TOD) which is the central accommodation of all the technologies, utilities, and operations of TOD and characterized by efficient and sustainable designs, user-friendliness, safety, ease of access (motorized and unmotorized), efficient energy usage, and mix usage directly relate to these SDGs. Thus, Fig. 7.1 conceptually illustrates the relationship among TOD characteristics, relevant SDGs, and TOB&U.

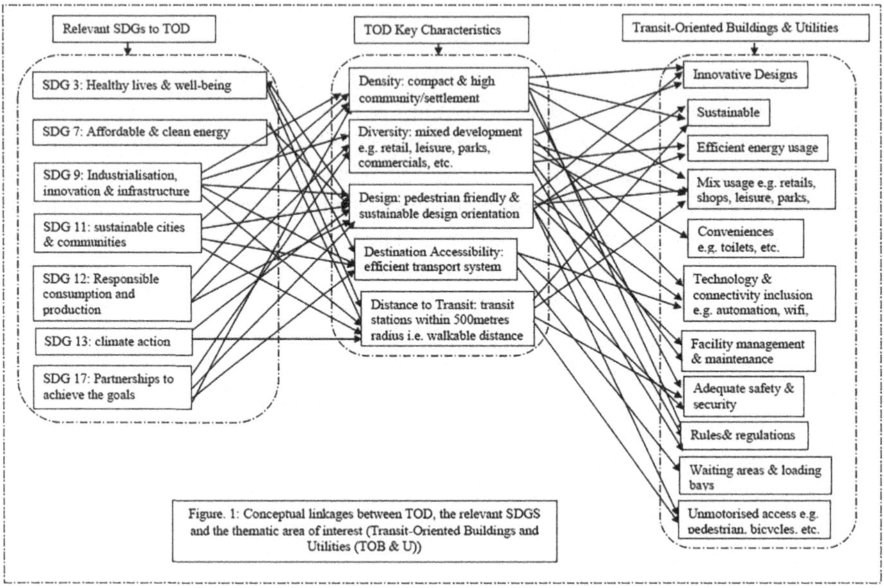

Fig. 7.1 Conceptual illustration of the relationship among TOD characteristics, relevant SDGs, and TOB&U

7.5 Case Studies from Cities Outside Lagos

Transit-oriented development (TOD) was made popular in the United States as a persuasive city planning idea by Calthorpe and others in the 1990s (Jacobson & Forsyth, 2008; Pennsylvania Department of Transportation, 2012) and later adopted across the western world. Today, the concept has become a benchmark for city planning and development across the world. Below are some cities with TOD as case studies.

7.5.1 Denver City TOD

The execution of TOD in Denver is hinged on connectivity, innovation, efficiency (closeness to transit station), compactness, mix (diversity, choice, and resilience), and shift (pedestrian, bikes, and reduced carbon emission) (Denver City, 2014). The development is mixed in nature (residential, retail, and office space) and congregates around transit stations with rail lines as its driver (Bardaka & Hersey, 2019; Ewing et al., 2017). For instance, the 50-acre revamped Denver Union Station was upgraded to an all-in-one intermodal system operating transit system such as Amtrak, Commuter Rail, LRT, Inter- city Bus, and City Bus (Denver City, 2014;

Fig. 7.2 Rail platform, Denver Union Station. (Source: Denver Union Station, 2023. https://help.g2rail.com/stations/denver-union-station-colorado)

Kidokoro, 2019). Urban rail system in Denver was commissioned on October 7 1994 with about 5 miles corridor and 13 stations (Bardaka & Hersey, 2019; Ratner & Goetz, 2013). However, with the rising inhabitants, the city diligently increased its rail line from 34.9 miles in 2006 to 81 miles in 2016, which is a huge leap on TOD implementation (Bardaka & Hersey, 2019; Denver City, 2014; Ratner & Goetz, 2013). Along these rail corridors are development of around 18,000 units of housing and over ten million square feet of retail, office, education, and medical spaces spread within the 800-meter distance to rail stations (Ratner & Goetz, 2013).

These rail stations are identical in terms of high-density pedestrian spaces that allow access to light rail bus lines, parking, bicycle ridership, access to retails, and other engagements by foot (Kwoka et al., 2015). Consequently, the city has been able to reduce vehicular activities significantly and increased unmotorized infrastructure that has led to increased walking and biking among the dwellers (Bardaka & Hersey, 2019). TOD development is still ongoing as the city's regional transportation district plans to launch additional three rail lines by the end of 2020 (Bardaka & Hersey, 2019). The implementation stretches across all the five settlement zones in the city which were achieved through policy implementation and research that today made it a success story of a city-wide TOD in the United States (Denver City, 2014) (Fig. 7.2).

7.5.2 Dallas TOD Through Dallas Area Rapid Transport (DART) System

Just as Denver, TOD implementation in Dallas is centered around the transit lines and stretched across the urban and suburban areas (DART, 2008; DART Reference Book, 2019; Denver City, 2014; Kidokoro, 2019). Denver light rail transit (LRT) system (commissioned on October 7 1994) covers 45 miles of transit lines and 35 LRT stations. Currently, DART services 13 cities (DART, 2008; DART Reference Book, 2019; Trepci et al., 2020) including Dallas, Plano, Addison, Carrollton, and Irving with over 600 fleets of buses operated by different operators, over 93 miles of light rail transit, and 64 stations (163 fleets). Among these developments are also paratransit for the person with mobility impairment, passenger transfer location (PTL), and park-and-ride. The light rail network is considered the longest in the United States (DART, 2008; DART Reference Book, 2019; Daqrouq & Anjomani, 2019) which has Mockingbird Station as one of its core transit stations. Mockingbird Station was the foremost TOD project to be launched in Texas (DART, 2008; DART Reference Book, 2019; Trepci et al., 2020). Mockingbird TOD was planned as a 10-acre urban settlement surrounding a DART station.

The station was commissioned in 2001 at a cost of $105 million. Its development span over 600,000 square feet with several loft apartments, over 90 shops, retail outlets, office spaces, direct platform access to DART trains, a cinema and café, underground parking space integrated into the development, and other surface parking for commuters and shoppers with a plan for future extension (DART, 2008; DART Reference Book, 2019). Its implementation success made it a nationwide model of TOD that other states emulate. However, Trepci et al. (2020) posited that Mockingbird is still lagging in being a superlative TOD due to shortfalls in housing units and frail pedestrian linkages within the development (Trepci et al., 2020) (Fig. 7.3).

Fig. 7.3 Mockingbird Rail platform. (Source: Mockingbird Station, 2023. https://www.bddeng.com/project/mockingbird-station/)

7.6 Thematic-Based Issues and Priorities for Lagos from a Transdisciplinary Lens

Transit-oriented buildings and utilities (TOB&U) are key components of transit-oriented development that is widely evident in most developed countries, while its existence is likely veiled in developing cities. TOB&U hinged on the SDGs (3, 7, 9, 11, and 13): healthy lives and well-being, sustainable cities and communities, clean energy, innovation and infrastructures, clean water and sanitation, clean environment, and climate actions (International Association of Public Transport, 2019; UN-Habitat et al., 2015; United Nation Development Programme [UNDP], 2015; United Nations, 2018b). Similar to TOD, TOB&U supports a mix of space and commercial uses that are accessible within a short distance from households, designed to ensure a successful public transit system and enhance the convenience and safety of walking, bicycling, and driving close to the public transit center within 500 m to provide livable cities (Institute for Public Administration, n.d.). Evidence has shown that there were limited TOB&U in rail and water forms of transport system operated in the cities, while the scarce TOB&U available in the road transport system are often associated with numerous problems in Lagos cities (Adeyinka, 2013).

These issues include and are not limited to a low standard of building design and layout; the poor physical condition of utilities such as retail outlets, shops, and canteens in the buildings; the lack of potable water; appalling state of sanitation and poor toilet facilities; wrong use of facilities; and poor operation and maintenance of transit-oriented buildings. Further, the absence of convenient waiting areas for users, inadequate loading bays, absence of multilevel parking space, lack of unmotorized access for pedestrians and riders, low level or no technology and connectivity like Internet facilities, the absence of efficient energy use to power appliances and services, inadequate safety and security measures, and the absence of rules and guidelines to guide the personnel managing relative transport activities in the buildings and utilities were apparent.

These aforementioned issues have serious implications for the healthy lives and well-being of the citizenry as well as halting a livable city. This could affect users' needs/priorities in terms of safety, security, and satisfaction. It is observed that the absence of clean energy comes to play because individual facilities and users within the transit-oriented buildings run a generating set to power appliances, thereby causing noise and air pollution. Rapid degradation of facilities or utilities evident in most of the transit-oriented buildings showcases an uncaring attitude of users of public facilities. Often, people give the impression that the maintenance of public facilities is the sole responsibility of the government. It becomes necessary that both the owners and the users of the public facilities give immediate attention to keep the facilities fit and curb the menace of facilities mismanagement.

Currently, the type of buildings and utilities supporting the transport system seems insufficient and incomplete compared to some development seen in other developed climes. Besides the paucity of supportive transit-oriented buildings evident in the road transport system, there is barely or nonexistence of supportive transit-oriented buildings for rail transport and water transport systems. Consequently, the present limited features and poor management of revenue generated in the existing transit-oriented buildings and utilities supporting the road transport system calls for more inclusion of integrated public transit system delivering, transparency and accountability of revenue as well as upgrade of facilities (Aleshinloye, 2014).

Largely, there are well-defined characteristics for TOD; these include a short-distance walk to a transit station (like a 5-min walk to a transit stop or 10-min walk to a major transit station offering frequent access to high-speed service area), the mix of uses of places (to live, to work, to learn, to shop for daily needs, and to relax), a fine-grained network of streets, dispersing traffic to allow for quiet and friendly thoroughfares, innovative and sustainable design comprising of parking spaces for four-wheel drives and bicycles, convenient walkway, stairway, ramp, escalators, and an efficient public transport service that is fast, regular, and comfortable within a short time interval (say 5–15 min) (City of Maryland Heights Community Development, n.d.). It is observed that transit-oriented development in this clime exhibits less of these characteristics.

Most of the major transit stations are at far distance to the majority of the users transiting to another destination. It is hard to find a transit station with a full mix of uses where communal could do varieties of activities such as parking of personal cars to join public transport, eating in a decent restaurant, toilet, relaxing, watch movies while on transit. Severe traffic jam and bottleneck have always been the order of our cities resulting in unfriendly thoroughfare which impede the health and lifestyles of city dwellers. Other problems facing the cities in Lagos include an unpleasant road network, an increase in privately owned vehicles, a large percentage of transit business being managed by a disorganized informal sector that is driven by a market-based, unregulated, and low-capacity service (Oshodi, 2016). However, the city of Lagos is close to having an efficient public transport system.

LAMATA law is pushing a stride toward achieving a more effective and efficient transit system by introducing and sustaining the BRT options (Edema, 2019). New infrastructure development of TOB&U should be innovative and all-encompassing, exhibiting features that are outlined in the Lagos State Development Planning Law. These include a bus station, security system, loading bay, waiting areas, multilevel car parks, toilet facilities, eatery, relaxation space, fire station, mini-mart, cinemas, administrative office, and fuel station, among others. The inclusion of these features would enhance a livable city, especially when owners of private cars can park their cars and join the public bus to get to their destination, ensures affordable and clean energy within the transit space, and maintains and sustains facilities and development that are climate-friendly. Besides the road transport system, more investment in sustainable TOB&U should be advanced in other transport systems such as rail and water transport system to ease the traffic bottleneck and meet the growing demand for affordable and suitable mass transit in the cities (Aleshinloye, 2014).

7.7 Data Analysis

The study focuses on Lagos. Lagos is the economic nerve center of Nigeria with an estimated population of 24.6 million inhabitants (Lagos State government, 2023). The map of Lagos State is shown in Fig. 7.4.

This section presents the analysis of the data collected from the panelist at the Sustainable Construction and Skill Development Research Cluster stakeholders' meeting that was virtually held on Tuesday, 28 July 2020. The meeting was closed to only the invited panelist and was conducted using the Chatham House rules. The Chatham House rules were observed so that more vulnerable stakeholders and politically exposed participants can freely express their views on the subject of discussion. The conversation at the stakeholders' meeting was framed around the following questions:

- What is the present state of transit-oriented buildings and utilities in Lagos State?
- Are there regulatory frameworks (such as standards and codes) for planning, executing, monitoring, and controlling (vis à vis QA/QC) and closing out on transit-oriented buildings and utilities?
- Are best practices followed in the commissioning, postconstruction use, operation, and maintenance of transit-oriented buildings and utilities?
- What measures could guarantee improved transit-oriented buildings and utilities' project delivery, hence, a liveable city in Lagos, Nigeria?

7.7.1 Demographics of Participants

Six panelists (tagged panelist 1 to panelist 6) comprising stakeholders with expertise and diverse experience on transit-oriented buildings and utilities within Lagos metropolis from both the public and private sectors participated in the virtual meeting.

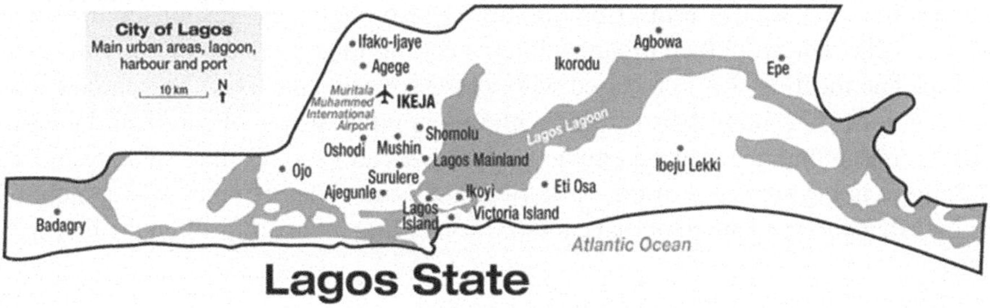

Fig. 7.4 Lagos State Map. (Source: PNGkit, 2023. https://www.pngkit.com/bigpic/u2e6e6w7o0a9r5t4/)

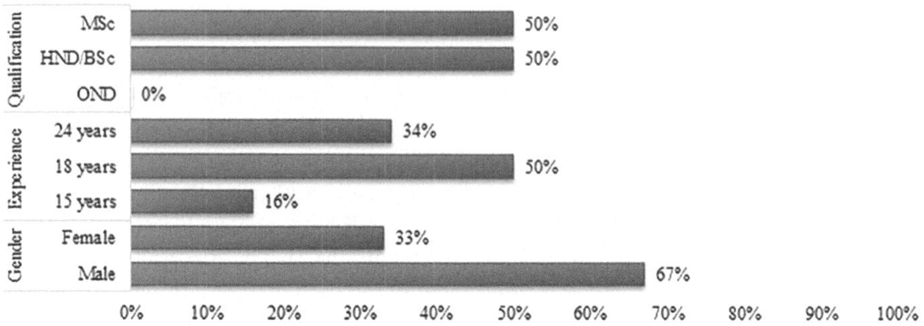

Fig. 7.5 Demographics of participants

Figure 7.5 shows that 67 percent of the panelists were males, while 33 percent were females. Moreover, 16 percent, 50 percent, and 34 percent of the panelists had 15 years, 18 years, and 24 years working experience respectively. The result further shows that the panelists had 50 percent at par with higher national diploma or Bachelor's degrees and master's degrees, respectively. The analysis implies that the perception of both male and female experts on transit-oriented development was captured during the data collection process for the study, thus implying fair trans-gender responses to the research questions. Furthermore, the result implies that the panelists are very suitable to have provided valuable contributions during the stakeholders' meeting owing to their adequate level of education and significant years of working experience.

7.7.2 Analysis of Stakeholders' Responses

Thematic analysis was adopted in analyzing the responses of the stakeholders that participated in the meeting. Thematic analysis refers to the process of identifying patterns or themes within qualitative data (American Psychological Association, 2010). The six-phase framework for doing thematic analysis as proposed by Braun and Clark (2006) was adopted for the analysis.

The first step was to transcribe verbatim the participants' responses from the meeting. The data were then organized in a meaningful and systematic way through coding. The meeting was concerned with addressing specific TOD's research questions, and the data analysis was done with this in mind; hence, theoretical thematic analysis rather than inductive thematic analysis was adopted for the coding exercise. Therefore, not every piece of the transcribed text was coded but the text segment of the data that was relevant to or that captured something interesting about the research question.

The analysis used open coding; this means that there were no preset codes, but codes were developed and modified as the data analysis progresses through code iteration during the coding process. The resulting codes were further organized into broader themes that align with the research questions. The themes are predominantly descriptive, as they describe patterns in the data that were relevant to the research questions. Table 7.1 shows the themes that are identified from the data along with the codes that are associated with them.

From Table 7.1, it could be seen that during the data analysis, it was discovered that some of the codes such as innovative designs, regulation and enforcements, and facility management and maintenance were associated with more than one theme. The interconnectivity and relationships between the codes are presented in the postscript plot as shown in Fig. 7.6.

The postscript plot in Fig. 7.6 is the final refinement of the themes and the associated codes. The significance of the postscripts plot is to identify the essence of what each of the theme is about. It provides the interaction between the codes and the themes and shows how the themes relate to each other.

7.7.3 Results from the Analysis Codes and Themes

The results of each of the themes from the qualitative data analysis in Table 7.1 and Figure 7.6 are discussed as follows.

Table 7.1 Themes that emerged from the qualitative data analysis

Codes	Themes
Conveniences and restrooms	Condition of transit-oriented buildings and utilities
Safety and security	
Innovative designs	
Efficient energy use	
Facility management and maintenance	Expectations of transit-oriented buildings and utilities
Sustainability	
Technology and connectivity inclusion	
Unmotorized access	
Waiting areas and loading bays	
Regulations and enforcement	A regulatory framework for transit-oriented buildings and utilities
Improved transit-oriented buildings and utilities	
Facility management and maintenance	
Innovative designs	Required intervention
Regulations and enforcement	

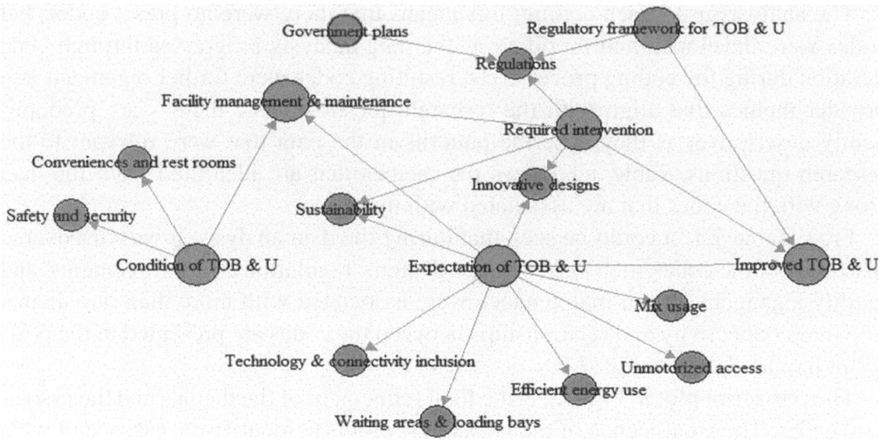

Fig. 7.6 Postscript plot of themes and codes for the study

7.7.4 Theme 1: Condition of Transit-Oriented Buildings and Utilities

The first theme—condition of transit-oriented buildings and utilities (Table 7.1)—seeks to appraise the current condition of buildings and utilities at major motor parks and garages within Lagos metropolis. All the panelists submitted that the buildings and allied utilities such as retail shops, restrooms, garage offices, and firefighting installations at the motor parks and garages are in poor physical and operational state. They explained that many of the motor parks and garages do not have toilet facilities, while the few existing toilet facilities are in a state of disrepair resulting in the practice of urinating and/or defecating in public open spaces around the park:

> *Some of the motor parks do not even have a toilet facility and you may be there at the park waiting for a bus that will take you to the next destination for the next hour. If someone wants to at least ease himself or whatever, at the end of the day the passenger will be looking for one corner to go and bend down and urinate and before you know it within a short time the whole place is messed up, and there are smells everywhere. (Panelist 1, male)*

> *Of course, the public toilet is almost non-existent because God help you, and you are at the park and something happens to you and you need to use the toilet, just keep praying because you just might be on your own. (Panelist 6, female)*

The stakeholders also lamented that many of the safety equipment that were installed at motor parks and garages are moribund which put transiting passengers and park operators at very high risk:

> *Some of the parks have firefighting equipment installation within them but they are not being maintained and may not function whenever the need arises. (Panelist 3, female)*

Taking the Ojota Park, for example, there used to be fire services in there, but it is no more there, so the issue of the post-construction operations are major issues across our motor parks and garages. (Panelist 4, male)

7.7.5 Theme 2: Expectation of Transit-Oriented Buildings and Utilities in Lagos State

The second theme—expectation of transit-oriented buildings and utilities—examined the expectation of stakeholders on what should be obtainable at the various public motor parks and garages within Lagos metropolis (Table 7.1). The stakeholders submitted that existing and future transit-oriented buildings and utilities across the various motor parks and garages are expected to facilitate a smooth and safe public transportation system. The buildings and utilities should be sustainable and they should be characterized by innovative designs, efficient facility and maintenance management system, provision for unmotorized access,and adequate waiting areas for passengers and goods:

Transit-oriented buildings, infrastructure, and utilities are supposed to be all-encompassing, in such a way that they are not just to be a park, but industry and lifestyle change are supposed to come along with it in line with the relevant SDGs (Sustainable Development Goals) vis-à-vis; sustainable buildings, healthier living, low poverty rate all together. (Panelist 4, male)

Our motor parks and garages should be operated in such a way that if you have your bus moving around six O'clock, you can go in the afternoon with your laptop and can stay there and connecting to the park public Wi-Fi, you could do all the work you want to do before you move on to your destination. (Panelist 1, male)

Let me go to the specifics, in the area of infrastructural provision, an effective or an ideal transit-oriented community is expected to have provisions for walking and cycling. (Panelist 5, male)

Although we have walking facility around some of the parks in Lagos, the question is, are they safe? In many cases, you have Okada using a pedestrian bridge or sharing walkways with pedestrians. Also, do we have the availability of space for cyclists on our roads without sharing the same carriageway with motorised vehicles? Do we have them in place? Of course, you and I will agree that the answer is NO. (Panelist 3, female)

Transit-oriented buildings and utilities are expected to encourage private-public investment and drive the economy of the places where they are situated. (Panelist 2, male)

What we have right now are touts and uniformed men manning the gates of our motor parks and garages. The touts and the policemen work hand in hand. The revenue that is generated from the motor parks are unaccounted for and thus constituting sustainability treat. Because it is when we can sustain these parks or garages, both economically and nvironmentally, that they would be attractive for private investors to come in. (Panelist 5, male)

Especially at a time like this when the revenue inflow from crude oil is dwindling, the government needs to think out of the box and make these investments viable to private investors to participate and drive sustainable developments through transit-oriented buildings and facilities. (Panelist 1, male)

7.7.6 Theme 3: Regulatory Framework for Transit-Oriented Buildings and Utilities

The third theme—regulatory framework for transit-oriented buildings and utilities—centers on regulatory frameworks for the construction of transit- oriented buildings and utilities. The stakeholders affirmed that there are government regulations concerning preconstruction, construction, and post construction of transit-oriented buildings and utilities in Lagos State. They, however, express concern on the lack of compliance by key players noting that approved designs are often varied during the construction or development phase of transit-oriented building projects:

Transit-oriented buildings especially garages and bus parks, are well captured in the Lagos State Urban Development Law (2010). The law stipulates required standards for bus stations, security system, loading bays, waiting areas, multi-level car parks, toilet and bathrooms, eateries, relaxation centres, fire stations, mini-marts, cinemas and administrative offices at the stations. (Panelist 3, female)

Taking the Ojota Park as an example, laws and regulations are governing the park operations, but the private operators there behave as if they don't know the law and they do as they like. (Panelist 2, male)

Also, the Federal Road Safety Corps (FRSC), in an attempt to regulate the public bus operators in 2007 initiated a programme called Road Transport Safety Standardisation Scheme which prescribes minimum safety standards for every operator in the industry especially the commuter bus and one of the requirements of the policy is that there must be adequate parking space that can accommodate their vehicle and mandate them to provide befitting space that can adequately accommodate their passengers. (Panelist 4, male)

7.7.7 Theme 4: Required Intervention

The fourth theme—required intervention—focuses on intervention exercise that could promote improvement in the physical and functional condition of transit- oriented buildings and utilities (Table 7.1). From the stakeholders' submissions on transit-oriented buildings and utilities, interventions are required in the areas of facility management and maintenance management, innovations in the designs for new and upgrade projects, and sustainable practices in the design, construction, and postconstruction operations of motor parks and garages:

The government should develop other means of transportation to ease the pressure on our roads. They should focus more on water transportation and rail transportation systems.

Public utilities like public toilets and lodging facilities should be provided alongside whenever motor parks and garages are being constructed. (Panelist 1, male)

More work needs to be done in the area of enforcement and control of transit-oriented buildings and facilities, I think we need to develop and apply technology. We need to deploy Electronic Transport System (ETS) like terminal management information system where we can control the inflow of vehicles; we need to bring in things like drones to monitor in real-time what happens in our terminals and the usage of this facilities, then, aside from that we need to build model parks and terminals through public-private participation and management. Above all, we must as a matter of responsibility strengthen community participation and collective identity, it is very, very key because those projects of government will never succeed without shared community participation and ownership. (Panelist 3, female)

7.8 Conclusion

The study investigated the conditions and the expectations of stakeholders on transit-oriented buildings and utilities with a view to developing a framework that could promote sustainability and improve the state of TOD&U in Lagos as it evolves into a smart city. The study sought the opinions of stakeholders who were experienced on the thematic areas. Each stakeholder's inputs were processed and analyzed.

Premised on the analyzed data, the study concludes that transit-oriented buildings and utilities in Lagos State are in deplorable state. Some of the challenges with TOD &U in Lagos State include poor physical and operational state of buildings and utilities, poor security, and low comfort level within various transit-oriented buildings. The majority of the TOD&U in Lagos State are socially obsolete, lacking state-of-the-art facilities such as free or open Wi- Fi, recreational facilities, and accommodation facilities for transiting users, among others. Moreover, there is a tradition of low-level compliance to physical planning and development regulations by developers and contractors that are awarded the contracts of constructing transit-oriented buildings which often results in premature failures and high level of defective systems and components during the use or operational phase of the facility. Furthermore, majority of the existing TOD&U in Lagos are operated by touts, thereby eroding the potential for maximizing the return on investment for the government which truncates revenue backflow and threatens economic sustainability of the facilities.

7.9 Recommendations for the Government, Investors, and the Community

The study recommends that government should be committed to routine maintenance of systems and services within TOD&U with a view to keeping the buildings and allied utilities such as retail shops, restrooms, garage offices, and firefighting installations in good physical and functional state. The government should consider upgrading existing TOD&U by way of refurbishment such that state-of-the-art facilities could be added to make the TOD&U best in class.

Moreover, government agencies that are saddled with construction and development control need to be more diligent in the area of enforcement and control during the construction of transit-oriented buildings and utilities. Government and partner operators of TOD&U should consider the development and deployment of ETS such as terminal management information systems. This could enhance the operations of the TODs, improve users' experience, and boost the economic performance of the facilities.

The government should consider investing in other modes of transportation within the state. Attention should particularly be paid to water and rail transportation systems in order to reduce the pressure on road transportation and allied infrastructure. The government should partner with private investors on such diversification. Such partnership could reduce the financial burden of such capital project on government and could also guarantee the project's economic sustainability.

Lastly, government should be committed to the consideration for facility management and maintenance management of proposed TOD projects right from the planning stage. Moreover, the government should encourage TOD stakeholders on innovations and best practices coupled with behavioral conduct and transparency with respect to design, construction, and upgrade of existing projects.

References

Abdullah, J., & Mazlan, M. H. (2016). Characteristics of and quality of life in a transit oriented development (TOD) of Bandar Sri Permaisuri, Kuala Lumpur. *Procedia: Social and Behavioral Sciences, 234*, 498–505. https://doi.org/10.1016/j.sbspro.2016.10.268

Adeyinka, A. M. (2013). Assessment of the quality of urban transport services in Nigeria. *Academic Journal of Interdisciplinary Studies, 2*(1), 49–58.

Aleshinloye, K. (2014). *What does the future of sustainable transport look like in Lagos? The city fix*. https://thecityfix.com/blog/future-sustainable-transport-lagos-nigeria-eko-rail-brt-cable-car-ferry-climate-kasope-aleshinloye/

American Psychological Association (Ed.). (2010). *Publication manual of the American Psychological Association* (6th ed.). American Psychological Association.

Bardaka, E., & Hersey, J. (2019). Comparing the travel behaviour of affordable and market-rate housing residents in the transit-rich neighbourhoods of Denver, CO. *Travel Behaviour and Society, 15*, 74–87. https://doi.org/10.1016/j.tbs.2019.01.001

Barter, P. A. (2011). Parking requirement in some major Asian cities. *Transportation Research Record, 2245*(1), 79–86.

Berawi, M. A., Saroji, G., Iskandar, F. A., Ibrahim, B. E., Miraj, P., & Sari, M. (2020). Optimizing land use allocation of transit-oriented development (TOD) to generate maximum ridership. *Sustainability (Switzerland), 12*, 1–20. https://doi.org/10.3390/su12093798

Braun, V., & Clark, V. (2006). Using thematic analysis in psychology. *Qualitative Research in Psychology, 3*(2), 77–101.

Centre for Neighbourhood Technology. (2013). *Transit-Oriented Development in the Chicago region: Efficient and resilient communities for the 21st century.*

City of Maryland Heights Community Development. (n.d.). *Characteristics of a successful Transit-Oriented Development.*

Currie, G. (2006). Bus transit-oriented development—Strengths and challenges relative to rail. *Journal of Public Transportation, 9*(4), 1–7.

Dallas Area Rapid Transit (DART). (2008). *Transit-Oriented Development (TOD) guidelines: Promoting TOD around DART transit facilities.*

Dallas Area Rapid Transit (DART) Reference Book. (2019). *Dallas area RapidTransit reference book* (10th ed.). Dallas Area Rapid Transit. https://doi.org/10.1016/j.ajodo.2019.01.008

Daqrouq, A., & Anjomani, A. (2019). Public transit ridership and car-oriented cities: The case of the Dallas region. *Economies, 7*(3), 1–17. https://doi.org/10.3390/economies7030086

De Vos, J., Van Acker, V., & Witlox, F. (2014). The influence of attitudes on transit-oriented development: An explorative analysis. *Transport Policy, 35*, 326–329. https://doi.org/10.1016/j.tranpol.2014.04.004

Denver City. (2014). *Transit-oriented Denver: Transit-Oriented Development strategic plan.*

Denver Union Station. (2023). *Denver Union Train and Bus Station*, Colorado [Online image]. https://help.g2rail.com/stations/denver-union-station-colorado

Dittmar, H., & Ohland, G. (2004). *Defining transit-oriented development: The new regional building block.* Island Press.

Economic and Social Commission for Asia and the Pacific. (2017). *Transport and sustainable development goals* (Issue 87).

Edema, J. (2019). *Poor public transport infrastructure in Lagos Nigeria, how sustainable improvement could enhance well-being of the people and provide environmental benefits.*

ESCAP. (2016). Sustainable development goals and transport. *Economic and Social Council for Asia and the Pacific: Ministerial Conference on Transport, 3*(2), 1–20. https://doi.org/10.1017/s0020818300034901

Ewing, R., Tian, G., Lyons, T., & Terzano, K. (2017). Trip and parking generation at transit-oriented developments: Five US case studies. *Landscape and Urban Planning, 160*, 69–78. https://doi.org/10.1016/j.landurbplan.2016.12.002

Food and Agriculture Organisation. (2017). *The 2030 agenda and the sustainable development goals: The challenge for aquaculture development and management* (Vol. 1141).

Gifford, J., Pelletiere, D., & Collura, J. (2001). Stakeholder requirements for traffic signal preemption and priority in the Washington, D.C., region. *Journal of the Transportation Research Board, 1748*, 1–7.

Gomez, C. P., Omar, M., & Nallusamy, R. (2019). A study on the benefits of transit-oriented development in Malaysia and incorporation of those benefits in planning. *MATEC Web of Conferences, 266*, 06016. https://doi.org/10.1051/matecconf/201926606016

Ibraeva, A., de Ameida-Correia, G. H., Silva, C., & Antunes, A. P. (2020). Transit-oriented development: A review of research achievements and challenges. *Transportation Research Part A: Policy and Practice, 132*, 110–130.

Institute for Public Administration. (n.d.). *What is transit-oriented design.* https://www.completecommunitiesde.org/planning/complete-streets/tod/

International Association of Public Transport. (2019). *Mobility and the SDGS: A safe, affordable, accessible and sustainable transportation system for all.*

Jacobson, J., & Forsyth, A. (2008). Seven American TODs: Good practices for urban design in transit-oriented development projects. *Journal of Transport and Land Use, 1*(2), 51–88. https://doi.org/10.5198/jtlu.v1i2.67

Jenks, C. W. (2005). *Transit-Oriented Development: Developing a strategy to measure success. Research Results Digest 294*. Transportation Research Board of the National Academies.

Kamruzzaman, M., Baker, D., Washington, S., & Turrell, G. (2014). Advance transit-oriented development typology: Case study in Brisbane, Australia. *Journal of Transport Geography, 34*, 54–70. https://doi.org/10.1016/j.jtrangeo.2013.11.002

Kidokoro, T. (2019). Transit-oriented development policies and station area development in Asian cities. *ADBI Working Paper 947*. Tokyo: Asian Development Bank Institute. https://doi.org/10.2139/ssrn.3470063.

Kwoka, G. J., Boschmann, E. E., & Goetz, A. R. (2015). The impact of transit station areas on the travel behaviours of workers in Denver, Colorado. *Transportation Research Part A: Policy and Practice, 80*, 277–287. https://doi.org/10.1016/j.tra.2015.08.004

Li, C. N., Lin, C., & Hsieh, T. K. (2016). TOD district planning based on residents' perspectives. *ISPRS International Journal of Geo- Information, 5*(4), 1–18. https://doi.org/10.3390/ijgi5040052

Mockingbird Station. (2023). *Mockingbird Train Station*, Dallas Texas [Online image]. https://www.bddeng.com/project/mockingbird-station/

Nawaz, M., Somenahalli, S., & Allan, A. (2017). Bus-based transit-oriented development (BTOD): Opportunities and challenges for low-density, car-dependent Australian cities. *Road & Transport Research: A Journal of Australian and New Zealand Research and Practice, 26*(2), 62–73.

OECD. (2013). *Recent development in rail transportation services*.

Oliveira, A., Calili, R., Almeida, M. F., & Sousa, M. (2019). A systemic and contextual framework to define a country's 2030 agenda from a foresight perspective. *Sustainability (Switzerland), 11*(22), 1–28. https://doi.org/10.3390/su11226360

Oshodi, L. (2016). *International development, urban infrastructure and governance: Transportation and mobility systems in Lagos*. https://oshlookman.wordpress.com/2016/08/12/transportation-and-mobility-system-in-lagos/

Paoli, A. D., & Addeo, F. (2019). Assessing SDGs: A methodology to measure sustainability. *Athens Journal of Social Sciences, 6*(3), 229–250. https://doi.org/10.30958/ajss.6-3-4

Pennsylvania Department of Transportation. (2012). *Pennsylvania Transit- Oriented Development: Toolkit for designing + building communities around transit*.

PNGkit. (2023). *The map of Lagos State* [Online image]. https://www.pngkit.com/bigpic/u2e6e6w7o0a9r5t4/

Ratner, K. A., & Goetz, A. R. (2013). The reshaping of land use and urban form in Denver through transit-oriented development. *Cities, 30*, 31–46. https://doi.org/10.1016/j.cities.2012.08.007

Renne, J. L. (2016). *Transit-Oriented Development*. https://doi.org/10.4324/9781315550008

Renne, J., & Wells, J. S. (2005). *Transit-Oriented Development: Developing a strategy to measure success*.

Rodriguez, D. A., & Targa, F. (2004). Value of accessibility to Bogatas bus rapid transit system. *Transport Reviews, 24*(5), 587–610.

Scharlemann, J. P. W., Brock, R. C., Balfour, N., Brown, C., Burgess, N. D., Guth, M. K., Ingram, D. J., Lane, R., Martin, J. G. C., Wicander, S., & Kapos, V. (2020). Towards understanding interactions between sustainable development goals: The role of environment–human linkages. *Sustainability Science*, 1–12. https://doi.org/10.1007/s11625-020-00799-6

Thilakshan, T., & Bandara, S. J. (2019). Identification of relevant sustainable transportation links to sustainable development goals (SDGs) in the national context. *Annual Sessions of IESL, 10*, 341–348.

Trepci, E., Maghelal, P., & Azar, E. (2020). Effect of densification and compactness on urban building energy consumption: Case of a transit-oriented development in Dallas, TX. *Sustainable Cities and Society, 56*, 101987. https://doi.org/10.1016/j.scs.2019.101987

UN-Habitat, UNEP & SLoCaT. (2015). *Analysis of the transport relevance of each of the 17 SDGs*.

United Nation Development Programme (UNDP). (2015). Sustainable development goals. *United Nations, 5*(2).

United Nations. (2018a). *The 2030 Agenda and the sustainable development goals: An opportunity for Latin America and the Caribbean*.

United Nations. (2018b). *World urbanisation prospects: The 2018 revision*.

Widyahari, N. L. A., & Indradjati, P. N. (2015). The potential of transit-oriented development (TOD) and its opportunity in Bandung metropolitan area. *Procedia Environmental Sciences, 28*, 474–482. https://doi.org/10.1016/j.proenv.2015.07.057

World Economic Forum. (2016). *Inspiring future cities & urban services shaping the future of urban development & services initiative*.

Xu, W. A., Guthrie, A., Fan, Y., & Li, Y. (2017). Transit-oriented development in China: Literature review and evaluation of TOD potential across 50 Chinese cities. *Journal of Transport and Land Use, 10*(1), 743–762. https://doi.org/10.5198/jtlu.2017.1217

Xueming, C. H. E. N. (2010). Prospect of the transit-oriented development in China. *Management Research and Practice, 2*(1), 83–93.

Chapter 8
Transit-Oriented Development and Pro-poor Development in African Cities

Taibat Lawanson, Richard Unuigboje, Olamide Udoma-Ejorh, Ademola Omoegun, Olufemi Saibu, and Taofiki Salau

8.1 Introduction

Transportation is a major challenge as the mass of residents and goods required to meet their needs have to be moved efficiently in order for cities to function effectively. Over time, various strategies have been proposed to develop sustainable urban mobility and transport systems. In the last two decades, transit-oriented development (TOD) has emerged globally as a critical approach to achieve this aim (Suzuki et al., 2013; Padeiro et al., 2019).

The concept originated in the United States as a response to prevailing mobility challenges in cities. TOD became prominent after Peter Calthorpe introduced it in his book, *The Next American Metropolis* in 1993. TOD was a contrast to conventional, auto-dependent development which was dominant in the United States following the Second World War which created negative societal outcomes such as

T. Lawanson (✉)
Department of Urban and Regional Planning/Centre for Housing and Sustainable Development, University of Lagos, Lagos, Nigeria
e-mail: tlawanson@unilag.edu.ng

R. Unuigboje
Department of Urban and Regional Planning, University of Lagos, Lagos, Nigeria

O. Udoma-Ejorh · T. Salau
Lagos Urban Development Initiative, Lagos, Nigeria

A. Omoegun
Department of Architecture, University of Lagos, Lagos, Nigeria

O. Saibu
Department of Economics, University of Lagos, Lagos, Nigeria

T. Nubi et al. (eds.), *Transit Oriented Development in West African Cities*, Urban Perspectives from the Global South,
https://doi.org/10.1007/978-3-031-58726-9_8

traffic congestion, air pollution, and urban sprawl. Consequently, American cities adopted TOD to improve local living conditions (Xu et al., 2017).

TOD is a model of urban development that promotes high-quality, compact, mixed-use (residential and commercial), multifunctional and pedestrian-friendly densification, situated in close proximity to transit nodes and corridors, which seeks to encourage people to live near and use public transit with the aim of reducing land consumption and car dependency as well as mitigating the negative effects of motorization (Suzuki et al., 2013; Feudo, 2014; Xu et al., 2017; Padeiro et al., 2019). However, implementing TOD is generally expensive and difficult to finance, thus requiring major infrastructure investments and various financing mechanisms (United States Environmental Protection Agency, 2013).

Haas (2020) opined that TOD significantly enhances integrated multimodal transport planning as it combines transport and land-use planning to ensure that different public transport options are in close proximity to passengers. He highlights the importance of integrating informal private sector transport providers alongside formal operators. Other researchers also argue for the integration of informal private sector transport providers, such as bikes, mini-buses, rickshaws, etc., and careful consideration of local contextual factors in the adoption of TOD principles in developing countries (IBI Group, 2016; Talamini & Ferreira, 2019), particularly in Africa (Klopp et al., 2017; Pieterse & Owens, 2018), where urbanization is characterized by a huge survivalist informal sector (Watson, 2007; Jenkins et al., 2007).

Informal actors are major but controversial participants (workers and customers) in transportation, providing a vital means of mobility and employment especially for the poor. Nevertheless, often unsuccessful efforts toward transport improvements seek their replacement with formal (high- capacity transport) services (Collier et al., 2019). Although TOD is relatively new in Africa, useful examples of the implementation of its principles through the BRT system in Accra, Lagos, and Johannesburg suggests the need for innovative collaboration with informal operators as being key for success (Pieterse & Owens, 2018; Collier et al., 2019).

In order to further explore the connection between TOD and the poor in African cities, this paper starts with a review of transportation, pro-poor development, and urban governance in Africa. Thereafter, conceptual linkages between TOD, the SDGs, and pro-poor development are explored. Some case studies of TOD are then reviewed with key issues and priorities for the implementation of TOD in Lagos subsequently elaborated. The final part of the paper provides key recommendations for improving transportation in Lagos with implications for other African cities.

8.2 Review of Relationship Between Transportation Challenges and Pro- poor Development and Governance in African Cities

As a result of migration and annual population growth rates at around 4.5%, urban areas in most parts of Africa are faced with major development challenges, with levels of access and mobility considered to be dramatically low for meeting current and future needs of the populations. It is projected that over the next 20 years, there will be an additional 300 million urban residents in Africa, and by 2050, 60% of all Africans will live in an urban area (United Nations, 2018), with many of them operating in the informal economy and residing in slums and informal settlements.

Congestion is already hampering the movement of people and goods in many cities and economic zones and causes environmental, health, and economic concerns which are challenges for transport. Crippling congestion in urban areas is aggravated by inadequate policy frameworks and a weak capacity to address the environmental, social, and safety risks of fast-growing motorization.

8.2.1 Transport and the Urban Poor

There are socioeconomic differences in how transit is used in African cities, as elsewhere globally. More than 75% of total daily trips made by Africa's poor are by walking, compared to 45% by the more affluent (Future Cape Town, 2018). Despite the large percent zage of the population relying on nonmotorized transport, pedestrian and bike-friendly city infrastructure is uncommon or in a poor state. Most cities in Sub-Saharan Africa depend on informal transportation as the collective mobility/mass transit that allows people to connect and engage with the city. Communal taxis and motor-taxis are popular, accounting for between 75% and 80% of total motorized trips.

This is consistent with the continent's low rates of car ownership: 30–70 vehicles per 1000 people, according to a 2011 survey of Kampala, Lagos, and Douala (Kumar, 2011).

Chatterton (2018) asserted that investment in developing cities has focused primarily on profit generation and globalized capital growth, rather than meeting the needs of the urban poor in cities. The transport needs of the poor have been neglected or invisible partly because walking itself has also been too often ignored and not even considered to be transport at all. People who are economically, physically, and socially disadvantaged are harmed by transport policies that focus on economic efficiency and automobile-focused transport priorities that do nothing to meet their travel needs. The poor cannot afford private cars. Nor can the very poor afford the commercial motorcycles that abound in many African cities. The prevalence of long walking trips is a key indicator of poor accessibility by the urban poor to facilities and affordable mobility.

The main elements related to conditions of transport and social disadvantages include limited access to travel (limited access to public transport or a car), prohibitive fare costs, and limited or nonexistent information. Several cities in Africa are faced with the effect of insufficient and flawed public transport, within the scope of journey time, affordability, quality, comfort, accessibility, and security (Badejo, 2009). Furthermore, gendered constraints regarding safety and security are also common occurrences.

8.2.2 Availability, Accessibility, and Affordability

On the social side, elements that influence conditions of disadvantage include the lack of jobs and qualifications, poor housing, and low income, which can adversely affect the ability of poorer people to access opportunities. The relationships between these types of disadvantages are straightforward and show that a lack of assets can be worsened by poor transport provision, increasing barriers to access activities and leading to processes of social exclusion. According to O'Hara (2015), "access to affordable and reliable public transport can be the difference between getting employed or being able to take advantage of education opportunities that might aid social mobility."

Affordability is seldom recognized as a transportation planning objective, and if it is, it is usually evaluated based simply on fuel costs. Conventional planning assumes that faster modes are more important than slower modes and that congestion reduction is the most important planning objective and so favors wider roads with higher design speeds and longer block lengths over narrower, lower-speed, more connected roads. Similarly, conventional planning assumes that it is desirable to require generous parking supply at new developments which stimulates sprawl and reduces housing affordability.

Conditions of social vulnerability force people in poverty to live in densely populated slums, with no access to jobs and where roads are generally unpaved, narrow, and poorly maintained. In fact, Kanter (2015) asserts that "the poor are often stuck in urban neighbourhoods with the fewest transportation options for getting to jobs, childcare and healthcare." The transport environment in many deprived areas and the high dependence on walking for people living in these areas mean that they are often exposed to the negative impacts of transport such as pedestrian fatalities and injuries, pollution, and isolation resulting from living near busy roads that are difficult to cross (Lawanson, 2019). These distributional effects increase the vulnerability of poor families to transport externalities. This is aside from the fact that transportation in Africa is expensive—with commuters spending as much as 40% of their income on transportation in cities like Lagos (Nagaba, 2019).

8.2.3 Quality

The lack of coordinated planning of land use and transport leads to inefficient cities, congestion, and unsatisfied transport demand, for the poorer segment of the population. Beard et al. (2016) reported that up to 70% of city residents in the Global South may be underserved, lacking access to one or more core services, such as housing, sanitation, energy, and transportation.

The poor and inadequate state of much of Africa's transport network is holding many countries back from competing effectively on the global market. Effective transport networks are a key component of the investment climate, enabling people and goods access to markets and reducing the costs of doing business. Good infrastructure has historically proven to be a key factor in economic development and is a precondition for, and an enabler of, growth for private sector development and for trade. While building and maintaining roads and expanding ports boost output and jobs, it must be said that the poor state of transport infrastructure has blocked the quick movement of goods and people in the continent and increased transport costs. Africa's transport costs, local, national, and international, are around twice as high as those for a typical Asian country (Teravaninthorn & Raballand, 2008).

8.2.4 Governance in Transportation

Transport governance is key to effective policy, planning, and implementation. The transport governance system has a unique set of characteristics because both the public and private sectors are actively involved, and in the case of African cities, it has an active informal economy. Furthermore, there is often cross-jurisdiction of infrastructure such as highway, rail, and telecommunication networks. A study by Christie, Smith, and Conroy identified key transport governance issues in African cities to include political interference in projects, limited or nonexistent integrated transport sector policy, dominance of roads, poor implementation of institutional and regulatory arrangements, unrealistic budgets, long-term underinvestment, and nonprioritization of maintenance across subsectors.

In Africa, billions of dollars have been spent on improving and rehabilitating transport infrastructure, but it has been long recognized that the poor performance of the transport sector is due to far more than merely inadequate finance or technical capacity constraints (Christie et al., 2013). For instance, the Lagos Strategic Transport Master Plan (STMP) was developed as a long-term plan focused on integration of the mass rapid transit system, developing the waterways and rail networks, installing a modern intelligent transport system, and introducing walking and cycling facilities (Oshodi et al., 2016). The implementation of the plan is hinged on extensive public sector borrowing (Saibu, 2018) that discountenances the value currently provided by the informal sector that currently operates over 90% of public transport services in the city.

8.2.5 Transit-Oriented Development, Sustainable Development Goals, and Pro-poor Development: The Conceptual Linkages

The Sustainable Development Goals (SDGs) were developed to guide economic growth, ensure environmentally sustainable development, and end poverty. Within the framework of the SDGs, cities and transport take prominent roles. UN-Habitat,

UNEP, and SLOCAT (2015) illustrate the correlation between transport and each of the SDGs; this has been highlighted as important because most SDGs are dependent on transport to be achieved. Out of the 17 SDGs, there are several that directly address transport:

- Target 3.6: By 2020, halve the number of global deaths and injuries from road traffic accidents.
- Target 7.3: By 2030, double the global rate of improvement in energy efficiency.
- Target 9.1: By 2030, develop quality, reliable, sustainable, and resilient infrastructure, including regional and transborder infrastructure, to support economic development and human well-being, with a focus on affordable and equitable access for all.
- Target 11.2: By 2030, provide access to safe, affordable, accessible, and sustainable transport systems for all, improving road safety, notably by expanding public transport, with special attention to the needs of those in vulnerable situations, women, children, persons with disabilities, and older persons.
- Target 12.c: Rationalize inefficient fossil-fuel subsidies that encourage wasteful consumption by removing market distortions.

In addition, transport is an enabler for achieving other targets such as Targets 2.3 (agricultural productivity), 3.9 (reduction in air pollution), 6.1 (improve access to safe water), 11.6 (sustainable cities), 12.3 (reduce food waste), and 13.1 (enhance climate resilience).

Transit-oriented development (TOD) has become prominent in the sustainable development discourse in recent times. Renne (2009) defines TOD succinctly as compact, mixed-use developments that facilitate walking, bicycling, and using public transport through its urban design. This description hinges on the idea of sustainable development from a land use and mobility perspective.

According to Renne (2009), this context of sustainable development seeks to "create an urban environment which maximises economic development and social equity, whilst minimising negative externalities upon the natural environment." According to Deloitte (2019), TOD has the potential to accelerate true city building and smart cities for the benefit of all.

TOD achieves economic development by introducing mass transit (such as rail or bus) in close proximity to residential and or commercial areas. This leads to increased and/or sustained property values. It also brings people closer to more opportunities and jobs due to better accessibility to other parts of a city. In addition, introducing public transport will decrease the cost of transportation, freeing up more household income for other purposes. Having public transport in close proximity to residential and or commercial areas reduces household driving which therefore decreases congestion, air pollution, and greenhouse gas emissions. There is a 30–50% reduction in per capita annual congestion delay and 20–40% less transportation energy consumption in TOD in comparison to auto-oriented cities (Jaiswal et al., 2012). TOD also encourages a healthier and more active lifestyle by having walkable and compact communities. There is also an element of safety improvements as studies have shown that residents of TOD communities have lower per-capita traffic fatality rate as residents of cities that rely on automobiles for the majority of their transportation needs (Jaiswal et al., 2012; Pongprasert & Kubota, 2019).

The economic, environmental, and lifestyle benefits of TOD all add to social equity; however, the rise in property value can have an adverse effect on residents and business owners as it can lead to gentrification and consequent displacement of low-income groups. Low-income communities are often characterized by a larger proportion of renters, who are more likely to be displaced by increasing rents. Displacement further exacerbates poverty because there is a big risk that the displaced population will be forced to move further away from jobs and families due to rent prices, increasing travel distances and therefore the cost of transportation. In addition, the threat and process of eviction can cause negative psychological issues, and displaced persons might accept more expensive, precarious, or overcrowded housing (Padeiro et al., 2019). In addition, other transit choices which the low-income population rely on may be pushed aside due to the new bus or rail line being implemented as part of TOD.

TOD in theory can achieve all transport-related SDGs. However, in practice, it often goes against SDG 1 (no poverty). TOD has the likelihood to exacerbate poverty. For TOD to counteract market forces and to ensure social equity, other policies and processes must be implemented in addition. Chief of this is social inclusion, for example, ensuring there is a diverse range of affordable housing in close proximity to the transport hub. Including higher-income housing will lead to benefits such as better education outcomes for low-income children as well as new and better services for the neighborhood (Padeiro et al., 2019).

For TOD to improve social equity, residents need to be part of the process of development. They need to agree on their goals for the community, and the policies need to be aligned with these. TOD needs to go hand in hand with policies that secure and increase the range of affordable housing available to low- and moderate-income residents and business owners (Lung-Amam et al., 2014). TOD must also align with already existing routes and sites that benefit low- income riders. In addition, TOD requires appropriate mitigation measures and fair compensation for those who are inconvenienced or displaced by construction. All levels of government have a huge role to play in not just developing the policies but supporting it at all levels and providing the required finances and subsidies where necessary.

If all these can be achieved with government buy-in and innovative land-use and housing policies, TOD can truly be what it says on paper, a transportation tool to deliver sustainable development.

Examples to illustrate how TOD has impacted on city structure and the people are discussed in subsequent sections.

8.3 Case Studies

Efforts to adopt transit-oriented development in Africa are recent. Dar-es-Salaam, Johannesburg, Addis-Ababa, and Lagos are some of the cities that have adopted this model. This includes the use of bus rapid transits (BRT) and light rails, all of which are modes under transit-oriented development (TOD). The pros and cons of this model in these cities and two others in developing countries are discussed below.

8.3.1 Addis-Ababa, Ethiopia

Addis-Ababa implemented the first light rail transit in Sub-Saharan Africa in 2015 as a strategic approach to rapid urbanization, growing automobile rate, and the transport demand and supply gap. Evidence reveals that the light rail has provided cheaper transport access and assisted the unemployed urban residents to search more intensively for new jobs. It is currently in operation at full capacity, serving up to 60,000 passengers across four lines, every hour (C40, 2016). However, there have also been growing complaints by motorists about the rails cutting off key turning points. It has also kept businesses on the wrong side of the tracks as they face economic losses due to customers' refusal to walk all the way down to the nearest pedestrian crossing to purchase a good/service (Alade, Edelenbos & Gianoli, 2020).

8.3.2 Dar-es-Salaam, Tanzania

The BRT system evolved as an economical transit option to cushion the pressure on existing public transport, create order in the hierarchy of roads and transit modes, transform the informal landscape of the city, and ensure low- income earners reach their job locations (Chengula & Kombe, 2017). The city implemented the Dar-es-Salaam Bus Rapid Transit (DBRT) system as a response to the transport challenges, the first East African country to possess such a system. In over a year of its operation, there is an observed increase in the strength of economic activities, mushrooming of developments and increased in land value along the corridor. Despite the completion of the first phase, there exists an exclusion of the urban poor. An origin and destination commute along the BRT corridor costs around 1300 shillings which amounts to 30% of the poor's daily earnings spent on transport as compared to the cost of commuting on minibuses (dala dala), which is 800 shillings. Furthermore, informal drivers are out of jobs due to the introduction of the BRT systems with no current plans of integrating these public transport workers into the new transit system (Rizzo, 2019).

8.3.3 Johannesburg, South Africa

Johannesburg introduced the Corridors of Freedom in 2013, a TOD initiative to address the challenges of spatial isolation of the urban poor, racial divides, and disconnect between housing and jobs (Pieterse & Owens, 2018). These challenges are recognized as impediments to the city's growth and development strategy for 2040. In Soweto, one of Johannesburg's black-populated areas with over 1.3 million residents, the BRT (Rea Vaya) has aided the commuters (low- income and informal workers) reach opportunities in the central business districts, minimized the spatial inequality between Soweto and Central Johannesburg, and garnered more investment on transit access and has contributed to urban regeneration of slum areas through the redevelopment of neighborhoods with mixed uses. However, the project has also come under criticism and controversy as it exempts the urban poor due to the concentration of public investment in wealthy communities along the corridor (Harrison et al., 2019).

8.3.4 Lagos, Nigeria

In Lagos, the Oshodi Transport Interchange sits within two urban development plans, namely, the Oshodi CBD Masterplan and the Bus Reform Initiative, kicking off the transformation of Oshodi into what is described as a world-class Central Business District (CBD). It is a $70 m project of three bus terminals and is expected to help one million Lagosians commute every day. The project compliments the expansion of the BRT lanes from Mile 12 to Ikorodu and links with the proposed Red Line (31-km railway project) (Ekechukwu, 2019). One terminal was opened in 2019, 3 years behind schedule, while the other two are yet to commence operations.

While the project's objective is to address the traffic, environmental safety, and transportation problems in and around Oshodi as well as to generate employment and enhance connectivity, it has also led to the spatial displacement of residents and traders (Omoegun, 2015). Apart from the street traders who were forcefully removed from the area, a large section of the Oshodi market was demolished to make way for the bus interchange without contingency plans for effective resettlement. There was also a massive demolition of shanties and the militarization of the area by the positioning of a special squad of police to patrol the area.

8.3.5 Curitiba, Brazil

The BRT system in Curitiba is regarded as a model of urban sustainability. For this, it is described as a "best practice city" (Lindau et al., 2010), an "eco-city" (Barth, 2014). Curitiba has a planned transportation system, which includes lanes on major streets devoted to a bus rapid transit system. The BRT system has a system length of 76.6 km with 123 bus stations carrying 619,500 persons daily. Spatially, the BRT system aided the successful regeneration of Curitiba as neighborhoods and schools were reorganized densely along the corridor. The economic benefit reflects that Curitibanos spend about 10% of their income on travel, which is low, compared to other regions in Brazil. Further evidence reveals that 46% of total trips are made by public transport (Global BRT Data, 2016). Despite the positives, the BRT system has failed to integrate its growing suburbs into a consistent regional plan. This has exempted the suburban and low-income residents from accessing the most livable parts of the city (Scruggs, 2013). Furthermore, the concentration on the formal city and inadequacy to access municipal housing has led to the growth of informal settlements at the periphery of the city.

8.3.6 Ahmedabad, India

In 2020, Ahmedabad was the fifth largest city in India with over 7 million inhabitants having a population density of 9900 per km^2 (World Population Prospect, 2022). The TOD initiative was centered on a mobility plan called *Accessible Ahmedabad*, focusing on economic, social, and physical accessibility. The corridor was designed to respond to peculiar challenges of the area in connection to busy places while avoiding busy roads. Despite its success in circulation improvement and moving people more swiftly, evidence reveals that it did not bring about the mix in urban activities. Prices of land near stations doubled between 2006 and 2011. With the absence of a land value capture mechanisms, private developers have been the major beneficiaries of the initiative. In the course of promoting TOD, there have been land fragmentations, informality in land transactions, degraded buildings, and urban sprawl (Suzuki et al., 2013). These case studies all reveal that transit-oriented development is a double-edged concept—providing value for economic development and compact city design, alongside with the unintended consequence of alienating the poor from the very opportunities they require for a better life. The examples of Johannesburg, Addis Ababa, and Dar es Salaam show that while transit-oriented development is a strategic planning approach for integrating the city, it also serves to increase social exclusion of low-income communities and job losses within the informal economy. It is important to examine the urban transport environment (infrastructure, policies, innovations, and people's need) from a transdisciplinary lens and ensure that the programs being developed are multisectoral, multiscalar, and inclusive.

8.4 Thematic-Based Issues and Priorities for Lagos from a Transdisciplinary Lens

This section seeks to identify the most pressing transportation challenges in Lagos and understand how recent TOD initiatives and transport innovations (BRT and ride hailing) impact the urban poor.

Unprecedented urban growth in Lagos, coupled with the lack of integration between urban planning and transportation planning systems (see Fig. 8.1), has led to several challenges in the city. This ranges from managerial/administrative, environmental, physical, and sociocultural challenges.

The Lagos transportation sector is multilayered, with both formal and informal actors providing various services to different categories of commuters. Transportation-related challenges include insufficient transportation modes and the lack of integration between modes (see Fig. 8.1), high transport costs, traffic congestion, weak freight sector management, poor traffic regulation and management, and inadequate transport infrastructure. Others are conceptualization/practice gap in transportation and mobility as well as poor transportation planning and the lack of coherent and consistent policies, weak linkages to urban planning and regional development frameworks, political interference and poor institutional structures, as well as dominance of informal sector and commuter behavioral challenges (see Table 8.1).

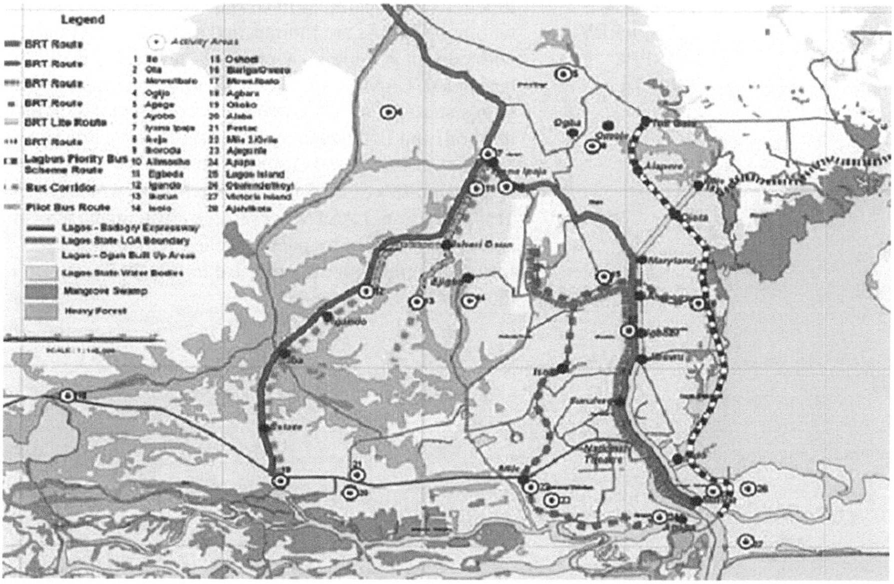

Fig. 8.1 Components of intermodal transport system in Lagos State. (Source: Lagos Strategic transport master plan)

Table 8.1 Overview of actors and challenges in Lagos transportation sector

Sector	Main actors	Major challenge
Private sector	– Public transport service operators such as BRT franchisee, LAGFERRY, and LAGBUS – Ride-hailing companies – Logistics companies	Public transport service businesses are unable to fund the extensive service structure required to support their businesses, especially under a regulated cost structure. The water transport sector is underexploited due to the cost and water apathy on the part of many Lagosians. Technology-enabled/ride-hailing companies and logistic companies are subject to changing government policies, while their riders face unfair labor practices
Private sector (informal businesses)	– Danfo – Keke Marwa – Commercial motorcyclist – Street traders	The yellow buses move over 80% of commuters in Lagos and are mostly poorly maintained, and their drivers were prone to reckless driving and violence. The commercial motorcyclists and tricycles are banned in many parts of the city, with their services subject to violent enforcement. These informal sectors are also subject to changing policies with little or no notice of changes
Policy maker	– Ministry of Transportation – Lagos house of assembly	The ministry of transportation has oversight on the various transport agencies and interfaces with various government arms on transport-related issues. There are many cases of agencies having overlapping functions and operating in silos
Policy implementation agency	– LAMATA. – LASTMA – LAGFERRY – LASEPA – LASWA	There are many cases of agencies having overlapping functions and operating in silos. Funding and capacity within the MDAs are limited, and projects need to align with current governor's agenda, or they are unlikely to be funded. LAMATA is saddled with implementing the Lagos strategic transportation master plan of which the light rail and BRT schemes are part of. There are frequent jurisdictional clashes with statutory responsibilities of LASWA and the Ministry of Transportation. LASTMA oversees traffic management and is encumbered by poorly trained staff who are too few to grapple with the extensive traffic challenges of the city

(continued)

Table 8.1 (continued)

Sector	Main actors	Major challenge
Commuter/ general public	– Private car owners – People living with disability (PLWD) – Cyclists – Elderly – Visitors to the city – School children – Market women – Corporate sector	Not part of urban development process and so are always playing a reactive role. In addition, there are a high number of accidents due to noncompliance with traffic laws, poorly maintained cars, the lack of signage, and traffic congestion. Private car owners are usually single-user occupants, who contribute to the traffic issue. They travel on poorly maintained roads and so spend more money on vehicle repairs. PLWD and the elderly are unable to navigate the city without assistance as inclusive road infrastructure and/or transport services are negligible. There are very few roads with dedicated cycle tracks; hence, cyclists have to contend with safety risks when moving around. Traffic and transportation information such as route mapping around the city is poor; hence, new visitors to Lagos are unable to navigate the city independently. There is also a clear gender gap in public transportation provisioning as the needs of women (especially those in the informal economy), and children are often not considered in planning and infrastructure

With these challenges showing no signs of abating, the public—especially the urban poor—are negatively impacted. This is due primarily to the fact that the urban poor rely on the transportation sector for daily commuting around the city (public transport) and also for livelihoods (street trading and para-transport).

The growing transport challenges have seen various initiatives and actions come to play, especially smart mobility and technology. Lagos has seen its fair share of smart mobility in action with the bus rapid transit (BRT) and transport network companies (Uber, Bolt, Opay, Max, among others). So far, these initiatives have impacted positively on the economy in terms of employment opportunities and improvement in access for commuters. With the introduction of the BRT, there was an instant and huge impact, with many corporate sector workers opting for it due to efficiency and cleanliness. Due to managerial inefficiency and poor maintenance of vehicles, the impact is fast dwindling. In addition, there are inconsistencies in the policy landscape governing the gig economy; hence, the initial optimism concerning the technology-enabled taxi service is fast dwindling.

The transport network companies have broadened transport options, enhanced accessibility, provided relatively standard service with a satisfactory level of convenience and security, reduced transport cost, created an accountable and transparent platform for commuters to interface with service providers, and have stabilized the economy for people. In reverse, it has thrown some people (yellow taxis) out of business. The yellow taxis are now exploring how to input technology into their business operations.

The policy environment for these ride-hailing businesses is also controversial—with many of them running into problems with licensing challenges and harassment by local transport union touts and even the police. The ride-hailing companies consider themselves technology companies and as such believe they are not bound by the state regulations for commercial transport operators. Therefore, they do not obtain the necessary licenses nor remit the requisite operational fees to the government. However, the state regulations consider the activities commercial transport; hence, the vehicle inspection officers (VIOs) are empowered to detain and impound their vehicles for lack of license especially hackney permit. In recent times, many drivers have exited the business due to the tensions, with commuters having to pay higher cost of movement.

The recent ban on motorcycles in 15 metropolitan local governments of the city has also hit hard on commuters who use it to navigate the interiors of the city and last-mile travel. The commercial cyclists are left redundant—as many have no alternative livelihood sources.

8.5 Effects of COVID-19 on TOD and Pro-poor Development

Since the COVID-19 lockdown was announced in March 2020, the urban poor has been negatively impacted. Apart from the fact that their dense communities and informal livelihoods are not conducive for government-recommended social distancing, their transportation patterns have also been affected. In Lagos, for example, the government recommended that informal commercial buses only carry 60% of their capacity. Apart from affecting the income of the drivers (informal workers), it has resulted in price hikes as commuters are charged more per trip on both informal and formal transports. The BRT instituted a 46% price hike (Olisah, 2020). This is particularly difficult as many are no longer earning full incomes due to the restrictions on movement and general economic downturn.

On the other hand, the pandemic is also seen as an opportunity for sustainable mobility. According to Unuigboje (2020), the pandemic lockdown has resulted in many commuters embracing non-motorized transport—walking and cycling.

8.6 Recommendations for the Government, Investors, and Communities

From the research and stakeholder engagement, the recommendations for the government, investors, and communities are described below.

8.6.1 Policy Makers and Implementers

- A more contextualized response in transport project design and implementation that involves the people, especially the poor.
- Intensive, collaborative, and transparent project conception and planning with an emphasis on the financial implications of long-term project maintenance and sustenance and the attainment of a self-sustaining model considering the financial capacity of the poor who are key stakeholders.
- Transport regulatory and supervisory bodies should leverage and collaborate with research institutions toward improving the transportation sector.
- The use of tracking systems on new transport initiatives to enhance commuters' safety and security on board and at terminals.
- The process of masterplan development should be extensively participatory and involve widespread public engagement especially among less influential segments of society who are often the most directly affected and most likely to compromise the plans if not adequately engaged.
- Masterplans should be in the public domain, so, people can critic and share ideas on it. This will enhance people's preparedness for any change. In addition, it will encourage the private sector to understand where it can come in.
- Fostering synergy between Lagos State Ministry of Transportation (including its agencies) and the Ministry of Physical Planning and Urban Development to bridge the gap between transport and urban planning.

Opportunities for private sector leveraging on public sector opportunities should be embraced. PPP initiatives can be developed in response to this.

- There is a need to understand how gaps (resulting from quick changes) can be filled by the informal and private sector. Hence, the government should create a playground (with rules and regulation) that regulates the system and creates an enabling environment that spurs proactive response.
- The implementation of the Lagos Bus Reform Initiative (BRI) to address the issues of formal-informal bus operators.
- Importance of last-mile travel so there are policies and development strategies to provide services that the majority of low-income commuters require. This should be done in such a way that links to the BRT service seamlessly.
- All levels of government (community, state, and national) need to work together to develop strategies that can be implemented at all levels.
- Urban planning policies and housing policies need to work side by side with transport policies, which means that there must be continuous collaboration with the relevant government ministries.
- Prioritization of walking and cycling as an alternative mode of transportation. Tree planting and construction of new/upgrade of existing walkways will bolster this alternative.

- Development of a gradual, comprehensive, and contextually suitable (short, medium, and long term) strategy to guide the implementation of TOD principles in the state cutting across small- and larger-scale projects citywide and how these links up into an integrated network with a careful consideration of the financial implications. This will then serve as a sustainable framework into which present and future stakeholders and investors can key into.

8.6.2 Investors

- Introduction of more transport services that respond to long- and short- term development plans.
- Funding approach to transit-oriented development in which a positive and mutually beneficial relationship with the informal transport sector should be explored.
- Development of innovative financial models that will ensure low-income communities are included in city-wide developments.
- Innovative deployment of technology to improve service delivery of existing informal transport operators (potentially formalizing them on the long run) and connecting their services with the formal high-capacity services.
- Deployment of seven-sitter buses to inner roads.

8.6.3 Communities

- Advocacy for better walking and cycling infrastructure.
- Better organization of associations and community groups so they can negotiate their needs with government or private sector officials.

This report has revealed the clear influence that transportation has on the lives and livelihoods of the urban poor. More importantly, it shows that urban transportation design and planning decisions have extensive impacts beyond mere mobility. The high cost of living within cities and the complicated access to land therein has triggered a large wave of peri-urbanization across African cities. This has resulted in the need for affordable transportation for residents of peri-urban areas who work in the city center. Many low-income earners have been able to bridge this gap and make out livelihoods in the largely informal transport sector as commercial minibus, tricycle, and motorcycle operators. In fact, the introduction of technology into the informal transportation system—through apps—provided some structure and established a value chain. The introduction of some principles of TOD through the establishment of government-sponsored/government-supported BRT systems has expanded access to public transport for longer routes but inherently threatens the existence of smaller scale operators.

It therefore behooves on government and investors to creatively seek out ways of integrating formal and informal transport systems as they often serve different classes of society and can contribute to addressing mobility challenges while building a safe, sustainable, inclusive, and resilient African city.

References

Alade, T., Edelenbos, J., & Gianoli, A. (2020). A sustainable approach to innovation adoption in light-rail transport. *Sustainability, 12*(3), 1–23.

Badejo, B. (2009). *Unbundling the challenges of transportation and development in Nigeria: The Lagos state example being the sixtieth anniversary lecture organised by the Department of Geography, Faculty of Social Sciences*. University of Ibadan.

Barth, B. (2014, March 15). Curitiba: The greenest city on earth. *The Ecologist*. http://www.theecologist.org/green_green_living/2299325/curitiba_the_greenest_city_on_earth.html.

Beard, V., Mahendra, A., & Westphal, M. (2016). Towards a more equal city: Framing the challenges and opportunities. *World Resources Institute Working Paper* (pp. 1–46). 25857.pdf (issuelab.org).

C40. (2016, November). *Addis Ababa light rail project*. https://www.c40.org/awards/2016-awards/profiles/107

Chatterton, P. (2018). Unlocking sustainable cities. A manifesto of real change. *Urbanities, 9*(1), 108–123. https://doi.org/10.2307/j.ctv7vctm4

Chengula, D., & Kombe, K. (2017). Assessment of the effectiveness of Dar es Salaam Bus rapid transit (DBRT) system in Tanzania. *International Journal of Sciences: Basic and Applied Research, 36*(8), 10–30.

Christie, A., Smith, D., & Conroy, K. (2013). Transport governance indicators for sub-Saharan Africa. *Working paper, 95*, sub-Saharan African Transport policy program. https://www.ssatp.org/sites/ssatp/files/publications/SSATPWP95-Governance-Indicators[1].pdf

Collier, P., Glaeser, E., Venables, T., Manwaring, P., & Blake, M. (2019). Urban mobility: policy decisions for connecting the city. *IGC Cities that Work Policy Brief* (pp. 1–12). CtW-urban--mobility-policy-brief-sept-19.pdf (theigc.org).

Deloitte. (2019). *The future of urbanisation and Transit-Oriented Development: Real estate predictions*. https://www2.deloitte.com/lu/en/pages/real-estate/articles/the-future-urbanization-transit-oriented-development.html

Ekechukwu, F. (2019). *Lagos residents to commute easily with Oshodi interchange operation*. https://www.thisdaylive.com/index.php/2019/04/27/lagos-residents-to-commute-easily-with-oshodi-interchange-operation/

Feudo, F. (2014). How to build an alternative to sprawl and auto-centric development model through a TOD scenario for the North-Pas-de-Calais region? Lessons from an integrated transportation-land use modelling. *Transportation Research Procedia, 4*, 154–177.

Future Cape Town. (2018). *The African commute: City transport trends analysis on African transportation by engineering for change in partnership with Yamaha ventures*. https://medium.com/impact-engineered/the-african-commute-city-transport-trends-cf369e5106bd

Global BRT Data. (2016). http://brtdata.org

Haas, A. (2020). Key considerations for integrated multimodal transport planning. United Kingdom: International Growth Centre. *Cities that Work* (pp. 1–26). Key considerations for integrated multimodal transport planning | Global Future Cities Programme.

Harrison, P., Rubin, M., Appelbaum, A., & Dittgen, R. (2019). Corridors of freedom: Analysing Johannesburg's ambitious inclusionary transit- oriented development. *Journal of Planning Education and Research, 39*(4), 456–468. https://doi.org/10.1177/0739456X19870312

IBI Group. (2016). *Transit-Oriented Development guidance document for Ministry of Urban Development*, Government of India. India: IBI Group. http://moud.gov.in/upload/uploadfiles/files/GuidanceDocument FINAL.pdf

Jaiswal, A., Rathore, K., & Jain, D. (2012). Sustainable transit-oriented development: Solving congestion problem at Delhi. *OIDA International Journal of Sustainable Development, 3*(8), 59–68.

Jenkins, P., Smith, H., & Wang, Y. (2007). *Planning and housing in the rapidly urbanising world.* Routledge. https://doi.org/10.4324/9780203003992

Kanter, R. (2015, July 25). Move: Putting America's infrastructure back in the lead. *GreenBiz.* Move: Putting America's infrastructure back in the lead | Greenbiz.

Klopp, J., Orwa, D., Wagacha, P., Williams, S., & White, A. (2017). Informal 2.0: Seeing and improving urban informal practices through digital technologies. The digital matatus case in Nairobi. *Field Actions Science Reports, 16*, 39–43.

Kumar, A. (2011). Understanding the emerging role of motorcycles in African cities: A political economy perspective. *Working Paper of the Sub-Saharan Africa Transport Policy Program.* http://documents1.worldbank.org/curated/en/391141468007199012/pdf/669410NWP0DP130IC00Role0Motorcycles.pdf

Lawanson, T. (2019). Transport is not gender blind in Lagos Nigeria. *Paper presented at Conference on Mobility and Access in African cities* organised by Volvo Research and Education Foundation and GIZ Transforming Urban Mobility Programme, September 2019.

Lindau, L., Hidalgo, D., & Facchini, D. (2010). Bus rapid transit in Curitiba, Brazil: A look at the outcome after 35 years of bus-oriented development. *Transportation Research Record, 36*(3), 17–27. https://doi.org/10.3141/2193-03

Lung-Amam, W., Pendall, R., Scott, M., & Knaap, E. (2014). *The promise and challenge of equitable Transit-Oriented Development in diverse suburbia. Chapter for Book on Transit-Oriented Development* (pp. 1–24). Washington, DC and Paris. (11) (PDF) The Promise and Challenge of Equitable Transit-Oriented Development in Diverse Suburbia. (researchgate.net).

Nagaba, T. (2019, March 14). The African commute: City transport trends. *Engineering for Change Report.* https://www.engineeringforchange.org/research/african-commute-city-transport-trends/.

O'Hara, M. (2015, May 26). When the poor people can't get on due to lack of public transport. *The Guardian.* When poor people can't get on due to lack of public transport | Mary O'Hara | The Guardian.

Olisah, C. (2020, July 21). *Lagos bus service limited announces 46% fare increase, to use BRT lanes.* https://nairametrics.com/2020/07/21/lagos-bus-service-limited-announce-46-fare-increase-to-use-brt-lanes/

Omoegun, A. (2015). *Street trader displacements and the relevance of the right to the city concept in a rapidly urbanising African city: Lagos, Nigeria* (Unpublished Ph.D. thesis). Cardiff University. https://orca.cf.ac.uk/72513/1/Ademola%20Omoegun%20Thesis.pdf

Oshodi, L., Salau, T., & Udoma, E. (2016). Urban mobility & transportation. In Heinrich Boll Foundation & Fabulous Urban (ed.), *Urban Planning Processes. Lagos* (pp. 83–84). 160206_urban_planning_processes_digital_new.pdf (boell.org).

Padeiro, M., Louro, A., & Marques da Costa, N. (2019). Transit-oriented development and gentrification: A systematic review. *Transport Reviews, 39*(6), 733–754.

Pieterse, E., & Owens, K. (2018). Towards a more equal city Johannesburg: Confronting spatial inequality. *World resources report* (pp. 1–28). https://files.wri.org/s3fs-public/johannesburg-confronting-spatial-Inequality.pdf.

Pongprasert, P., & Kubota, H. (2019). TOD residents' attitudes toward walking to transit station: A case study of transit-oriented developments (TODs) in Bangkok, Thailand. *Journal of Modern Transportation, 27*, 39–51. https://doi.org/10.1007/s40534-018-0170-1

Renne, J. (2009). *Evaluating Transit-Oriented Development using a sustainability framework: Lessons from Perth's Network City* (pp. 1–26). Microsoft Word- FRONT-PAGES-1.doc (vtpi.org).

Rizzo, M. (2019, January 22). *Dar es Salaam's new rapid bus system won international acclaim–But it excludes the poor.* https://theconversation.com/dar-es-salaams-new-rapid-bus-system-won-international-acclaim-but-it-excludes-the-poor-109987.

Saibu, O. (2018). Public spending, fiscal sustainability and macroeconomic performance in Nigeria. *NDIC Quarterly, 33*(3 & 4), 33–48.

Scruggs, G. (2013, November 1). Cracks in the Curitiba myth. *Next City.* https://nextcity.org/daily/entry/cracks-in-the-curitiba-myth

Suzuki, H., Cervero, R., & Iuchi, K. (2013). Transforming cities with transit: Transit and land-use integration for sustainable urban development. *Urban development* (pp. 1–233). World Bank. http://hdl.handle.net/10986/12233

Talamini, G., & Ferreira, D. (2019). An informal transportation as a feeder of the rapid transit system. Spatial analysis of the E-bike taxi service in Shenzhen, China. *Transportation Research Interdisciplinary Perspectives, 1*, 1–12. https://doi.org/10.1016/J.TRIP.2019.100002

Teravaninthorn, S., & Raballand, G. (2008). Transport prices and costs in Africa: A review of the main international corridors. *Working Paper 14.* Africa Infrastructure Country Diagnostic. https://devpolicy.org/transport-costs-in-africa-why-are-they-so-high-and-what-can-be-done-about-them20120302/

UN-Habitat, UNEP, SLoCaT. (2015). *Analysis of the transport relevance of each of the 17 SDGs.* https://sustainabledevelopment.un.org/content/documents/8656Analysis%20of%20transport%20relevance%20of%20SDGs.pdf

United Nations (2018). *World urbanisation prospects*, https://esa.un.org/unpd/wup/.

United States Environmental Protection Agency. (2013). *Infrastructure financing options for transit-oriented development* (pp. 1–251). http://www.epa.gov/smartgrowth

Unuigboje, R. (2020). *Coronavirus in Lagos: Reflections on mobility and transport poverty.* http://chsunilag.com/blog/coronavirus-in-lagos-reflections-on-mobility-and-transport-poverty

Watson, V. (2007). Urban planning and twenty-first century cities: Can it meet the challenge. In A. Garland, M. Massoumi, & B. Ruble (Eds.), *Global urban poverty: Setting the agenda.* Woodrow Wilson International Centre for Scholars.

World Population Prospects. (2022). Ahmedabad population. Ahmedabad Population 2022/2023 (populationu.com).

Xu, W., Guthrie, A., Fan, Y., & Li, Y. (2017). Transit-oriented development in China literature review and evaluation of TOD potential across 50 Chinese cities. *Journal of Transport and Land Use, 10*(1), 743–762. https://doi.org/10.5198/jtlu.2017.1217

Chapter 9
Transit-Oriented Development and Spatial Data Infrastructure in African Cities

Muyiwa Agunbiade, Oluyemi Akindeju, Peter Elias, Oluwafemi Olajide, and Franca Agama

9.1 Introduction

Infrastructural deficiency remains a major challenge for Africa in achieving its full potential. As one of the global hubs growing at unprecedented rate into economic centres, attending to the demand for key infrastructure has become an exigency based on the fact that inadequate transport infrastructure can add up to 40% to the costs of goods traded among African countries (Export- Import Bank of India, 2018). In an increasingly urbanised, technology-driven and overstretched world with its attendant challenges, the need to attain prosperity, frantically combat environmental degradation due to transportation and ensure sustainable growth contends with digitisation, which is driving new patterns of transport for goods, services and people (Smith et al., 2017). As transport needs and challenges are increasing world over, the unavailability of a data infrastructure framework needed to make optimal use of different data sets has further made prevailing challenges more overwhelming for major stakeholders in the transportation sector such as the urban/transport planners, investors, policy makers and commuters.

Improvements in transport technology and transport networks, through its effects on transport costs, access and connectivity have brought about more complexities and volume to data and its management especially in developing economies such as in sub-Saharan Africa, but the uniqueness in infrastructure requirements which varies from region to region and across the globe requires integration of localised data

M. Agunbiade (✉) · O. Akindeju · O. Olajide
Department of Urban and Regional Planning, University of Lagos, Lagos, Nigeria
e-mail: magunbiade@unilag.edu.ng

P. Elias
Department of Geography, University of Lagos, Akoka, Lagos, Nigeria

F. Agama
Department of Urban and Regional Planning, Yaba College of Technology, Lagos, Nigeria

T. Nubi et al. (eds.), *Transit Oriented Development in West African Cities*, Urban Perspectives from the Global South,
https://doi.org/10.1007/978-3-031-58726-9_9

as an umbrella approach of bridging the gap. As demographic and geospatial attributes of the society or economy are location specific and peculiar, the need to engender and promote in the long run local knowledge and workable solutions to seemingly intractable problems impeding the functionality and sustainability of African cities becomes imperative.

Urban transport data infrastructures are usually characterised by multiple sources, heterogeneity, large quantity, disparity, and independence, therefore efficiently classifying, integrating, and interconnecting such as multi-source data constitutes a major challenge and cause for concern for African cities of the twenty-first century (Page et al., 2020). Based on the premise that economic growth, urban transport and land use can be managed more efficiently if inclusively planned together, the transit-oriented development (TOD) model links residential, retail, commercial and community service developments to frequent, accessible mass/rail transit services to stimulate sustainable development in the form of decreased land use and transport integration (Zhao et al., 2014).

The key towards realising data interconnection of transportation infrastructure for a sustainable TOD lies in a spatial data infrastructure framework which combines data activation and multimedia annotation technologies to process and integrate multiple sources and heterogeneous, massive and decentralised data. This must accommodate both real-time signal processing and scalability supporting data analysis on a large scale using batch and stream processing (Luckow et al., 2015).

As an archetypal African city, the current diverse mix of transport services in relation to the urban morphology of Lagos unsuccessfully seeks to meet critical mobility needs of the city but also creates myriad of problems ranging from pollution, economic loss to challenging environment for walking and cycling, among others. A data-enabled platform and infrastructural framework that will synchronise multiple data for effective designs and decisions-making can effectively integrate land use and transport developments to check horizontal sprawl and promote transit-oriented policies and developments, thereby improving health, safety and environmental quality and enhancing social equity and economic activities (Li et al., 2019).

9.2 Transportation and Spatial Data Infrastructure in African Cities

One of the critical ills confronting rapidly growing cities is the overburdened transport system (World Bank Group, 2019), where loss of man-hour due to traffic gridlocks and air and noise pollutions represent a major constraint to continued economic growth and access to opportunities. The complexity of other numerous challenges arising from high population, diverse socio-economic or demographic configurations, prevailing socio-cultural milieu and political atmosphere have in recent times been further aggravated by advanced technological applications resulting in complex multitudinous information. Despite growth projection of global transport-induced emission from 25% to 33% by 2050, automobile-centred developments are still prevalent in most African cities (ICLEI, 2016) (World Bank, 2018). As a result of these trends, urban sprawl, segregated land use and uncoordinated urban development are having direct impact on the environment, social equity, economic growth and quality of life.

Urban form, either horizontal or vertical, linear, compact or clustered, influences the accessibility and integration between various land uses, which are interrelated and interdependent (Cilliers & Schoeman, 2008). Land use generates urban activities for which transport provides platforms for interaction. In recent years, cities have evolved as spatial structure that increases reliance on motorised transportation, particularly private cars. Furthermore, development patterns and the demand for movement of goods and services within and between intense activity areas vary making it almost impossible to rely on one mode or means of movement.

Transportation is therefore influenced by urban patterns, types, scale and location of activities; this is because they all have varied mobility demands (Aderamo, 2012). With increasing population, expanding settlements impose further travel and intensive use of transport facilities with unsatisfactory traffic situation in major cities, as demand and volume of traffic generated surpasses road infrastructure resulting in wide variety of impacts, consequences and failing carrying capacities (Agamah, 2019). Interventions and approaches include the application of advanced and emerging technologies, communication and electronic devices for advanced transport management, intelligent transport, traffic management systems, etc., all within the ambit of relevant and usable data and appropriate data management framework. This invariably saves lives, time, money, energy and the environment and requires real-time interdisciplinary data. Although mass transit and land use can be examined at different levels or scales, micro, meta and macro scales, ranging from the neighbourhood, metropolitan and regional levels, each has its own unique challenges and complexities which would require tailored approaches as amplified by the location-specific data. Integration of these in an SDI, i.e. a relevant base collection of technologies, policies and institutional arrangements that facilitate the availability of and access to spatial data, is important to achieving sustainable urban form and mobility. The disconnection between land use and transportation in Nigerian cities has been attributed to lack of comprehensive geographical databases.

As transport needs and challenges are increasing, the unavailability of reliable and usable data and a data infrastructure framework needed to make optimal use of different data sets has further made the challenges more overwhelming for transport policy makers, planners and operators around the world (Smith et al., 2017). As knowledge consists of data and information that have been organised and processed to convey understanding, experience, accumulated learning and expertise as they apply to a current problem or activity (Vandervalk, 2012), an understanding of local specificity in terms of demographic composition, travel needs, lifestyles, choices and preferences, spatial composition and configuration can only be addressed by a robust data structure for informed policy statements and decisions. The integration of data, people, platforms, standards and institutional framework for a complete SDI deployment (Kitahara et al., 2018) in the transportation sector is as depicted in

Fig. 9.1. This integration should be the base framework on which transport policies and development should be carried out, the TOD more importantly.

Transit-oriented development (TOD) as an urban development approach emphasises reliable public transport and prioritises the maintenance of a closely knit urban fabric through high-density, mixed-use and human-scale design that ensures residents are within walking distance from transit stations, with the capability of decreasing CO2 emissions by up to 43%, and can reduce private vehicle use by 50% (ICLEI, 2016).

Therefore, the development of transport policies must as a matter of essence involve trans- and inter-agency collaboration for inputs from all allied and respective agencies of the environment, physical planning, departments of works and infrastructure, transport and traffic management (Agamah, 2019). To ensure seamless and successful design/construction, impact assessments, implementation, traffic guidelines, regulations and enforcements, compliance, safety, public education and other aspects of managing and sustaining transport, the spatial data infrastructure provides the needed framework for better planning and implementation (Agunbiade et al., 2019). For any successful transport planning, there is need for data, secondary baseline and or real-time data on which decision will be taken for solutions. In this era, when the Internet of Things (IoT) and other digital data are transforming social infrastructure and businesses, the development of platforms (Kitahara et al., 2018) that can integrate diverse forms of transportation data in ways

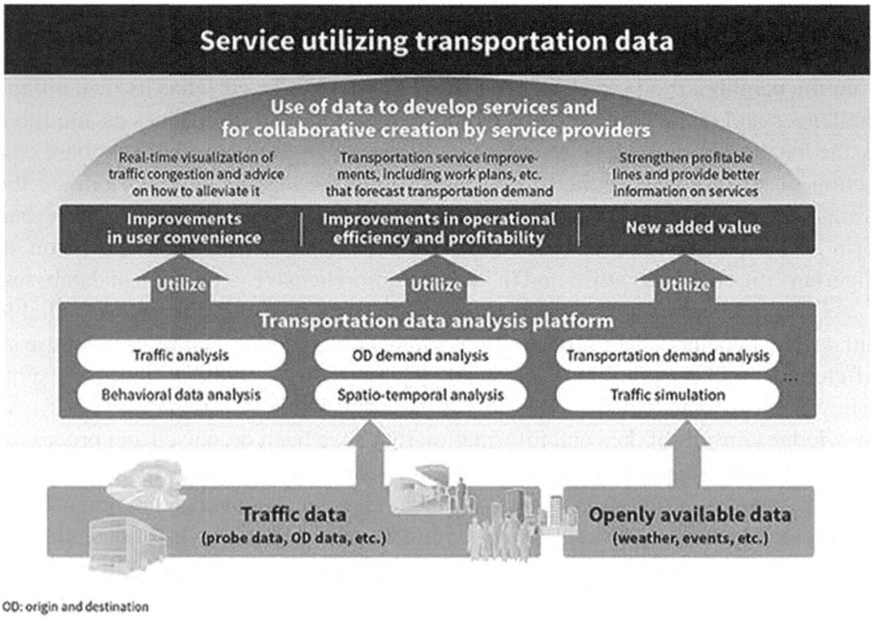

Fig. 9.1 Transport data integration framework. (Adapted from Kitahara et al., 2018)

that users require and the delivery of services that best meet their needs are becoming crucial challenges.

Without a dedicated spatial data infrastructure or repository, it will be difficult and almost impossible to access baseline data upon which real-time data can be derived. TOD, as an ideal option, harmonises all existing initiatives and adapts to urban forms and activities, enhancing mobility, convenience and liveability while reducing demand for personal vehicular travels with serious traffic challenges. Rather than having fragmented laws, investing in an integrated spatial data infrastructure and transport policy is more desirable.

9.3 Transit-Oriented Development, SDGs and Spatial Data Infrastructure: Conceptual Linkages

Transportation is instrumental to the economic and social development of any society by providing access to economic and social opportunities (Anin et al., 2013; Ewing & Cervero, 2010). It ensures the movement of people, goods, services and information over specific locations in space. The configuration of transport infrastructure and connectivity can greatly influence the well-being of the people and the quality of their environment (Solanke, 2013). Like the other side of a coin, transport has negatively impacted issues of global concern such as climate change, fatalities, health and well-being, etc., thereby making actualising the Sustainable Development Goals, especially in developing regions such as sub-Saharan Africa, burdensome. One of the long-standing issues, in this regard, is the lack of reliable data in appropriate format to aid decision-making and monitor progress. Where data are available, they are fragmented among different institutions and agencies (private and public) and as such not integrated and accessible. In addition, the available data are not disaggregated in a usable format. Specifically, it is established that accurate and reliable geospatial data, integrated across sectors, are central to the implementation and monitoring of progress towards the attainment of SDGs (Agunbiade et al., 2019).

Within the context of the SDGs, as a global roadmap towards achieving a balanced, sustainable and interconnected society (UNDP, 2016), transportation also conversely plays a pivotal role in actualising the global agenda. Transport development can contribute immensely towards the achievements of the SDGs where its critical components of planet, people, prosperity, peace and partnership are integrated into transport policies, planning and operation in an information society that is spatially enabled. One way to address the major challenge bedevilling the city is through transit-oriented development (TOD), which is a multidisciplinary planning and design strategy to promote sustainable urban development (World Bank Group, 2019).

Through a framework facilitating the management of information assets, TOD can effectively contribute towards the actualisation of the Sustainable Development Goals (SDGs) (Berawi et al., 2019). The improved connectivity and mobility increase employment opportunities that are very instrumental to alleviating poverty and putting an end to hunger (SDG 1 & SDG 2) and increase access to education and health services (SDG 3 & SDG 4) as well (Tilahun & Fan, 2014). Rail transit, one of the key elements in TOD, can reduce private vehicle dependence, hence improving energy efficiency in transportation (SDG 7) and reducing the impacts of pollution and GHG emissions on the climate (SDG 13) (Renee & Ashish, 2019). Moreover, the increased mobility offered by TOD is also crucial to the achievement of sustainable cities (SDG 11) (Simon et al., 2016; Valencia et al., 2019). Furthermore, TOD is also a solution to achieving the New Urban Agenda (NUA), which emphasises the need for well-integrated transport-land-use planning to reduce travel and transport needs and enhance connectivity between areas in cities and its surroundings. However, to achieve these goals, the fundamental premise of TOD, i.e. the application of context-sensitive and interdependent design standards, including building densities based on variables surrounding the station area such as transit capacity, plot sizes, street widths and infrastructure capacities; innovative real estate negotiations; affordable housing near transit stations; or public space design with high-quality public realm, demographic details, land use pattern, mobility needs, preferences and patterns and other numerous data ranging from micro scale to national or regional level, becomes very essential. Therefore, to arrive at an optimal intermodal mix of transportation that justifiably caters for the social, economic and environmental dimensions of development, an integrated approach of spatially enabling information and data aggregation towards arriving at specific solution to a local mobility need is essential. A spatial data infrastructure platform at the appropriate scales and levels provides the needed framework in accomplishing this.

9.4 Transit-Oriented Development Case Studies

The TOD as a transport policy identifies urban development projects that are located within walking distance of a high-capacity transit station and present specific urban design and land use characteristics known to support, facilitate and prioritise the use of public transport, walking, cycling and other non-motorised modes (Otunola et al., 2019) (LAMATA, 2018). TODs encourage developments that are proactively oriented towards, rather than simply adjacent to, public transport and invariably capable of optimising the nexus between the land use and the transit needed to maximise the potentials. For example, easy access to public facilities and commercial buildings along the road (to the station) can encourage people to utilise transit more often. Although TOD principles cannot be applied uniformly across an entire city or transit network due to local specificity and peculiarities, they can serve as a tool to initiate a paradigm shift in transforming the future of the growing number of cities in the world; appraisal of two adjudged successful TOD projects are hereby undertaken as case studies.

9.4.1 Curitiba, Brazil

The city of Curitiba, the capital of the Brazilian state of Paraná, is inhabited by over 1.8 million people as of 2015, living within an area of 430 km^2, having grown from 350,000 to 940,000 inhabitants between 1955 and 1975 (IBGE, 2010). The continuous influx of inhabitants forced the city to prioritise a strategy for long-term, sustainable growth. The city turned the challenges of increased urbanisation and private vehicle use into opportunities by adopting a transit-oriented development (TOD) approach through integrated land use and transport planning of vibrant, high-density, mixed-use communities along mass public transport corridors in a series of strategic incremental steps.

With 85% of Curitiba's population using the bus rapid transit (BRT), being a very high-capacity vehicle with the capability of conveying 10,000–20,000 passengers at one time, with a capacity of 40,000 passengers on busy roads, it is a necessary element given the size and expansion of the city. This in the long run improved quality of life and access to public space for the city's residents and also helped make the streets of Curitiba pedestrian friendly (ICLEI, 2016; Fig. 9.2).

As a result, the TOD efforts have made Curitiba a global model in low-carbon urban development practices with numerous other advantages. Through coupling land use and transport planning, the city has substantially reduced its carbon footprint while boosting the quality of urban life. It is one of the most sophisticated low-cost, high-capacity public transit networks to date which supports resource-efficient and long-term investments in urban infrastructure. With reduced travel times and private vehicle use, there has been tremendous improvement in residents' quality of life. Between 1970 and 2000, the Human Development Index (HDI) for the city rose from 0.7 to 0.8, placing it above the.

HDI levels of the entire state of Paraná (both of which stood at 0.4 in 1970 and rose to 0.7 in 2000) (ICLEI, 2016). Curitiba's TOD phase-based approach to low-carbon urban development, emphasising structured urban development and sound transport policy, guaranteed compact, liveable communities and helped to develop the infrastructure needed to support the growth of the system over time, making it a huge success and a transport policy model for Africa.

9.4.2 Cape Town, South Africa

Cape Town is the second largest city in South Africa, after Johannesburg, with a population of over 3.7 million people (Atlas of Urban Expansion, 2016). Due to high infrastructural deficit not commensurate with a high population figure, the 2010 World Cup in South Africa offered a very valuable opportunity of interventions of which improvement in public transportation specifically for the low-income groups was a priority (Statistics South Africa, 2015; Fig. 9.3).

Fig. 9.2 Map of Curitiba City and its regional districts (Parana, Brazil). (Adapted from Krause et al., 2009)

MyCiTi, Cape Town's version of bus rapid transit (BRT), is a public transport strategy for the city to provide greater mobility to the majority of the population (Tennant, 2015) and also contribute to economic development through integrated transportation. The MyCiTi Integrated Rapid Transport system uniquely incorporated all the other motorised and non-motorised transport methods that had already existed in the area into one cohesive new system instead of a total replacement

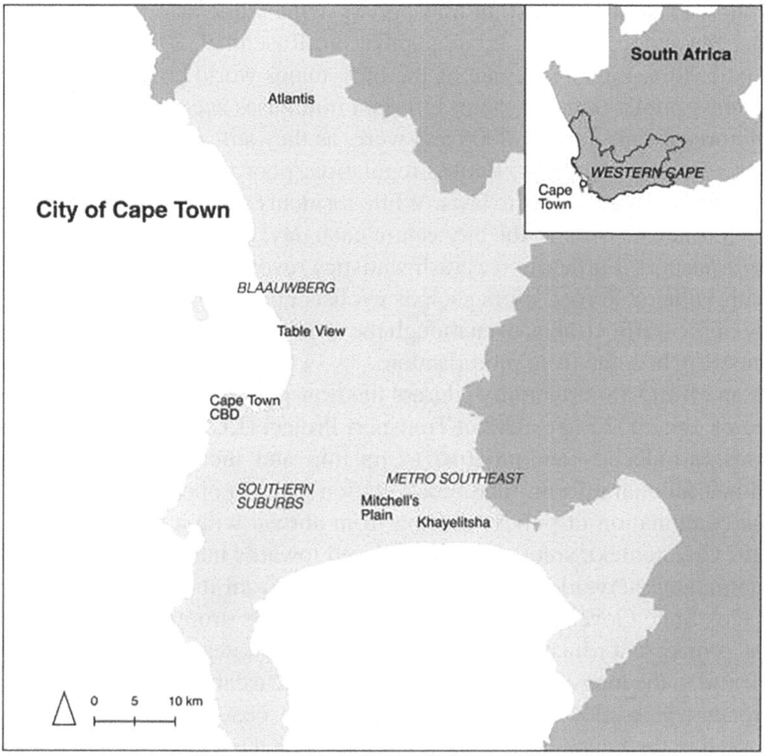

Fig. 9.3 Map of the city of Cape Town. (Adapted from Lisa Rayle, 2017. PhD Thesis on *Bus rapid transit as formalisation: Accessibility impacts of transport reform in Cape Town, South Africa*)

(Tennant, 2015). This transport system developed the urban areas around the transit-oriented principles, and to attain a considerable level of impact, measures such as zoning land for dense, high-rise developments around rapid transit corridors and discouraging development further away from them were ensured to increase the ridership needed for BRT in the city of Cape Town.

9.5 Trans-disciplinary Perspectives of SDI and TOD Issues for Lagos

Rapidly growing cities in World Bank client countries like Lagos are, more often than not, densely populated even before the introduction of public transit. With a population estimated between 18 and 20 million and an annual growth rate at 6%, Lagos is set to become the third largest megacity in the world. As the second largest city in Africa, and the fastest growing megacity in the world, it has contended with improving the quality of existing transport systems while also providing new services that can support a larger urban population (Rosenthal, 2012).

Up until 2008, Lagos was one of the only major world cities without any formalised mass public transit system. Informal minibuses accounted for around 69% of all motorised trips back in 2008 and were, as they still remain, characterised by their lack of safety standards, limited regulation, poor maintenance, poor security and frequent violation of traffic laws. While residents spend over 2 h in traffic on a one-way journey to work in the city centre each day, about 40% of their income is spent on transport. Furthermore, crash statistics revealed an increasing trend over time, with vulnerable road users such as cyclists and pedestrians accounting for a majority of the traffic deaths, even though these groups often have little or no access to the mobility benefits from motorisation.

With an effort to substantially address the transport challenge and its associated negativities, the 2002 Lagos Urban Transport Project (LUTP) funded by the World Bank was introduced—an initiative to revamp and modernise public transport through institutional reform, road rehabilitation and bus operations improvements. With the combination of policy learning from abroad with a strong knowledge of the Lagos City context, solutions were tailored towards integrating society and the danfo-dominated network rather than excluding it from the decision-making process (Lagos State Government, 2013). Furthermore, a strong institutional foundation for project coordination and setting of the strategic plan which remains fundamental to the long-term success of reform were established, and the incremental approach which allowed the state to make quick, cost-effective and visible wins was adopted in the Lagos BRT project. As a fairly TOD option, the seeming coordination borders on limited and isolated spatial data management.

The Lagos BRT and the wider transport reforms of Lagos have helped to transform the sprawling and unplanned city, characterised by fragmentation and heavy traffic, to a much better example of transport development for liveable and productive cities (Otunola et al., 2019). As the first ever bus rapid transit (BRT) system on the African continent in 2008, it currently boasts of two different lines which cover over 35.5 km of track and transport over 350,000 commuters on a daily basis (Mobereola, 2009). Since the implementation of the BRT, the city has seen remarkable improvements across a number of indicators that impact upon the everyday lives of citizens. An evaluation by the World Bank in 2009 suggested that within a year of the operation of the BRT system, over 200,000 commuters were using it daily, with passengers enjoying a reduction of fares, journey time and an overall increase in quality of service. While it can be said that the BRT has captured significant ridership, the corridor still constitutes a single route, which is by far too little to attend to the city's massive urbanised area (BRT Lite corridor in Lagos stretches 35 km from CMS to Ikorodu and carries an estimated 120,000 passengers per day) (LAMATA, 2018).

Towards this end, the option initiated by the Lagos Metropolitan Area Transport Authority (LAMATA) of the Lagos State Government through the creation of a non-motorised transport policy to guide the implementation of transport systems that prioritise the needs of pedestrians and cyclists within the context of integrated land use, public transportation and transit zones stems from the underpinning principles of transit-oriented development. In adopting a NMT-PT-based approach, Lagos aims to discourage the use of personal motor vehicles, thereby improving health, safety and environmental quality and enhancing social equity and economic activity. As a viable transport system, more accessibility per unit of mobility is paramount. Walking, cycling and the use of public transport will greatly enhance accessibility and improve mobility. Unfortunately, recent and reliable data on walking and cycling are not available, but non-motorised modes likely account for upwards of a third of all trips in the city of Lagos.

As submitted in the Lagos State Development Plan (2012), to achieve reasonable access for public transport and NMT users and stabilise the use of personal motor vehicles, Lagos will need to build at least 470 km of MRT, 900 km of footpaths and 300 km of cycle tracks. Equally important, Lagos will need to adopt measures to check horizontal sprawl and promote transit-oriented land uses.

Going forward, it needs be established that mobility planning should focus more on people and not on vehicles. All decisions require data, and as data becomes more volatile, human issues of data sharing, security, accuracy and access forge the need for more defined relationships between people and data. The rights, restrictions and responsibilities influencing the relationship of people to data become increasingly complex, through compelling and often competing issues of social, environmental and economic management. Facilitating the role of people and data in governance that appropriately supports decision-making and sustainable development objectives is central to the tenets of SDI. The motorisation extent in the transport reform efforts revealed this gap in data and its management in addressing spatial issues. Passengers, transport providers, infrastructure operators and planners and regulators alike can make better-informed decisions if public transport systems and key trip generators, such as shopping malls and parking lots, share usage and availability data. Due to the contexts in which spatial data is produced, analysed, circulated and used, SDIs contribute immensely to the complexity of the 'spatial information universe' (Ballatore, 2014) and meet the goal of critical data studies (Kitchin & Lauriault, 2014).

9.6 Recommendations

It has become evident that land uses and transport are *Siamese twins* being conjoined entities of the urban space. An attempt to isolate one from the other mostly results in dysfunctional urban settings characterised by congestion, pollution, insecurity, exclusion and health issues. Emanating from a transdisciplinary meeting interrogating the place of spatial data infrastructure in transportation to achieve sustainable transit-oriented development were diverse perspectives from various stakeholders in the society. Considering the place of data and technology in the scheme of things of modern society, for African cities to achieve sustainability for the benefit of all, opinions and views as expressed by the different stakeholder groups must form the basis of policy formulation.

For the purpose of achieving compact transit-oriented development, while the accuracy of available data for the desired purposes cannot be ascertained, the need to extend information gathering on travel pattern, trip behaviour, etc., to the grass-root/local constituencies at the neighbourhood level remains pivotal to the success of any policy engagement. To a large extent, conscious efforts should be made to synchronise the available data (both longitudinally and laterally) domiciled in different units and levels for consistency and operationalisation. Notable effort on the part of the Lagos State Government in this regard is the mandate that all ministries, departments and agencies must, on a monthly basis, forward data at their disposal to the Lagos State Government Ministry of Information and Strategy. This, however, must be undertaken within a National Spatial Data Infrastructure framework to make meaningful decisions.

Summarily, for African cities to optimally explore the benefits inherent in urbanisation through transit-oriented developments within the context of sustainable development and the spatial data infrastructure framework, conscious efforts should be taken towards acquiring the right and reliable data; finding the right balance between density, building design and development; establishing a central coordinating body saddled with the responsibility of implementation; and engaging a wide range of relevant multi-sectoral stakeholders through inclusive participation and co-production.

Applying automation to transit-oriented development using technology will help streamline and overhaul complete business processes in a number of ways where technology gives everyone involved in the transportation spectrum from traffic agencies to commuters the power to do more and achieve more. It will give commuters real-time information and options to know the conditions of roads and alternate routes and choose the best mode of transportation at that particular moment so that they can make informed transportation decisions that are not based solely on uncertainties. All these will not be possible without data coming from sensors, traffic lights and every source of IoT–/cloud-connected devices. This results in information becoming the biggest currency in the technology of transit-oriented developments, helping road authorities and planners on the operational side of things to make transit more efficient and more attractive for transit-oriented developments. It is noteworthy that in some parts of Europe like the Netherlands, transit-preferential traffic signals and lane-preemption technology are reducing bus headways.

Road authorities, transit entities and transportation companies can then provide unified and comprehensive information which can be accessed online via social media and web platforms or apps, via USSD and via Bluetooth or beacon devices to reach users more effectively and efficiently; this will make transit even more engaging and participatory, amplifying the value of transit-oriented developments, and just as with the new economy, the data used by these platforms can be crowdsourced from road users to get a win-win situation for everyone. The above would not be possible without the technology that connects all of these components. From Bluetooth technology to USSD, apps and platforms that can easily be accessed via a PC, mobile or tablet, you can order for a product and pick it up or have it delivered or have the ability to access transit information in the cloud that can show you exactly where your bus or train is and how long until it arrives to get you to wherever you need to go quickly and inexpensively. Technology with available data like fuel in a car will help transit-oriented development advance and work best to the advantage of everyone who uses and is connected to it in new and exciting ways.

9.7 Conclusion

It is sacrosanct that transport and land uses are inseparable components of human settlements. However, the prevailing global health challenge posed by the COVID-19 pandemic and the already evolving complexity of the info-digital era have put the transport systems of cities in more dilemma and are further aggravating the crisis and challenging its management. Having the right data, from the right sources, at the right time, in the right environment, in the custody of the right people can sustainably guarantee a more prosperous, socially inclusive and environmentally balanced city system fit for all.

References

Aderamo, A. J. (2012). Urban transportation problems and challenges in Nigeria: A planner's view. *Prime Research on Education (Pre)*, 198–203.

Agamah, F. U. (2019). *Integrated transport planning and policy towards sustainable urban development – A case study of Lagos metropolis*. A paper presented at the 6th International Conference on Water, Energy and Environmental Research (ICWER 2019) 2019 Held between 22nd -24th September 2019 in Abuja.

Agunbiade, M., Oluwafemi, O., & Akindeju, O. (2019). SDGs and geospatial information perspective from Nigeria-Africa. *Sustainable Development Goals Connectivity Dilemma*, 297–312. https://doi.org/10.1201/9780429290626-19

Anin, E. K., Annan, J., & Otchere, F. A. (2013). Assessing the causes of urban transportation challenges in the Kumasi Metropolis of Ghana. *American Based Research Journal, 2*(6).

Atlas of Urban Expansion. (2016). *Cape Town*. http://www.atlasofurbanexpansion.org/cities/view/CapeTown

Ballatore, A. (2014). Defacing the map: Cartographic vandalism in the digital commons. *The Cartographic Journal, 51*(3), 214–224.

Berawi, M. A., Saroji, G., Iskandar, F. A., Ibrahim, B. E., Miraj, P., & Sari, M. (2019). Optimising land use allocation of Transit-Oriented Development (TOD) to generate maximum ridership. *Sustainability*. Article 11.

Cilliers, E. J., & Schoeman, C. B. (2008). *The urban development boundary as a planning tool for sustainable urban form. Transactions on ecology and the environment* (Vol. 117, pp. 85–94)., 10p; ref:13. The sustainable city V (urban regeneration and sustainability). A. Gospodini, (Ed.) University of Thessaly, Greece, C.A. Brebbia, Wessex Institute of Technology, UK and E. Tiezzi, University of Siena, Italy.

Ewing, R., & Cervero, R. (2010). Travel and the built environment: A meta- analysis. *Journal of the American Planning Association, 76*(3), 265–294.

Export-Import Bank of India. (2018). *Working Paper No. 72 connecting Africa: Role of transport infrastructure* (pp. 10–11). www.eximbankindia.in
ICLEI. (2016). Curitiba, Brazil: A model for transit-oriented development, 1–8. https://doi.org/10.4135/9781412973816.n42
Instituto Brasileiro de Geografia e Estatística (IBGE). (2010). Brazilian Institute of Geography and Statistics.
Kitahara, R., Mizutani, A., Hirao, A., Sakamoto, T., & Taniguchi, N. (2018). How data will transform future urban transportation. *Hitachi Review, 67*(5), 568–569.
Kitchin, R., & Lauriault, T. (2014). *Towards critical data studies: Charting and unpacking data assemblages and their work*. University of Nebraska Press.
Krause, M., Januário, R., Hallage, T., Haile, L., Miculis, C., Gama, M., Goss, F., & da Silva, S. (2009). A comparison of functional fitness of older Brazilian and American women. *Journal of Aging and Physical Activity, 17*, 387–397. https://doi.org/10.1123/japa.17.4.387
Lagos State Development Plan. (2012). *Lagos State Development Plan 2012–2025*. https://www.Scribd.Com/Document/271150413/Lagos-State-Development-Plan-2012-2025
LAMATA (2018, April). *Lagos non-motorised transport policy*. https://wedocs.unep.org/bitstream/handle/20.500.11822/25415/Lagos_ NMTPolicy.pdf?sequence=3.
Li, W., Luo, Q., & Cai, Q. (2019). Coordination of last train transfers using potential passenger demand from public transport modes. *IEEE Access, 7*, 126037–112605.
Luckow, A., Kennedy, K., Manhardt, F., Djerekarov, E., Vorster, B., & Apon, A. (2015). Automotive big data: Applications, workloads and infrastructures. In *2015 IEEE International Conference on Big Data (Big Data)* (pp. 1201–1210). https://doi.org/10.1109/BigData.2015.7363874
Mobereola, D. (2009). *Africa's first Bus Rapid Transit Scheme*. Sub-Saharan Africa Transport Policy Programme.
Otunola, B., Kriticos, S., & Harman, O. (2019). The BRT and the danfo: A case study of Lagos' Transport Reforms from 1999–2019. *IGC, Cities that Work, Case Study*. https://www.theigc.org/wp-content/uploads/2019/10/lagos-BRT-case-study.pdf
Page, J., Gutman, J., Madden, P., & Gandhi, D. (2020, September 24). *Urban economic growth in Africa; A framework for analysing constraints to agglomeration*. Africa Growth Initiative at Brookings. Working Paper.
Lisa Rayle, (2017). *Bus rapid transit as formalisation: Accessibility impacts of transport reform in Cape Town, South Africa* [Ph.D Thesis] for Doctor of Philosophy in City & Regional Planning University of California, Berkeley.
Renee N. S., & Ashish, K. J. (2019, September). Climate change threatens the achievement of effective universal healthcare. *BMJ* (Clinical Research ed.). 366:l5302. https://doi.org/10.1136/bmj.l5302. PMID: 31548271; PMCID: PMC6753637.
Rosenthal, E. (2012). Nigeria tested by rapid rise in population. *The New YorkTimes*. https://www.nytimes.com/2012/04/15/world/africa/in-nigeria-a-preview-of-anovercrowded-planet.html
Simon, D., Arfvidsson, H., Anand, G., Bazaz, A., Fenna, G., Foster, K., Jain, G., Hansson, S., Evans, L. M., Moodley, N., Nyambuga, C., Oloko, M., Ombara, D. C., Patel, Z., Perry, B., Primo, N., Revi, A., Niekerk, B. V., Wharton, A., & Wright, C. (2016). Developing and testing the urban sustainable development goal's targets and indicators—A five-city study. *Environment and Urbanization, 28*(1), 49–63. https://doi.org/10.1177/0956247815619865
Smith, J., Clayton, E., & Hanson, D. (2017). *Building sustainable, inclusive transportation systems: A framework for the future*. Price Waterhouse Coopers. https://www.strategyand.pwc.com/reports/building-sustainable-transport-systems
Solanke, M. O. (2013). Challenges of urban transportation in Nigeria. *International Journal of Development and Sustainability, 2*(2), 891–901.
Statistics South Africa. (2015). *Measuring household expenditure on public transport*. Statssa.gov.za. http://www.statssa.gov.za/?p=5943
Tennant, M. (2015). *Cape Town's MyCiti BRT – Four years into its 'democracy'*. UrbanAfrica.Net. http://www.urbanafrica.net/urban-voices/cape- towns-myciti-brt-four-years-into-its-democracy/

Tilahun, N., & Fan, Y. (2014). Transit and job accessibility: An empirical study of access to competitive clusters and regional growth strategies for enhancing transit accessibility. *Transport Policy, 33*(C), 17–25.

UNDP. (2016). Sustainable urbanisation strategy. UNDP's support to sustainable, inclusive and resilient cities in the developing. *World*, 1–54. http://www.undp.org/content/undp/en/home/librarypage/poverty-reduction/sustainable-urbanization-strategy.html

Valencia, S. C., Simon, D., Croese, S., Nordqvist, J., Oloko, M., Sharma, T., Buck, N. T., & Versace, I. (2019). Adapting the sustainable development goals and the new urban agenda to the city level: Initial reflections from a comparative research project. *International Journal of Urban Sustainable Development, 11*, 4. https://doi.org/10.1080/19463138.2019.1573172

Vandervalk, A. (2012). *Turning data into information for transport decision making*. Association for European Transport and Contributors.

World Bank. (2018). *Logistics performance index*. https://lpi.worldbank.org

World Bank Group. (2019). *Pandemic preparedness financing – Status update*. https://apps.who.int/gpmb/assets/thematic_papers/tr-4.pdf

Zhao, J., Deng, W., Song, Y., & Zhu, Y. (2014). Analysis of metro ridership at station level and station-to-station level in Nanjing: An approach based on direct demand models. *Transportation, 41*, 133–155.

Part III
City Study

Chapter 10
Urban Infrastructure and Transit-Oriented Development in Banjul, Gambia

Olawale Fabiyi, Umaru Tanimu Saminaka, Benjamin Ozirigbo, and Francis Mfobujong

10.1 Background

The Gambia, the smallest country in West Africa, is geographically surrounded by Senegal, bordering the North Atlantic Ocean. The name 'Gambia' is derived from the Mandinka term Kambra/Kambaa, meaning Gambia River, and for the Gambia River that flows through the heart of the country, it formed a short- lived Confederation of Senegambia between 1982 and 1989. It occupies a long narrow strip of land that surrounds the Gambia River. The land is flat and is dominated by the river, which is navigable throughout the length of the country. Between 2000 and 2020, the annual population growth rate was reduced from 2.2% to about 1.87% (2020 est.) (Anadolu Agency, 2020).

The urban population is at an all-time high of 62.6% of the total population (2020), and the rate of urbanisation has an annual rate of change of 4.07% (2015–2020 est.) The middle range, projected total population for 2000 is 898,000, and current population estimates stand at just over 2,173,999 (July 2020 est.). The major urban areas include Banjul and stretches into Kanifing with a population of 481,000 (2023) which include the local government areas of Banjul and Kanifing. In The Gambia, settlements are scattered along the Gambia River, with the largest communities, including the capital, Banjul, and the country's largest city, Serekunda, located near the river's mouth on the Atlantic coast (Central Intelligence Agency [CIA.gov], 2024).

O. Fabiyi (✉)
Department of Computer Science and Technology, American International University West Africa, Serekunda, The Gambia
e-mail: ofabiyi@aiu.edu.gm

U. T. Saminaka · B. Ozirigbo · F. Mfobujong
Department of Accounting & Finance, American International University West Africa, Serekunda, The Gambia

T. Nubi et al. (eds.), *Transit Oriented Development in West African Cities*, Urban Perspectives from the Global South,
https://doi.org/10.1007/978-3-031-58726-9_10

Almost 60% of the population is under the age of 25 and is likely to persist because the country's total fertility rate remains strong at nearly four children per woman. The Gambia's small economy relies primarily on tourism, rain- dependent agriculture and remittances and is vulnerable to external shocks. Real gross domestic product (GDP) growth was estimated at 6.5% in 2018 (from 4.8% in 2017); it also thrives on fishing because of its rich coastline The Gambia has sparse natural resource deposits, some of which include clay, silica sand, titanium (rutile and ilmenite), tin and zircon. It relies heavily on tourism, with a large influx of tourists from Europe and the Americas, and on remittances from citizens overseas with inflows amounting to about one-fifth of the country's GDP. The Gambia's location on the ocean and proximity to Europe has made it one of the most frequented tourist destinations in West Africa, boosted by private sector investments in eco-tourism and facilities. Tourism normally brings in about 20% of GDP, but it suffered in 2014 from tourists' fears of Ebola virus in neighbouring West African countries (World Bank, 2022).

10.2 City Profile

Banjul, officially the City of Banjul (former name Bathurst until 1973), is the capital of the Republic of The Gambia, in West Africa, as well as the administrative centre of the country and the seat of government. The port city has a population of about 31,000 and is divided into 3 districts. Its land area is 12 sq. km (4.6 sq. mi). Banjul is located on Saint Mary's Island at the southern part and mouth of the Gambia River; the Mandinka used to gather fibrous plants on the island for the manufacture of ropes; 'bang julo' is Mandinka for 'rope fibre'; mispronunciation over time caused the term to become the word Banjul. People have inhabited the area outside of Banjul Island for many millennia before the Portuguese navigators Antoniotto Usodimare and Luiz de Cadamosto entered the mouth of the Gambia River in 1455 before being ejected by locals. They returned in 1456 and managed to reach James Island further upstream. Banjul's commercial centre is around the docks at the city's east end. Despite the port's small size, it is a critical entryway for imports as well as exports, and the roads near the waterfront are often jammed with trucks and lorries waiting to load newly arrived consignments such as sugar, rice and cooking oil as they are offloaded from container ships. Some merchandise are destined for stores on Liberation and Ecowas Avenue as well as the Royal Albert Market; others get transported inland and to other countries in West Africa via entrepot (Access Gambia Ltd., 2009; Fig. 10.1).

Banjul as a port town, capital city and seat of power makes it very important to The Gambia as a whole. The quality of infrastructure is generally perceived as being above the low-income developing country (LIDC) average. The World Economic Forum conducts surveys of business leaders' impressions of overall infrastructure quality on a seven-point scale. Based on this survey, perceived quality in Banjul exceeds the average of the cities in sub-Saharan Africa and of LIDCs, though the

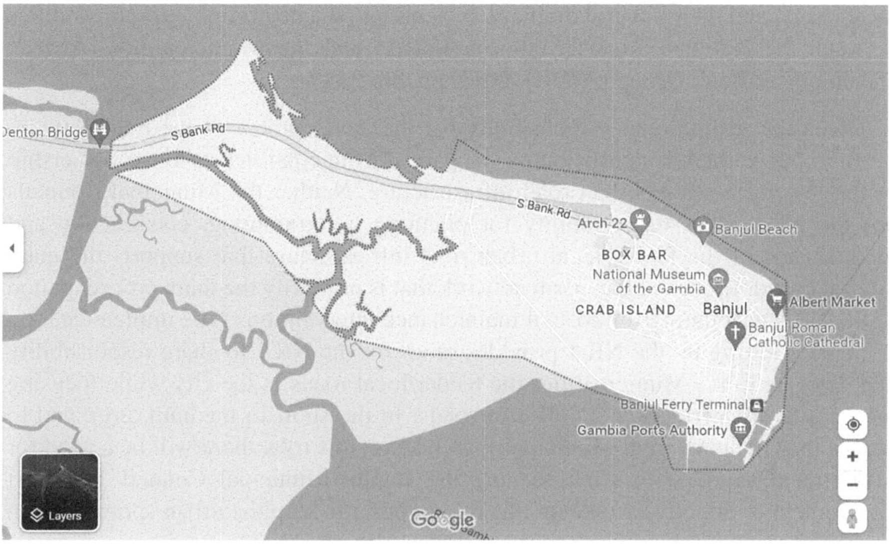

Fig. 10.1 The map showing the city of Banjul. (Source: Google Maps, 2023)

gap narrowed over the period from 2010 to 2015 (The World Economic Forum [WEF], 2022). Infrastructure quality rose sharply in the period from about 2006 to 2010 on the back of significant investment in areas such as roads and energy, but subsequently has fallen back, compared to the broadly flat trend across the comparator groups. Access to physical infrastructure is relatively good in the electricity sector, but worse in the areas of roads, education and health. Public infrastructure in health (measured by access to hospital beds) and education services (measured by access to secondary school teachers) fall below the averages for cities in sub-Saharan Africa and low-income developing cities. The infrastructure in the city of Banjul remains a challenge as with the whole country, with unreliable electricity supply and no proper drainage coupled with bad roads, which makes it very difficult for those using transport services and even for the residents of the city (Manneh, 2020).

10.3 Broad Infrastructure Challenges

Road Institutional responsibility for the hierarchy of roads contributes to the poor condition of the urban road network. Urban road infrastructure provision and management calls for the definition of hierarchy of roads in the city and the shared institutional responsibilities for their construction and maintenance. The poor state of urban road infrastructure in the Banjul area is impeding the flow of traffic, leading to loss of transit time for urban dwellers and higher vehicle operating costs which are a real cost to the economy. So, in the case of the Greater Banjul Area, there is no definition of the hierarchy of road infrastructure (arterial, collector and local roads) and their integral drainage systems and the institutional responsibilities between the National Roads Authority (NRA) and the Municipalities (African Development Bank Group [AfDB], 2013, pp. 30–35).

Given the institutional responsibility for the development of the road network given to NRA, there is confusion on the part of Municipalities as to the ownership of the lower classes of urban road infrastructure. Neither the Municipal Councils nor the NRA takes responsibility for planning, programming, construction and maintenance of the feeder/local urban road infrastructure that supports the main arterial corridor roads in the urban network that is primarily the main preoccupation of the NRA accounting for 60% of maintenance interventions. The implementation of the Act setting up the NRA provides room for the NRA to share responsibility and delegate to the Municipalities the feeder/local roads in the city while focusing on the arterial corridors and collector roads. In the short to medium term, and to enable the Greater Banjul Municipality to take on this role, there will be a need for institutional support for strengthening the Banjul Municipal Council Transport Department to effectively take on this role in order to stop the urban sprawl, which is a major issue for the city planners (AfDB, 2013, pp. 32–34).

Waterways A number of issues need to be addressed with respect to maritime and river transport. These issues require urgent actions on the part of the Gambia Ports Authority (GPA) and the Government in the face of globalisation of shipping, port and logistic industry. The role of the private sector in the further development of the sector calls for institutional restructuring of the GPA from a public service model to that of a landlord model in order to improve operational efficiency and hold its competitive position in the fierce competition with other regional ports (Organisation for Economic Co-operation and Development [OECD], 2007). Globalisation has affected the port and shipping industry in a way that the port management model has moved from public service ports to landlord model in order to improve efficiency in the face of increasing containerisation of world trade that is growing and has replaced break bulk cargo. Banjul is a port city, and the urban sprawl has led to significant urban development in areas adjacent to the port, leaving little room for port expansion or port service industries to develop in the future. This will call for the development of inland transport connection and effective coordination between the Banjul Port Master Plan, the Banjul Urban Development Master Plan and the National Transportation Master Plan, which require intermodal coordination of surface transport that may necessitate the use of container barges to off port ICD on the bank of the Gambia River outside the urban boundary (AfDB, 2022).

Airways There is an institutional issue with the dual role of the Gambia Civil Aviation Authority as regulator and at the same time operator of Banjul International Airport and its facilities. There is a perceived concern of possible conflict of interest in the two roles which may compromise safety standards. The Gambia Civil Aviation Authority (GCAA) explained that it has dealt with this perceived conflict with the definition of roles under an unwritten rule of shared responsibility between the Director General and the Deputy. The Deputy is directly responsible for managing

Banjul International Airport and its facilities, while the Director General handles the regulatory issues. It is necessary to move beyond this as traffic matures for better commercial management of Banjul International Airport. The GCAA should then take on only regulatory functions, and an airport authority should be created for the management of the airport and its facilities (OECD, 2007).

10.4 Transportation Infrastructure Overview

The development of the transport sector in Banjul as a model city had its roots in the colonial period because it was the entry and exit point which is the natural port of Banjul, supported by the Gambia River. It provided the international gateway and the main transport artery for The Gambia and its neighbouring countries of Senegal, Mali, Guinea Bissau and Guinea Conakry external trade. The Transport systems is currently made up of a road transport system consisting of 37 km of urban roads, an air transport system provided by Banjul International Airport at Yundum, and a maritime and river transport system consisting of Banjul Port and transport along The Gambia River (Ministry of Transport Works & Infrastructure [MOTWI], 2017).

10.4.1 Road

The urban transport infrastructure of Banjul is about 37 km, and 150 km is in the Greater Banjul Area. The primary focus in the urban transport subsector is to address the emerging transport crisis in the Greater Banjul Metropolis, which is facing the challenge of rapid urbanisation. A key deficiency in managing urban transport is the lack of institutional authority due to the multiplicity of agencies involved. Currently, there is a blur with respect to institutional responsibilities for the hierarchy of urban road network and regulation of urban transport services on the arterial, collector and local roads in Banjul. This would need to be rationalised and strengthened in order to improve the efficiency of the urban transport system and accountability for system failure particularly for the interrelated plans covering road network, bus network and truck routes and traffic management plan for the central business district (CBD) in the city environment (MOTWI, 2017).

Road Infrastructure Development

The current stock of urban road infrastructure in Banjul as highlighted earlier is about 37 km, and most of the urban road network is in very poor condition. This has been caused by a lack of proper drainage infrastructure not just in the capital city but in the country as a whole. The institutional responsibility for construction, maintenance and management of urban roads and drainage infrastructure lies with the

National Roads Authority that also has the power under the enabling Act to delegate the construction and management of some roads to municipal authorities or the local administration. The Greater Banjul Metropolis requires a long-term transport needs study within an urban framework based on the future land use plan in order to contain the current sprawl of the Banjul-Brikama area. But any further development of urban transport infrastructure in the Greater Banjul area should be based on integrated land use and transport planning approach. This will enable effective planning between the central business district and residential areas and the transport system that would respond efficiently to urban mobility needs. There is no hierarchical road system, a network of arterial, collector and local roads, established for the Greater Banjul Area and other emerging cities, which will form the basis on which the National Roads Authority can delegate some local roads to the lower agencies such as the municipal authorities and local councils. The arterial and collector roads can then be the responsibility of the National Roads Authority, while the local roads are left to the municipal councils. Institutional responsibility is currently blurred both for the road infrastructure and the leadership for urban transport services (AfDB, 2013, pp. ix–x).

10.4.2 Waterways

The advantage of geography, with Banjul situated on the mouth of the Gambia River, had made the port of Banjul a regional trading centre, with river transport providing deep links hinterland and servicing the transit traffic to Senegal, Guinea Bissau, Mali and Guinea Conakry. Up till the 1960s, Banjul Port and its integral River Gambia was the backbone of The Gambia transport system. Ocean-going vessels of up to 300 gross tonnage can go up the river for about 240 km, and smaller ships could go up to 500 km from Banjul. Principal river ports were at Kaur chanelling up to about 193 km upstream from Banjul and Kuntaureon the river for handling the traffic that then consisted mainly of groundnuts and cotton for export. The river is still navigable up to Fatoto, which is 477 km from Banjul. However, this strategic role of the subsector has weakened over the years due to competition from road transport. This has led to the collapse of the river transport system, and most of the river port infrastructure is currently in disrepair. As a result, the reinstatement of the competitive position of the port of Banjul and the river transport mode for intermodal and multimodal transport continues to be a challenge for The Gambia transport sector (AfDB, 2013, pp. 35–37).

The Port of Banjul is the only maritime port managed by the Gambia Ports Authority and is the gateway for the export/import trade of The Gambia accounting for over 80% of the total international trade. The Gambia Maritime Administration is responsible for the regulation of the maritime transport sector. Banjul Port presently has two jetties, the Banjul Wharf and the New Banjul Jetty, with a combined quay length of 428 metres. The two installations comprise five berths at the inner and outer sides and a Ro-Ro ramp facility at the northern end of the New Banjul

Jetty. Depth at approach channel and at berth sides is a constraint to the bigger vessels in the changing trends in the port and shipping industry. The port throughput was recorded as 1,402,135 metric tonnes in 2010 and increased to 1,718,938 metric tonnes in 2011, representing an increase of 9.0% over the previous years.

Based on the berth occupancy ratio of 65% applied in the 2008 Ports Master Plan Study as trigger for additional berth capacity, the Government of The Gambia is planning for a port expansion with a 200 metres wharf extension with capacity to handle larger vessels of over 20,000 DWT. But this is on the proviso that there is concessioning of the container terminals which will lead to increased efficiency to restore the ports' competitive edge to participate in transit and re-exports trade currently threatened by increased competition from the neighbouring ports of Dakar and Conakry that have moved to landlord model ports. Banjul Port is still a public service port model. This has affected its efficiency compared to its neighbouring ports, resulting in erosion of its competitive position for the same hinterland. Other infrastructure constraints include additional container storage area needed to improve the capacity and efficiency of the Port of Banjul, inter-modality in terms of inland transport connections particularly on regional road networks with trade and transport facilitation instruments and use of the Gambia River complemented with inland container depots. A major investment in infrastructure, however, needs to be preceded by port reforms (AfDB, 2022).

Due to the collapse of river transport, river port infrastructures are not in use and have fallen into disrepair. Thus, infrastructure supply alone would not revive the river transport industry, as an analysis of flow of freight based on origin/destination survey on the road network indicated weekly movements of construction aggregates, cement and building materials and other bulky freight on the road that can be shifted to river transport given a level playing field for all modes. The participation of the private sector in river transport services therefore would require a holistic transport sector multi–/intermodal planning and appropriate infrastructure pricing policies and incentives, including axle load control on the road network, increase in road access fees, fuel levy charges on road user vehicles and tax holidays to water transport operators for bulk haulage by barges. Besides, trade technologies have changed, and the use of roll on-roll off ships and intermodal terminals to back up policy changes and other tax incentives is needed to get the private sector operators back into the river transport business in The Gambia (Canadian Pacific Consulting Services [CPCS], 2020).

Banjul Port Infrastructure

Banjul Port presently has two jetties, the Banjul Wharf and the New Banjul Jetty, with a combined quay length of 428 metres. The two installations comprise of an inner and outer sides, five berths and a roll on-roll off ramp facility at the northern end of the New Banjul Jetty. Out of these five berths, only the three outer berths (1, 3A and 3B) have safe and sufficient water depth (9.5 metres) for vessels with more than 6 metres draught. Access to berths is limited in many cases for large vessels (large vessels are defined to be bigger than 15,000 tons dead weight). Both inner berths can be approached or left only under certain tide conditions, which can vary slightly depending on the intensity of tide. The approach channel water depth is 9.5 metres at mean low water spring tide (MLWST). With trends towards larger vessels, a study financed under the AfDB Bilateral Funds has revealed that dredging a 10.5 metres channel would be possible in line with meeting the objectives of the Government's freeport strategy. Sixty percent of the berths are in good condition, while 40% are in need of rehabilitation. Other facilities at the port include a container freight station, 2.4 hectares of open storage for containers, Sheds A and B for general cargo storage and a workshop for maintenance of port handling. The African Development Bank Group has been one of the main financiers of the port expansions (Banjul Port I approved with Bank Group participation of UA 6.99 out of the total UA 17.47 million co-financed with IDA, KWF and the Government of The Gambia in 1982 and Banjul Port II with an ADF loan of UA 16.0 million approved in 1993) (MOTWI, 2017; Fig. 10.2).

River Transport Infrastructure

There are terminals at nine strategic locations for the ferry services (Banjul-Barra, Farafenni-Mansakonko, Janjanbureh (former Georgetown), Bansang, Basse and Fatoto) which are an integral part of the north-south road network and are maintained by the GPA. The Government in 1984 constructed 12 concrete jetties at strategic locations, 3 of these at Kaur, Kuntaur and Basse. The inland port of Kaur has the facilities to accommodate minor sea-going vessels and had been used in the past mainly for transshipment of groundnut and cotton to Banjul. Due to the collapse of river transport, these infrastructures are not in use and have fallen into disrepair.

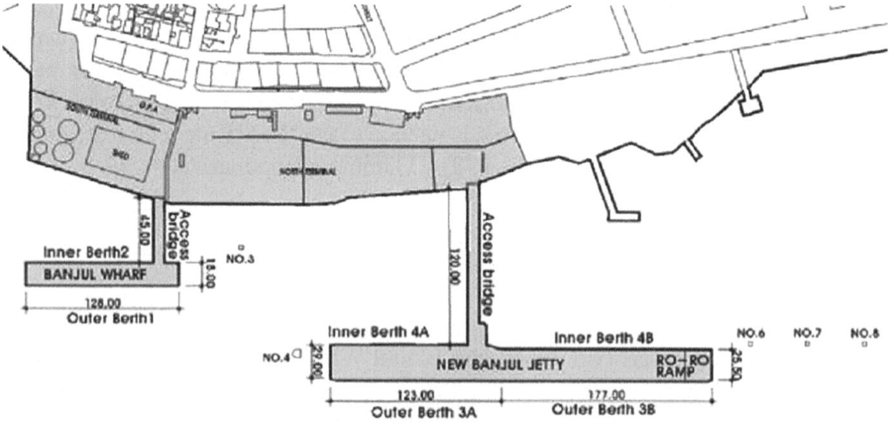

Fig. 10.2 Plan for the New Banjul Jetty. (Source: MOTWI, 2017)

Thus, infrastructure supply alone would not revive the river transport industry, as an analysis of flow of freight based on origin/destination survey on the road network indicated weekly movements of construction aggregates, cement and building materials and other bulky freight on the road that can be shifted to river transport given a level playing field for all modes (OECD, 2007).

10.4.3 Airways

Air transport plays a major role within The Gambia transport system in providing international gateways for the business community with the rest of the world and is of critical importance for The Gambia tourist industry. Because of the small size of the country, domestic air transport has not been part of the internal transport system. Transport by air into and out of The Gambia is via Banjul International Airport (BIA), situated at Yundum, about 24 kilometres to the south-east of Banjul, and it is the only airport. The Gambia Civil Aviation Authority (GCAA), established in 1991, operates the airport, manages and regulates the civil aviation industry and is responsible for ensuring compliance with the set Standards and Recommended Practices (SARPs) of the International Civil Aviation Organisation (ICAO) (AfDB, 2013, pp. 47–50).

The Gambia International Airlines Limited (GIA) was incorporated under the Companies Act 1955 as a flag-bearer airline to engage in air transportation and related activities and commenced operation in the same year. Following operational losses on regional routes using wet lease and chartered aircrafts, it has discontinued flight operations since 2005 and is only involved in ground and cargo handling, travel agency and Hajj operations. GIA is currently in search of a strategic partner for the development of BIA as a hub in the sub-region. The current capacity of Banjul International Airport (BIA) is only optimally utilised during the tourist season from November to April each year, waiting for a new business case for the rest of the year. The Gambia Civil Aviation Authority is vigorously marketing the airport for capacity utilisation during the off-tourist season; an emerging strategy is that of promoting it as an air transport hub in West Africa. This vision for an aerotropolis at Yundum requires a lot of buy-in from stakeholders both within and outside of The Gambia to justify any capacity expansion both for the passenger terminal and runway. Assessment under the sector diagnostic indicates that the exception in the medium term is with respect to apron and taxiway that are constrained during the peak tourist season and for upgrade of navigational aids for safety reasons. Given the current traffic level of about 300,000 passengers a year, the BIA may not be attractive for private sector investment, except for a long-term management contract (AfDB, 2013, pp. 47–47).

Airport Infrastructure and Facilities

The Banjul International Airport (BIA) Master Plan is supply driven in its development strategy of a hub airport for the ECOWAS sub-region. In forecasting and designing the facilities, the binding factor is the peak tourist traffic. The vision of positioning the airport as a hub and its progression from a traditional airport model to a full-fledged airport city has informed the implementation of the Banjul International Airport Master Plan in Phases. The disadvantage is that the infrastructure and facilities needed for peak tourist traffic are very much comparable to what is required by larger airports receiving much more traffic. The BIA is composed of a single 3600 metres long runway, a terminal building with capacity to handle one million passengers annually, a control tower and an international freight centre (approximately 550 sq. metres).

The air traffic control and navigational aid facilities at the airport, including VHF transceivers, VSAT, Doppler very high frequency omni-directional radio range (DVOR), distance measuring equipment (DME), instrument landing system (ILS), and an aeronautical fixed telecommunication network (AFTN)/voice link, were upgraded following the studies carried out by Belgian Airport Consultants (BEAC) in 1994, the economic part of which was updated by DHV Consultants in 1998 (AfDB, 2013, pp. 47–49).

10.5 Governance Framework for City Transportation Sector

The policy objectives of the Government were focused on ensuring each element of transport is provided in the appropriate quantity, quality and form; combining all transport components in a technologically optimal way so each mode carries the traffic it is best suited for at the least cost to the economy; operating each mode efficiently in a level policy playing field environment; and establishing appropriate institutional mechanisms to ensure effective intermodal coordination and communication between the user, operator, regulator, and the government (MOTWI, 2017).

The institutional framework for implementing the directives of the National Transport Policy and Transport Sector Programme includes the Ministry of Works, Construction and Infrastructure (MOWCI). It has the responsibility for overall policy direction and oversight (including regulation of the road subsector), and the other parastatals and agencies are in charge of the operation and regulation of the various modes. The National Roads Agency (NRA) with a Road Fund under the same Board is responsible for the planning, construction and maintenance of the national road system; The Gambia Civil Aviation Authority (GCAA) has operational and regulatory responsibilities for the civil aviation subsector; The Gambia Ports Authority (GPA) is responsible for the planning, construction, operation and maintenance of maritime/inland waterways infrastructure; and the nascent Maritime Transport Administration created in 2010 is the subsector regulator.

Deepening of institutional reform and improved governance for the civil aviation subsector is recommended. This is for the separation of operational and regulatory functions of the GCAA as traffic increases to a level of commercial profitability for Banjul International Airport (BIA) (AfDB, 2022). Also, for the road subsector, the creation of a separate Road Fund Board, apart from the Board for the NRA, to improve governance and for checks and balances to enable the Road Fund to carry out the required technical and financial audit of the subsector projects for value for money is recommended for further sector reform. The institutional mechanism for sector-wide regulation and integration of all the transport modes as a basis for prioritisation and in determining the appropriate quantity, quality and form of infrastructure recommended in the National Transport Policy (1998–2006) document by upgrading the Transport Planning Unit within the MOWCI to a National Transport Agency (NTA) has led to the recurring problem of uncoordinated transport infrastructure planning and lack of a transport sector database to inform sector-wide policy formulation and planning, so, to improve sector efficiency, this reform needs to be carried through and the NTA established as proposed in the Policy document, comprising two units: one for policy/regulation and another for planning.

These should be increased to three to include a unit for public-private partnership to drive the privatisation policy of government in the sector. The road sector is the weak link in the enforcement of existing traffic laws and regulations. Besides, some of the traffic laws and the penalties specified therein have become obsolete and irrelevant as deterrents. Some regulations in terms of level of service and technical standards of vehicles needed to be reviewed and updated for the efficiency of the subsector (AfDB, 2013, p. 7).

10.5.1 National Transport Policy

The policy and regulatory framework in transport for the city of Banjul stems from the National Transport Plan that was prepared in 1997 which is focused on the maintenance and development of transport infrastructure. The Government, noting that infrastructure alone was not sufficient for the smooth functioning of the transport system, decided to put in place a comprehensive policy framework to guide the development of the sector. This led to the preparation of the National Transport Policy 1998–2006 that limited the role of the Government ministry in charge of the sector to policy formulation and oversight, while regulatory, operational and management functions were delegated to the parastatals and other agencies in the sector. The policy objective was to ensure the application of commercial principles in the management of the sector and contribute to the national objectives of facilitating the optimal development of the economy (AfDB, 2013, p. 7–9).

The policy emphasised integration of the modes as a basis for prioritisation and in determining the appropriate size and quality of infrastructure, which must be based on a long-term and integrated transport infrastructure planning. In the institutional restructuring recommended in the National Transport Policy (1998–2006) document, the upgrading of the Transport Planning Unit to National Transport Agency (NTA) was recommended to function as a unit within the Ministry and be staffed with high-level professionals and organised in Transport Regulatory Office (TRO) for the enforcement of policies, legislations and rules and their update and Transport Policy Office (TPO), responsible for the preparation and update of the National Transport.

The idea behind policy and intermodal coordination include the preparation of an overall Transport Master Plan consistent with the Transport Policy and Strategy. Though the Transport Planning Unit was not upgraded to a National Transport Agency (NTA) as recommended, due to being understaffed, with only one professional who does not have the training and capacity to perform the functions of the NTA recommended under the policy (AfDB, 2013, p. 9).

10.5.2 Institutional and Regulatory Framework

For The Gambia, the legal basis and regulations pertaining to transport are summarised in the publication 'The Laws of the Gambia' which collated the relevant laws in the area of transport following the Law Revision Act of 1990, particularly Section 24 of the document that covers matters of relevance to the sector. The MOWCI is responsible for setting the overall policy objective and strategic role for the sector and overseeing the performance of the institutions under its mandate, including the regulatory and operator institutions in the transport sector. While enforcement is a major concern, some of the existing regulations are not adequate, not sufficiently comprehensive or in some cases outdated to meet the current realities of the sector; all these form a parent-child relationship as it pertains to Banjul City (AfDB, 2013, p. 6–8).

The Gambia Civil Aviation Authority (GCAA) was established as an autonomous body in July 1991 (under the Public Enterprise Act of 1989) for monitoring and regulating the aviation industry in The Gambia and was also entrusted with the management of Banjul International Airport (BIA). The GCAA is overseen by a Board of Directors, and the management is structured into 8 directorates under the leadership of the Director General supported by the Deputy and had an established staff position of 548 in 2011. The GCAA plays the dual role of both regulator and operator of Banjul International Airport and the navigational services for the industry; but it is only a regulator of air transport services in which The Gambia International Airlines (GIA) is the operator with room for private sector operators. In its regulatory role, the GCAA strives to fully comply with ICAO Standards and Recommended Practices (SARPS) and secondly the acquisition of International Aviation Safety Assessment Programme (IASA), which are based on regular audits respectively by ICAO and the US Federal Aviation Administration (FAA).

The GCAA as a regulatory authority has technical units for civil aviation including Air Safety Operations (ASO), Aircraft Safety (AS), Aviation Security (AVSEC) and Air Safety Infrastructure (ASI) in ensuring compliance with ICAO standards (AfDB, 2013, pp. 7–9).

For the maritime transport subsector, The Gambia Ports Authority (GPA) is in charge of construction, maintenance and operation of the infrastructure and ferry services, while the nascent Maritime Transport Administration created in 2010 is the regulator. The GPA is overseen by a Board of Directors, and the management is structured into seven directorates under the Managing Director assisted by the Deputy with an establishment summary of 759 staff in 2011 (AfDB, 2013, p. 8).

Also, for the road subsector, the MOWCI is responsible of the policy formulation and regulation of roads/road transport, while the National Roads Authority (NRA) is in charge with the operational responsibilities of construction, maintenance and management of the road network in a sustainable way. The NRA is overseen by a Board of Directors, while the management is structured in four directorates overseen by the Director General assisted by the Deputy. The NRA Board also oversees the Road Fund attached to the NRA. To deepen sector reform and improve sector governance, it is necessary that a separate Board is set up for the Road Fund. The Road Fund Board will appropriate checks on the NRA to ensure value for money generated from tax pays and road users. As of December 2012, the National Roads Authority (NRA) had an established staff structure at the management and directorate levels. The public transportation system for roads has been liberalized, with services intended to be provided by both private sector operators as well as the formerly state-run Gambia Public Transport Corporation, which is now defunct (AfDB, 2013, p. 10).

10.6 Mobility and Access in the City

In addressing urban mobility challenges in Banjul, a major concern is accessibility and affordability in transport use and transport infrastructure that facilitates and supports economic and social interactions which leads to continued and sustained economic growth. An effective and efficient transport system will boost private sector production by reducing production costs and increasing productivity and also increase employment opportunities. It also impacts on the lives of people both directly, by enabling them to attain their basic needs, like access to school, health centres, hospitals, food and energy supplies, and indirectly, through economic growth by enhancing the free movement of people, goods and services, to and from the urban and rural communities and to the local and regional markets (Bass & Tamba, 2022).

Employment in the transport sector is created in both the construction and operational phases of the transport system. Therefore, improving and modernising the transport infrastructure will contribute immensely to economic growth and employment generation. The transport sector has the potential to support the productive capacity and long-term expansion of the economy. It is the Government's objective to improve the transport sector's adaptation to climate change and mitigation of greenhouse gas emissions through the improvement of public transport, a better road maintenance regime, better maintained vehicles and the use of more water-based transport. The Government has taken significant steps towards the attainment of this goal through investments and projects to improve road transport, maritime transport, river transport and air transport (Cities Alliance Knowledge, 2021).

Transport infrastructure is one of the most important factors for a country's progress. It is no gainsaying that the availability of a good road network system is essential, as it facilitates the smooth flow of people, goods and services. In a nutshell, good road infrastructure has become an essential part of daily life. There is no doubt that the city is witnessing a rapid surge in the number of commuters, which seems to outnumber the available cars plying on our road daily. However, that should be an excuse as good road infrastructure is critical to any meaningful development. The road infrastructure in the country really needs a critical rethink. During peak rush hours or closing times, commuters often encounter extremely frustrating situations that can be likened to their worst nightmares (Cities Alliance Knowledge, 2021).

In Banjul as well as other places in The Gambia, cars and buses became a major mode of transportation. But the issue of a good road network is becoming a concern, especially with the increasing number of people buying cars for private and commercial use, thus increasing traffic congestion. The current problem being faced in the road network started many years ago. From the onset, those who were involved in the initial design of road infrastructures did not put in place long-term measures, considering the fact that future expansion is inevitable. The city is developing, as we need road connectivity that will meet international standards for the safety of passengers and even socio-economic growth. The deplorable road condition in the city with potholes in some streets shows that there is an increased need for maintenance or rehabilitation of the aging road system especially during the rainy season where mobility becomes a nightmare for many road users (Cities Alliance Knowledge, 2021).

Commuters battle high traffic congestion along Banjul and other connecting areas causing huge traffic congestion for people and vehicle users especially during and after working hours where commuters spend hours before reaching their destinations. A diverging diamond interchange also called the 'double crossover diamond interchange' is an innovative strategy, designed to guide motorists through the pathways. It minimises conflict points, vehicle delay, decreased speed and reduced environmental impact. The initiatives also ease traffic congestion and minimise accidents. Recent incidents in The Gambia show an increase in the rate of fatal car accidents in roads across the country and especially in Banjul. It is believed that using the latest engineering technology and techniques is better than adding lanes. The economic benefits however, combined with improved safety and mobility, will also contribute to enhanced competitiveness and quality of life for businesses and communities (Bass & Tamba, 2022).

10.7 Transportation Investment

10.7.1 Road Planning

Investment in the road network has not been based on a prioritised plan, where individual links on the network are ranked based on their economic internal rate of return (EIRR) or net present value (NPV) and consistency with the macro-economic and fiscal framework and the medium-term expenditure framework. Based on the strategy, recommended in the National Transport Policy 1998–2006, investment in the subsector had concentrated on inter-urban roads in the Greater Banjul area and moving eastwards thereafter with the focus on addressing bottlenecks on the most trafficked links eastward from Banjul. While this focus has been right, in addressing the investment needs on the highly trafficked part of the network, there remains a gap in ascertaining how the road sector investment has contributed to national and local development needs by using network approaches that take into account connectivity and the opportunity cost to the economy. Roads are enablers of other opportunities in an economy, and the issue of connectivity and opening up of areas with development potentials would demand a network approach in investment planning, which is not the case under the current approach to investment in the sector (AfDB, 2013, pp. 24–25).

Up until now, development and investment in the road network have been driven by feasibility studies on individual road links, rather than taking a sector-wide approach considering the entire classified road network as a whole. Such holistic approach will take into account new construction/upgrading to bitumen, rehabilitation, periodic maintenance, routine maintenance and capacity buildings/studies, all competing for the limited funds that take into cognisance the connectivity of the network to support overall socio-economic development and inclusive growth process. The EU is currently preparing an intervention in feeder roads where all the links are taken into account and prioritised for a feeder roads rehabilitation programme. For evaluation purposes, all the links were evaluated based on parameters for social, economic and technical features to arrive at a ranking determined by cost-effectiveness and multi-criteria analysis, and in cases where the ADT is between 50 and 100, the parameters are combined with economic analysis criteria using the RED software (AfDB, 2013, p. 24).

10.7.2 Capital Investment Financing

Significant investments have been made in The Gambia's primary road network from 2017 to 2020. The total investment on sections of the primary road network during this period has been estimated at a sizeable €69.99 million, including costs for ongoing road works and construction projects. The majority of the funding for these road infrastructure improvements has come from development partners and foreign sources. The European Union and Arab development funds have been the main providers, contributing large portions of the €69.99 million invested in upgrading and expanding the country's primary road arteries between 2017–2020. This influx of funds from international partners has allowed The Gambia to undertake major road enhancement projects across its core national highway system over the last few years, addressing key bottlenecks and modernizing vital transportation links. However, significant financing gaps likely still remain to comprehensively upgrade and maintain the full primary road network to meet the country's developmental needs (MOTWI, 2017).

10.7.3 Periodic Maintenance Funding

On abolishing the force account system and coming into effect of the NRA with the Road Fund in 2006, the Government of The Gambia's strategy in managing the primary road network has been the use of long-term maintenance contracts for a minimum period of 5 years awarded in lots to domestically registered foreign firms. This has led to higher quality of maintenance work as contractors are committed to a level of service specified as per contract. For the primary network between 2009 and 2011, a total amount of EUR 13.54 million had been spent. The long-term maintenance contract approach will ensure sustainability of the investment and value for money. The periodic maintenance funding is through subvention from the Government budget in spite of the fact that the Road Fund has been established since 2006. However, it is also important that the Road Fund does technical audits to ensure that road users are fully protected particularly when the fuel levy is introduced as one of the sources of revenue for the Road Fund. There has been a push and pull between the Banjul Area Council and the Government via the Ministry of Works for the management of roads in Banjul City (MOTWI, 2017).

10.7.4 Urban Transport Subsector Investment

Over the horizon of 2012–2015, only the Greater Banjul Area drainage system has found its way into the Government's investment plan with a total indicative allocation of USD 1.11 million. The deplorable situation of urban road infrastructure has not been explicitly stated, though its maintenance might be presumed under the allocation to the National Road Authority for maintenance of the road network, which is not even enough to meet the annual cost of maintenance needs. If there is definition of clear responsibilities for hierarchy of roads in the Greater Banjul area, there should be in the short term a capacity building support to the Banjul City Transport Department on planning, programming, construction and maintenance of its local road network, and this could be included as a capacity building activity for the sector and its stakeholders. The Government should take a holistic view of urban land use and urban transport master plan and commission a study to guide future urban development and improve the livelihood of urban dwellers and the productivity of the urban economy that are engine of growth for not just Banjul but the country at large (MOTWI, 2017).

10.7.5 Port and Maritime Transport Subsector Investment

An investment plan of EUR 118.5 million to reverse the decline in the competitive position of Banjul Port and improve ferry services has been proposed under the port and maritime transport subsector by the Gambia Ports Authority under Pillar 2: Improving and Modernising Infrastructure of 2012–2015. It is a known fact that the Government is not interested in the privatisation of Banjul Port and restructuring of its management to a landlord port model. This may be a serious constraint in raising funds in the capital market or borrowing from the private sector window of development partners given their credit policies. The construction of a new quay of 200 metres recommended by the recent study financed by the Korean Trust Fund hosted by the AfDB has made the privatisation of the container terminal a condition for the berth expansion. As per the AfDB Group credit policy, The Gambia is an ADF country and grants only; thus, the expansion of berth capacity through construction of a new quay of 200 metres cannot be financed through concessionary resources. The AfDB may however undertake a study on the privatisation of Banjul Port through the bilateral funds hosted to provide the Government an informed basis for dialogue to undertake the necessary reform of its port sector. Other maritime transportation investment projects under active consideration by GPA relate to ferry services and rehabilitation of the dockyard at a total cost of EUR 36.0 million. The rehabilitation of the dockyard will be important for the rival of the river transport if the private sector is incentivised to enter the river transport industry.

10.7.6 Airways Financing and Investment

A total investment of about USD 37.14 million has been made for improving the Banjul International Airport infrastructure and facilities between 2007 and 2020. Except for the construction of the new terminal building, the investment in the airport infrastructure and facilities has been mainly financed by Development Partners, with the Government handling some counterpart funds. The AfDB Group through its Nigeria Trust Fund window financed the construction of the runway at the airport in the 1970s, but since then the Bank has been absent from the subsector. The two main donors in the sector since the 1990s have been the Kuwait Fund for Arab Economic Development (KFAED) and the Arab Bank for Economic Development in Africa (BADEA) that have co-financed the Fast Track and Phase I—Airport Improvement Projects with the Government (AfDB, 2022).

10.8 Actions Taken in the Last 20 Years

The key issues that were addressed in the road subsector from both operational and investment perspectives relate to the following for which necessary action plans have been and are still being taken:

- Development of a prioritised five-year road development programme for the primary and secondary road network consistent with the macro-economic and fiscal framework and medium-term expenditure framework for the road sector under the;
- Development of a prioritised five-year feeder roads rehabilitation programme based on multi-criteria analysis to support the agricultural sector and all-inclusive growth under the 244 km of rural feeder roads rehabilitation programme prepared under the EU/TA can be a basis for donor involvement as it meets the sector planning criteria;
- Establishment of sustainable funding for road maintenance by introducing fuel levy charge on road users as main source of road maintenance funding for the network and revision of current schedule of user charges and road access fees to reflect damage caused to pavements by the vehicle type and introduction of transit fees;
- Development of the Five-Year Road Maintenance Action Plan (R-MAP) based on a revitalised Road Fund with main revenue source coming from fuel levy;
- Development of heavy vehicle management programme to prevent premature damage to road pavements;
- Initiation of a road safety programme with creation of necessary institutional framework—National Road Safety Council—with a secretariat to develop its work programme;
- Institutional capacity building for the NRA to be able to exercise its mandate for planning, programming and execution of road works and managing contractors and consultants;

- There is a need for the Government to explore options and strategies for introducing civil engineering and other relevant engineering courses in the university to meet the nation's medium-term and long-term manpower needs;
- Programme for the development of the local road construction industry; starting with labour-based method in collaboration with ILO and other development partners and GAMWORKS and private sector management of heavily trafficked road corridors;
- Review and update of road and road traffic laws and regulations to respond to the current realities of road use and operation and to improve road transport services (passenger and freight) and to improve road safety;
- Creation of corridor management committees essential for transit and trade facilitation improvement on Gambia Port-based regional road corridors should be implemented as follow-up action on the ISRT and Bilateral and Tripartite Agreements entered into by the GoTG (AfDB, 2013; MOTWI, 2017; AfDB, 2022).

10.8.1 Waterways

The key issues that were addressed with respect to maritime and river transport require urgent actions on the part of the GPA and the GoTG in the face of globalisation of shipping, port and logistic industry. The role of the private sector in the further development of the sector calls for institutional restructuring of the GPA from a public service model to that of a landlord model in order to improve operational efficiency and hold its competitive position in the fierce competition with other regional ports. Globalisation has affected the port and shipping industry in a way that the port management model has moved from public service ports to landlord model in order to improve efficiency in the face of increasing containerisation of world trade that is growing and has replaced break bulk cargo (Bojang, 2000).

Banjul is a port city, and the urban sprawl has led to significant urban development in areas adjacent to the port, leaving little room for port expansion or port service industries to develop in the future. This will call for the development of inland transport connection and effective coordination between the Banjul Port Master Plan, the Banjul Urban Development Master Plan and National Transportation Master Plan, which require intermodal coordination of surface transport that may necessitate the use of container barges to off port ICD on the bank of the Gambia River outside the urban boundary; better handling capacity for bulk cargo and container logistics to improve current operational performance; and improved regional road connections with trade and transport facilitation measures to the hinterland countries of the port (The Point, 2021).

10.8.2 Airways

There is an institutional issue of the dual role of GCAA as regulator and at the same time operator of Banjul International Airport and its facilities. There is a perceived concern of possible conflict of interest in the two roles which may compromise safety standards. The GCAA explained that it has dealt with this perceived conflict with the definition of roles under an unwritten rule of shared responsibility between the Director General and the Deputy. The Deputy is directly responsible for managing Banjul International Airport and its facilities, while the Director General handles the regulatory issues. It is necessary to move beyond this as traffic matures for better commercial management of Banjul International Airport. The GCAA should then take on only regulatory functions, and an airport authority should be created for the management of the airport and its facilities (GCAA, 2022).

The peak traffic generated by the tourists during the tourist season is the main determinant of the capacity requirements of Banjul International Airport. Thus, there is idle capacity during 6 months of the year after the tourist season from November to April. The dimensioning of airport infrastructure and facilities in this context calls for marketing the airport for business during the off-tourist season. With small business travel given the low per capita income of The Gambia and the small population, other export promotion opportunities are being explored including marketing the airport as a hub in the subregion. However, a hub must have the necessary domestic traffic density and a functional Gambian-based airline both of which are not the case. Getting all stakeholders to get the business case to make the BIA viable is a major challenge for the subsector. Given the current traffic level of about 300,000 passengers a year, the BIA may not be attractive for private sector investment, except for a long-term management contract which can be reviewed in detail by the PPP Unit proposed to be in the proposed NTA of MOWCI (AfDB, 2022; MOTWI, 2017).

10.9 Transportation Infrastructure GAPS

Since infrastructure is primarily implemented and maintained by The Government of The Gambia, all the failures and lack of dutiful interventions not done in the city is attributed to the Government, and they have agreed that infrastructure and interconnectivity, particularly transport, represent a major challenge to achieve most of its development goals. The Government has already made substantial efforts in the sector: 70% of the primary road network is in good condition (most of it financed through the EDF). The National Road Authority (NRA) and the Road Fund have been operational since 2006 and are partially operational with EU's support under the 9th and 10th EDF. However, the Government, through its road transport policy, identifies major issues in the governance of the sector to be able to take over efficiently and sustainably its management and in particular maintenance. These issues would request both additional support and strong commitment from the Government. No other development partner is involved in the transport sector in The Gambia (except AfDB and Arabic funds for some investments). Therefore, it is proposed that the programme provides support to build up and strengthen the governance of the transport sector, excluding additional investment in the trunk roads. It will be a short-term exit strategy to consolidate the work already done through EDF funding both with regard to investments and governance of the transport sector with:

(i) Technical assistance to the NRA.
(ii) Capacity building of the Road Fund enabling a full implementation of the maintenance programme, under the condition that the road fund is progressively fully financed by a reliable and sustainable source of funding as foreseen by the 2006 Road Fund Act.
(iii) Accompanying measures to local small-scale contractors currently not able to perform maintenance.
(iv) Additional support in implementing ECOWAS' rules on axle load control (AfDB, 2013, p. 29).

10.9.1 Main Expected Results (ERs)

ER1: An annual maintenance plan, including the required budget, is to be submitted to the Minister of Transport, Works and Infrastructure, as well as a rolling multi-annual maintenance plan. The NRA will implement these plans by tendering and awarding the required road maintenance contracts in accordance with the annual maintenance plan.

ER2: The Government will demonstrate its commitment by ensuring that the required funds are provided either through the levies specified in The Gambia Roads and Technical Services Authority Act and/or, if this is not sufficient, by a subvention from the general budget to finance the required maintenance budget.

ER3: An institution to be established or upgraded to cater for the training of small-scale contractors in labour-based intensive road construction works, including elements of environmental management.

ER4: A legal framework to be established to enable the Government to implement the ECOWAS' rules on axle load control in accordance with the agreement signed by the members of ECOWAS (MOTWI, 2017).

10.9.2 Main Indicators for Each ER

The main indicators for measuring the aforementioned results are contained in the sector intervention framework.

Donor coordination and policy dialogue: In the transport sector, there are only a few other donors, AFDB (with a high investment capital for the construction of the bridge over the Gambia River), The Kuwait Fund and the Arab Bank for Economic Development in Africa (BADEA) with limited amounts for investment in the road sector.

Government's financial and policy commitments: The Government of the Gambia shows that the levies meant for the Road Fund are actually going to the Road Fund. If these are less, based on needs assessment, the Government will top up the missing amount from the general budget. The Government has recently approved that the fuel levy will be included in the levies meant for the Road Fund. The Government will have to show a proactive approach of preparing and submitting new axle load laws in accordance with the ECOWAS' axle load rules to enable the Government to enforce the axle load control to decrease the overloading.

Overall risk assessment of the sector intervention maintenance funding risk: Lack of enough revenue to cater for the required maintenance.

Mitigation measures: The Government's commitment to include in the levies meant for the Road Fund also a levy on the fuel (MOTWI, 2017).

Regulatory enforcement: Non-enforcement of existing regulations on driver behaviour and traffic management has resulted in chaotic traffic situations in the Greater Banjul Area. The lack of effective control on parking and commercial activities on sidewalks and even on pavements are commonplace, reducing the capacity of the urban network. Much could be done in this respect in a shorter time and at low cost to improve traffic circulation (MOTWI, 2017).

Transition to regulation of small informal sector and bus rapid transit: There is no bus rapid transit system in the Banjul Metropolitan Area with priority bus lanes. There is currently a laissez-faire approach with a paratransit system consisting mainly of the small informal sector that now controls the provision of public transport in the city following the demise of GPTC that calls for urgent regulation of current chaotic service provision. This could be done through route allocation that confines them to feeder routes. This should however be accompanied by incentives to the private sector for big bus operators to run on the arterial routes with high-volume, uncongested routes with priority bus lanes, where higher productivity can be achieved and protected from mini buses that are confined to the collector and feeder routes (Sanneh, 2020).

Fleet renewal is a challenge for the urban transport operators and has resulted in an ageing fleet of public service vehicles on the urban road system. This has implications for road safety. This may require more franchising agreements that could provide more secured streams of future income and help facilitate bank financing. The Government can also provide incentives such as reduction of import duties on buses/spare parts and tax holidays for bus operators that operate the main arterial corridors. Managing the traffic in the central business district of the Greater Banjul Metropolitan Area will require complementing a bus rapid transit system that maximises accessibility to the CBD destinations with high-impact solutions, such as bus priority lanes/exclusive bus lanes, removal of on-street parking, signalisation at intersections, one-way streets and widening of road infrastructure for improve capacity. This would require the introduction of a road hierarchy that identifies primary distributor roads, linking arterial roads for interzonal traffic and collector roads used for intrazonal traffic movements (Sanneh, 2020).

The improvement of Greater Banjul Urban transportation operations will rely on implementing four key strategies: improving vehicular traffic control mechanisms, regulating vehicular flow through effective parking policies, establishing or upgrading traffic control centers to enhance monitoring and management, and optimizing intersection traffic management systems along with signal coordination to ensure smooth traffic flow and safety (Committee for Economic and Commercial Cooperation of the Organization of the Islamic Cooperation [COMCEC], 2015).

It is important to consider all these issues as vital components of the Greater Banjul Land Use/Transportation Plan. There should also be policy measures to mitigate urban environmental degradation. This should include enhancement of gas emission control through the establishment of vehicle inspection stations and policy for traffic demand management that favours public transportation service over the use of private vehicles. The above-stated policy streams and measures are to be implemented over the short, medium and long term with respect to infrastructure, traffic management, service quality and route network development. In the long term, following the commissioning of Greater Banjul Land Use and Transportation Plan Study, there may be institutional development leading to the establishment of a metropolitan transport authority in charge of both infrastructure and public transport (COMCEC, 2015).

10.10 Recommendation

To improve the performance of urban transport in the Greater Banjul Metropolis, priority should be given to urban transport infrastructure that protects public transport movements and non-motorized traffic (NMT) from the unrestricted expansion of private motorized trips. An incentivised private sector may need to be drawn into public transportation activities particularly those that can operate big buses on the arterial and collector roads and invest in terminals combined with route allocation on the local and secondary roads based on an agreed level of service. Enhanced traffic laws and regulations, however, need strict enforcement by the police to complement traffic management measures at intersections and thereby improve urban traffic circulation in Banjul and its peri-urban areas. In the long term, it is necessary with the increasing rate of urbanisation to plan ahead for a bus rapid transit system which could be the core of the transport plan for the Greater Banjul Area, but it would need building appropriate public sector institutions for urban transport regulation and partnering with the private sector concessionaires for operation. The policy stream with respect to infrastructure, traffic management, service quality and route network development has been discussed as the way forward. The Greater Banjul Land Use and Transportation Plan Study to be based on the bus rapid transit system has been recommended to inform a medium- to long-term solution, with institutional development leading to the establishment of a Banjul Metropolitan transport authority in charge of both infrastructure and public transport. Some key issues to improve the current poor road infrastructure and chaotic situation of unregulated paratransit mode have been recommended for action through:

- Definition of responsibility for the hierarchy of road infrastructure (of arterial, collector and local).
- Funding of urban road maintenance, traffic regulation and enforcement.
- Introduction of traffic demand management measures and signalisation at intersections.

10.11 Lessons for Other African Cities

Banjul, being the capital city and seat of power, is highly valued due to its strategic geography, proximity, central role in governance, and significance as a hub for administrative, economic, and cultural activities in the country. The city, as it has been noted earlier, relies on resources from the Central Government which in turn is reliant on donor agencies across the world. The city's compact size could have positioned it as a model for emulation by other cities, but the fact that the Central Government is in charge of infrastructure and the Banjul Area Council not being fully autonomous makes it difficult to take quick decisions. Still, since the Council has a good rapport with the Central Government and the Central Government and the seat of power are in the city, it is somewhat easier to liaise with them. In conclusion, we have been able to see the state of transport systems and infrastructure in Banjul. The Government just needs to put in efforts for it to become more of a model city that other countries can use as a reference point.

References

Access Gambia Ltd. (2009, April 21). *Banjul – Gambia*. Access Gambia. http://www.accessgambia.com/information/banjul-capital.html

African Development Bank Group. (2013, August 14). *Transport sector diagnostic study [The Gambia, transport sector diagnostic study]* [PDF]. African Development Bank Group. https://www.afdb.org/fileadmin/uploads/afdb/Documents/Publications/AfDB_-_Gambia_-_Transport_sector_diagnostic_study.pdf

African Development Bank Group. (2022, November 15). *Gambia – ADF/PPF advance Banjul port 4th expansion*. AfDB Data Portal. https://projectsportal.afdb.org/dataportal/VProject/show/P-GM-D00-008?cur=ua

Anadolu Agency. (2020, January 28). *Gambia, the smallest country in West Africa*. Anadolu Agency. https://www.aa.com.tr/en/africa/gambia-the-smallest-country-in-west-africa/1716845

Bass, M. S., & Tamba, A. (2022, July 26). The urban transport initiative:—The Gambia | standard news from The Gambia. *The Standard Newspaper* (The Gambia). https://standard.gm/the-urban-transport-initiative/

Bojang, I. M. (2000, February 20). *Proposals to transform the port of Banjul into a transhipment and distribution Centre with special emphasis on feedering*. Maritime Commons. https://commons.wmu.se/all_dissertations/61/

Central Intelligence Agency. (2024). *The Gambia, – The world Factbook*. CIA. https://www.cia.gov/the-world-factbook/countries/gambia-the/

Cities Alliance Knowledge. (2021, August 20). *Planning for inclusive greater Banjul*. Cities Alliance. https://www.citiesalliance.org/resources/publications/cities-alliance-knowledge/planning-inclusive-greater-banjul

Canadian Pacific Consulting Services. (2020, June 20). *The Gambia: Unlocking river transport to new destinations*. CPCS. https://cpcs.ca/projects/the-gambia-unlocking-river-transport-to-new-destinations/

Committee for Economic and Commercial Cooperation of the Organization of the Islamic Cooperation. (2015, October 22). *Urban transport in the OIC megacities*. COMCEC. https://www.comcec.org/wp-content/uploads/2021/07/The-Gambia.pdf

Gambia Civil Aviation Authority. (2022). *Gambia civil aviation authority – MoFEA*. Ministry of Finance and Economic Affairs. https://mofea.gm/soe/gcaa

Google Maps. (2023, January 26). *Banjul – Google Maps*. Map of Banjul. https://www.google.com/maps/place/Banjul/@13.4594539,-16.6164676,14z/data=!3m1!4b1!4m5!3m4!1s0xec29c2965af3807:0xa1e31751fbae2c52!8m2!3d13.454375!4d-16.5753186

Manneh, M. (2020). *Challenges and possible solutions to electricity generation, transmission and distribution in The Gambia*.

Ministry of Transport Works and Infrastructure. (2017, December 19). *The Gambia NATIONAL TRANSPORT Policy (2018–2027)*. Ministry of Transport works and infrastructure. https://www.motwi.gov.gm/sites/default/files/2020-03/THE%20GAMBIA%20NATIONAL%20TRANSPORT%20POLICY-%202018%20-%202027%20FINALPRINT_0.pdf

Organisation for Economic Co-operation and Development. (2007, July 3). *The Gambia*. OECD. https://www.oecd.org/aidfortrade/countryprofiles/dtis/Gambia-DTIS-2007.pdf

Sanneh, A. (2020, March 10). *The Gambia National Development Plan (2018–2021)*. United Nations in The Gambia. https://gambia.un.org/sites/default/files/2020-10/1.-The-Gambia-National-Development-Plan-2018-2021-Full-Version.pdf

The Point. (2021, November 2). *Gambia: The River Gambia and river transportation – Panacea for accelerated socio-economic transformation*. allAfrica.com. https://allafrica.com/stories/202111030241.html

World Economic Forum. (2022, September 7). *Annual report 2021–2022 | world economic forum*. The world Economic Forum. https://www.weforum.org/reports/annual-report-2021-2022/in-full

World Bank. (2022). *The Gambia overview: Development news, research, data*. World Bank. https://www.worldbank.org/en/country/gambia/overview

Chapter 11
Urban Infrastructure and Transit-Oriented Development in Freetown, Sierra Leone

Braima Koroma, Joseph Mustapha Macarthy, and Yasmina Yusuf

11.1 Freetown: A General Overview

Freetown, the capital of Sierra Leone, is in the western area of the country. The city is located on a mountainous peninsula, making the topography extremely challenging (Allen et al., 2020; Koroma et al., 2018). It is the largest city in the country and has a total land area of 74 square kilometres with a coastline of about 40 kilometres. Despite its difficult topography, it has experienced an incredible urban sprawl. In part, this unstructured development is due to its rapidly growing population (MLCPE/FCC, 2014; World Bank, 2019).

According to the 2015 census, the population of Freetown is about 1.1million (approx. 21.1% of the country's total population) (Fig. 11.1). The city has a population density of 8450 persons per km^2 compared to 97.2 persons per km^2 nationwide (Statistics Sierra Leone, 2017; World Bank, 2019). The population size of Freetown has increased with a corresponding increase in population density from 3.8 to 28.8 per square kilometre (km^2). This is expected to rise to 2.3 million and 44.8 per km^2 in 2030 if current trends persist (MLCPE & FCC, 2014; Tarawally et al., 2017). Freetown's average population growth rate (2010–2020) of 4.22%[1] largely fuelled by internal displacement from the civil war and rural-urban migration in search of employment opportunities has continued since 1963 (Statistics Sierra Leone, 2016). Increasing population growth has placed significant pressure on the environment, housing and sanitation infrastructure (Fig. 11.2, Table 11.1).

[1] World Urbanisation Prospects: the 2011 Revision, UNDESA, New York, 2012.

B. Koroma
Institute of Geography and Development Studies, Njala University, Freetown, Sierra Leone

J. M. Macarthy (✉) · Y. Yusuf
Sierra Leone Urban Research Centre, Njala University, Freetown, Sierra Leone
e-mail: jmacarthy@slurc.org

T. Nubi et al. (eds.), *Transit Oriented Development in West African Cities*, Urban Perspectives from the Global South,
https://doi.org/10.1007/978-3-031-58726-9_11

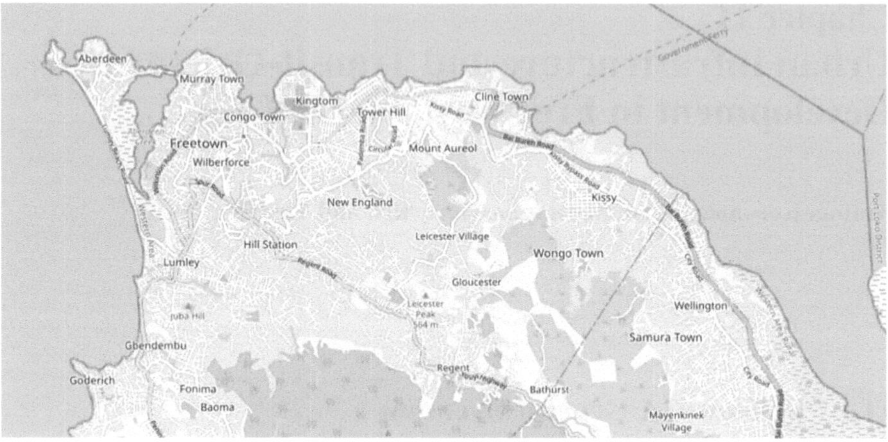

Fig. 11.1 Map showing Freetown, the capital of Sierra Leone. (Source: Open Street Map, https://www.openstreetmap.org/#map=13/8.4531/-13.2468)

As the main economic centre of the country, Freetown also has the most economic opportunities and social amenities. Recent studies show that Freetown contributes 30% of the country's GDP despite housing only 21.1% of its population and occupying less than 0.5% of the national land mass, indicating its economic potential to power economic transformation for Sierra Leone (World Bank Group, 2018a). About, 95% of Freetown's residents depend on the informal sector for their livelihood, which puts this group in a vulnerable position (World Bank Group, 2018b), as only 5% are employees in the public sector. This means that the majority have low and irregular incomes to afford their basic needs.

Like in many other sub-Saharan African countries, Sierra Leone has a high level of primacy; the capital city, Freetown, defined narrowly, is more than four times the size of the second biggest city, Bo. The city is the major urban, economic, financial, cultural, educational and political power house of the central and local government. The city's economy revolves largely around the natural harbour, which is the largest natural harbour on the continent of Africa. Freetown has the potential to attract productive firms and skilled workers to cluster together as a hub for exports and to support growing income and living standards across the country (World Bank, 2018a).

11.2 Broad Infrastructural Challenges Facing Freetown

Freetown has been unable to provide adequate housing, social infrastructure and service provision to keep pace with population growth. Most of this growth in population comes from unplanned development of informal settlements that have sprouted especially along the hill slopes and the shorelines of the Atlantic (Allen

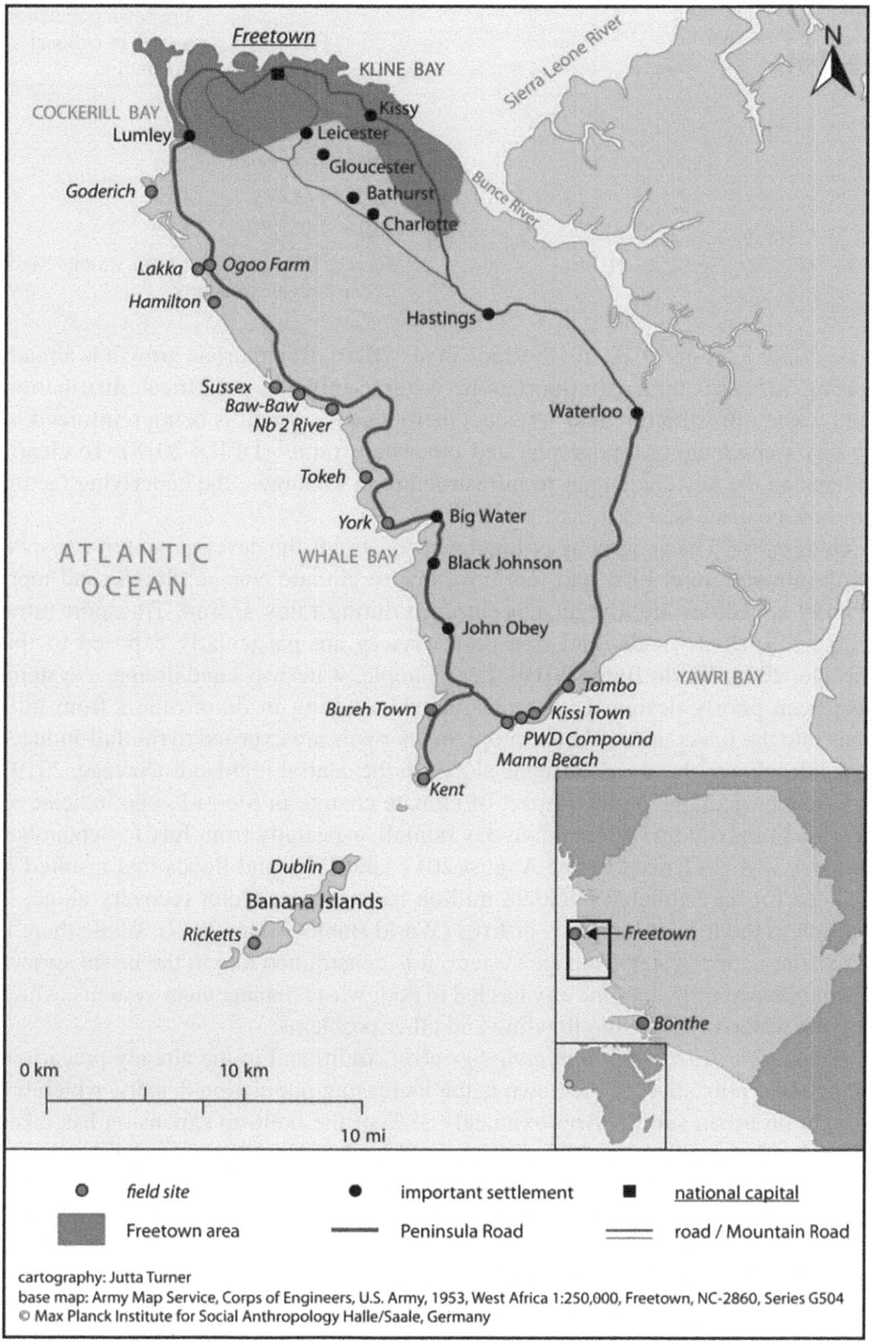

Fig. 11.2 The Freetown peninsula (1.5-column image). (Source: Ménard (2017))

Table 11.1 Population growth of Freetown (1963–2015)

Year	Freetown population	Freetown population as a % of national population
1963	127,917	5.9
1974	276,247	10.1
1985	469,776	13.3
2004	772,873	15.5
2015	1,055,964	14.9

Source: Data compiled from various Sierra Leone Statistics Reports

et al., 2020; Koroma et al., 2018; Rigon et al., 2020). Rapid urban growth is already causing infrastructural (transportation, water, sanitation, electrical distribution, solid waste infrastructure and services) inefficiencies which is being reinforced by the city's challenging topography and other constraints (DFID, 2018). To clearly understand the key challenges to infrastructure in Freetown, the underlying factors will now be discussed in turns.

Geography: The geography of Freetown constrains the development of transportation infrastructure. First and foremost, due to climate change, floods, and more recently landslides are a regular occurrence during rainy season. Transport infrastructure, such as roads, and transport services are particularly exposed to this (Savage, 2019; World Bank, 2019). For example, waterways and drainage systems have been poorly designed and constructed, resulting in drain runoff from hilly areas into the lower areas. Furthermore, many roads are exposed to rainfall-induced landslides due to the steep, unstable slopes in the central highlands (Savage, 2019).

Latest projections on the impact of climate change in Sierra Leone indicate an increase in maximum one- and five-day rainfall, especially from July to September. This risk was evidenced by the August 2017 landslides and floods that resulted in the need for an estimated US$0.98 million for transport sector recovery alone, in addition to the loss of hundreds of lives (World Bank Group, 2017). While there is an existing storm water drainage system, it is constrained due to the urban sprawl. The unplanned growth of the city has led to poor waste management systems, which clog the waterways causing flooding and other problems.

Population density and unplanned growth: Additional to the already precarious topographic allocation of Freetown is the increasing population density, which has brought on urban sprawl. Approximately 38% of the built-up expansion has taken place in either medium- or high-risk areas (World Bank, 2019). Though laws exist to prevent illegal construction in high-risk areas, these are often sidestepped or ignored. In addition to this, the shortage of affordable housing has further exasperated the situation (World Bank, 2019). For transport, this has resulted in a series of challenges: (1) Urban sprawl has meant that there is not enough space for transport infrastructure without displacing households and communities. (2) The development of the transport sector needs to be in line with population growth; otherwise, informal transport operators will fill this gap.

The road network: Despite the reliance on the road network for mobility in the city, Padrosa (2009) notes that the roads and streets in Freetown have deteriorated over the past decades. The demand for road space has increased dramatically in Freetown over the last 30 years, with consequent traffic problems caused by the high presence of street traders and pedestrians, multiple modes of low ridership transportation and poor traffic management.

11.3 Transportation Infrastructure Overview

Rapid population growth in Freetown has caused increasing pressure on the existing transport infrastructure. Until recently, most of the roads in Freetown were in a serious state of disrepair. However, even with the recent improvements, the demand for road space is still high owing to rising populations and the unsustainable urban transport and mobility situation resulting in traffic problems caused by uncontrolled street trading, poor traffic management and enforcement of rules, congestion, inadequate road maintenance and vehicular pollutions.

Central Freetown has benefitted from an orderly road layout of paved streets and a compact development pattern (Fig. 11.3), while development of the rest of the city has been shaped to a significant extent of the city's topography. The major road corridors in the city have not been designed and constructed using urban road standards and specifications with the provision of basic facilities to enhance safe and efficient movement of people and vehicles in an urban setting.

In addition to the physical condition of the road system, traffic management and control of the use of road space are also generally ineffective. The road network itself is inefficient, with no proper hierarchy of roads and poor connectivity; furthermore, vehicle capacity on some roads is inadequate for existing traffic volumes, while other roads have excess capacity. With the exception of recently constructed and rehabilitated roads, most roads, beyond the primary network in the city, are in very poor condition, the bituminous surfacing is often severely eroded and in many cases the surfacing has disappeared completely, resulting in the road being reduced to gravel standards, reduced traffic speeds, damages to vehicles and in some cases danger to road users and pedestrians. The design of some intersections creates bottlenecks and serious delays to traffic. Facilities for pedestrians are particularly poor, with footpaths of inadequate width and often in dangerous condition and some roads lacking footpaths altogether; this problem is more severe in the central area of the city, where pedestrian volumes are the highest. Furthermore, studies indicate that only 5% of the existing Freetown urban road network is covered by side drains. Most of the drainage is failing, erosion is apparent in many areas, which has become very dangerous to the pedestrians and vehicles. The drains are in many cases blocked or completely broken down. According to existing data, only 10% of the streets in the non-CBD roads and 40% of the CBD roads are provided with sidewalks

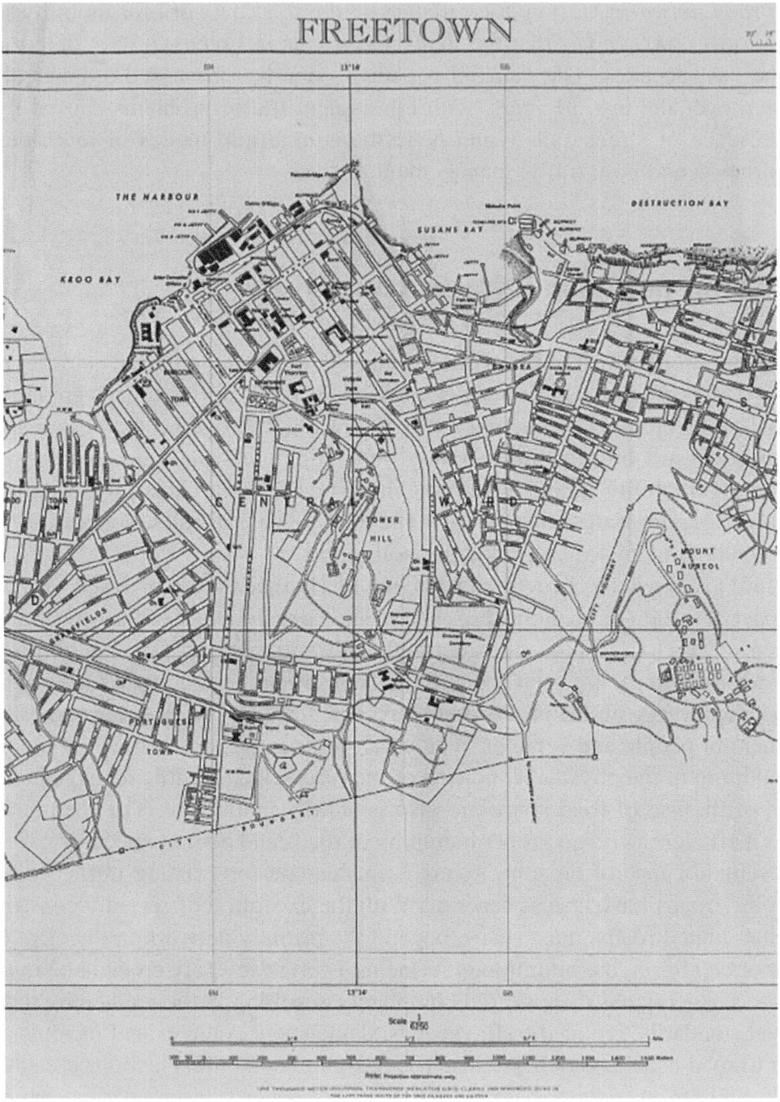

Fig. 11.3 Central business district in 1947. (Source: Padrosa, 2009)

(Padrosa, 2009). There are no off-street parking facilities available to the general public, and on-street parking is uncontrolled which further restricts road capacity. Existing sidewalks and pedestrian areas are often eroded or overgrown, thus forcing pedestrians to walk in the road. In addition, individual market stalls, street traders or parked vehicles occupy much of the footpaths in the town centre.

11.4 Governance Framework for City Transportation Sector

Table 11.2 below provides an overview of the names, roles and responsibilities of the different actors in the transportation sector. As the table shows, multiple government ministries and departments are responsible for planning, financing, running, implementation and enforcement in the urban transport sector. However, the institutional arrangements for the delivery of infrastructure and services are complex and inadequate, with several agencies having responsibility for various aspects of urban transport, often with inconsistent objectives and priorities, overlapping mandate and associated tensions between agencies. The main bodies responsible for urban transport in Freetown are the SLRSA, SLRA, SLRTC and the Sierra Leone Police Traffic Division. To date, the FCC is yet to exercise any major role. Although the different organisations have undergone several reforms over the years, there is still a lack of capacity, in terms of manpower, expertise, equipment and funding. Most of these agencies support the MoTA in the development of laws, regulations and framework for sustainable delivery of transport services in the country. Nevertheless, there is still no clear-cut institutional framework for urban transport in Sierra Leone. There is, however, a plan to establish an autonomous Metropolitan Transport Authority (MTA), which would take responsibility for all aspects of transport management in the Greater Freetown area. It is proposed that the MTA would be headed by a panel of representatives from the two local councils in the Greater Freetown area, relevant government agencies, police, transport operators' organisations, consumer groups and other relevant stakeholder organisations.

Table 11.2 Transport stakeholders in Freetown

Stakeholder	Role
Government	
Ministry of Transport and aviation (MoTA)	Overall responsibility for planning and policy
Ministry of Finance	Provides budget and policy guidance on road-user charging
The Ministry of Works and Public Assets (MoWPA)	Responsible for construction and maintenance of the main road network through the Sierra Leone roads authority (SLRA)
Ministry of Planning and Economic Development	Sets national agenda
The Sierra Leone roads authority (SLRA)	See above
The Sierra Leone road safety authority (SLRSA)	Responsible for testing and licensing all vehicles and drivers and for traffic management
The Sierra Leone road transport corporation (SLRTC)	Responsible for regulating public transport operations as well as operate and maintain the public transport services of the government
The Freetown City Council (FCC)	Responsible for urban planning and management, parking, enforcement, street management
The Sierra Leone traffic police (SLTP)	Responsible for controlling operations and enforcing regulations

(continued)

Table 11.2 (continued)

Stakeholder	Role
The road maintenance fund administration (RMFA)	Oversees the management of road funds for periodic and routine maintenance of road assets. The SLRA and RMFA collectively manage all classified roads; the latter is primarily responsible for financing the maintenance of the core road network
Paratransit associations	
Sierra Leone traders union (SLeTU)	Represents the members of the country's main five traders' unions. A trader's council leads it. There are several unions that represent the owners and drivers of commercial passenger and freight transport vehicles in Sierra Leone
Sierra Leone tricycle commercial motorbike riders association (SLTCMA)	Representing the needs of kekeh drivers in the city
Bike riders association	Representing the needs of okada drivers in the city
Passenger welfare association	Represents passenger needs in transportation
Indigenous transport owners association	Represents the indigenous transport owners in the city
The Sierra Leone motor drivers and general transport workers union (SLMDGWU)	Represents the general transport workers and drivers
Sierra Leone institutions of engineers	A network of engineers who work on all engineering projects, including transport
Universities and research institutions	SLURC, Fourah Bay college
International organisations	
World Bank	Funder
European Union	Funder
African development Bank	Funder
Kuwait fund	Funder
Islamic Development Bank	Funder
UNDP	Funder
Chinese international development cooperation agency	Funder

In addition, there is an active informal transport which is based on associations or unions – representing the interests of specific paratransit modes. As described by the World Bank (2021), the paratransit associations make up the informal transport sector. To provide service, they organise themselves into branches that divide the network. These branches typically function at suitable off-street or on-street parking locations. Routes are operated from or between these points. Membership in a branch is dependent on access to a suitable vehicle, with most of these vehicles sourced through operating leases from non-participant owners. Each route is then operated on the principle of fill-in-turn prior to departure. This process is managed by branch officials, and a departure levy is raised for this purpose. The Government, in negotiation with the associations, to protect the public interest, generally controls the fare charged for each route. Where fares have been kept too low, the associations have responded by shortening routes to the extent that revenues can exceed operating costs.

There is a strong commitment for the government to build resilient sustainable transport infrastructure as reflected in the UN's Sustainable Development Goal 11 (sustainable cities and communities), which includes in one of its targets an aim to: 'by 2030, provide access to safe, affordable, accessible and sustainable transport systems for all… with special attention on the needs of those in vulnerable situations, women, children, persons with disabilities and older persons' (UN, 2015: 21).

The current approach used by the government is to work with development partners and the private sector in designing an integrated, scaled, measurable, accountable and sustainable mobility plan. It is hoped that this will ensure inclusive development and the protection of women and girls from all forms of gender-based and sexual violence, thereby creating a safe commuter matters through the span of the daily movement of women and girls. The clusters of infrastructure projects contribute towards meeting the Sustainable Development Goal 11 by making the cities and communities more productive, safer, healthier, more accessible and therefore more sustainable and resilient. This they hope to achieve through the Integrated Resilient Urban Mobility Project (IRUMP) by strengthening the resilience of commuter thorough fares and major roads during adverse climate conditions by improving drainage capacity and slope stability (World Bank Group, 2018a; GoSL, 2019).

However, transit-oriented development has not been adopted as a major policy decision-making process by both national and city authorities. This reflects the fact that urban transport infrastructures are highly underdeveloped; the sector is largely informal, with no mass transit and increasing use of private vehicles, which has serious spatial, environmental and economic implications.

11.5 Mobility and Accessibility in Freetown

There is virtually no public transport in Freetown, with journeys mainly made by mini buses (poda-poda), three-wheel motorcycles/tuk-tuks (kekeh), motorcycles (okada) and shared taxis, contributing to traffic congestion. For different parts of the city, prices paid for transport services range from 2,000 Le (USD$0.21) up to 5,000 Le (USD$0.51). In effect, the average cost of routed public transport is 1,500 Le, which is considerably lower than the cheapest trip registered by kekeh (three wheelers). Generally, operators offer lower prices to men compared to women (particularly young women) indicating a gender-based pricing gap. This difference is influenced by both the amount paid and the willingness to negotiate. The quality of the road is also linked with the increase in the prices of transport services, e.g. okada and kekeh are more expensive along unpaved routes. However, the ability of users to negotiate down price on these less-regulated services can contribute to the increased affordability of certain trips and the homogenisation of travel cost for specific destination (Oviedo et al., 2021).

Combined, these factors are constraining mobility of both people and goods and the irrespective access to work and markets – particularly for people with disabilities and women in Freetown due to their different transport characteristics. For people with disabilities, basic mobility around the city is challenging with poor paving and inaccessible buildings, often lacking elevators (DFID, 2018). In addition to service quality, design considerations, social dynamics and enforcement are crucial for an inclusive public transportation system. However, there have been limited efforts to address the specific needs of PWD, who constitute about 1.3% of the country's total population (Statistics Sierra Leone, 2016). Previous research shows how 18% of women often suffer sexual harassment in public transportation (World Bank Group, 2018a). Besides sexual harassment, transport services and infrastructure are not designed to respond to either the women's mobility patterns linked with the economy of care (e.g. travelling with children, bags, in a multimodal way) (ibid).

Safety and security constitute a major concern for women in influencing their choice of transport mode, time and place of travel (Oviedo et al., 2021). In a user survey (World Bank, 2018b), 50% of interviewees responded that physical violence is one of the main barriers to access public transport services, and more than 10% mentioned theft and robbery. The study further shows that the current informal system challenges the possibility of institutionalising measures for enhancing women's safety.

Poor people's inability to access jobs and services is an important element of social exclusion that defines urban poverty. In the Western Urban, 25% of the population lacks access because they live farther than 500 metres from bus stops. Accessibility to transport shows major disparities among neighbourhoods, with over 40% of the population of York (in the west) and Waterloo (in the east) western area rural lacking easy access to transport services. Access to transport services is even more challenging for women and persons with disabilities in Freetown due to their different transport characteristics.

11.6 Transportation Investment

Padrosa (2009) indicates that the Government of Sierra Leone often obtain funding from different sources, both external and internal, to be able to provide infrastructure solutions. More generally, the major sources of funding for infrastructure development (road) have come mainly from the World Bank Group, European Union and African Development Bank. Other notable sources include DFID, Kuwait Fund, BADEA, OPEC Fund for International Development (OFID), Islamic Development Bank, the Arab Bank and the Saudi Fund for Economic Development. More recently, the Chinese through bilateral funding arrangements have funded not only new road projects but also the rehabilitation of some existing ones. This has included the widening of roads and repair of pavements (see Annex 11.1) as in the cases of the Jui to Regent and Hill Station to Regent roads. However, domestic public investment in infrastructure development is limited because of the level of debt already owed to donors and the under-collection of revenues.

Much of the Government's investment has often targeted feeder roads and the provision of counterpart contribution for donor-funded road projects. Government investment for the improvement of road networks has increased over time, more recently from Le 174.0 billion in 2019 to Le 217.0 billion in 2020. The government raises revenue for road improvement through the Road Maintenance Fund (RMF) which collects direct payment of road levy component of the fuel price to the SLRA by oil companies. More recently, the government tends to have started opening up to public-private partnership which will give private sector actors the opportunity to start operating in the sector.

As part of the IRUMP, the government seeks to pursue public-private partnership model as part of its desire to reform the Sierra Leone Road Transport Corporation to act as a regulatory agency. The Private-Public Partnership (PPP) Act introduced in 2010 will allow the government to promote and facilitate the active role of private sector actors whose involvement is still limited to only a handful of activities, such as the management of the container terminal at the country's main seaport in Freetown (DFID, 2018). While several incentives have already been outlined to steer private sector engagement which includes tax exemption and tax holidays, the limited capacity of the PPP Unit (PPPU) and the National Commission of Privatisation (NCP) in the Ministry of Finance (MoF) will likely constrain the mainstreaming of this priority agenda.

There is a strong commitment for the government to build resilient sustainable transport infrastructure (National Development Plan 2019-2023). Recent government efforts in Freetown have been the investment of US$ 250,000 for key transport infrastructure development such as the use of traffic signals including ongoing work to build new flyovers at key intersections in the city of Freetown. This will complement the $50 million World Bank-funded Integrated and Resilient Urban Mobility Project implemented by the MoTA, which was launched in November 2019. The project has three components: comprehensive corridor improvements (including a transport operator's terminal, improving road conditions and pedestrian ways), institutional capacity building and improvements in ferry terminals and services.

11.7 Actions Taken in the Last 20 Years and Impacts

Table 11.3 provides an analysis of the concluded and ongoing infrastructure projects that have been undertaken in the last 20 years (see Annex 11.1 for details).

As noted from the previous sections, much of the funds have come from bilateral and multilateral sources. However, mindful of the implications unplanned growth and climate change will have on Freetown, the municipal government is already in action to address the likely effects these will have on residents and the city infrastructure. In 2019, the FCC introduced the Transform Freetown plan as a three-year vision for the development of the city. The plan seeks to address a range of challenges in Freetown including poor housing, unplanned growth, poor waste management and disaster risks. There is also the World Bank-funded IRUMP which is

Table 11.3 Ongoing and concluded projects over the last 20 years in Freetown

Project description	Distance (km)	Date
Hillcot road-Lumley	3.4	2013
City roads	34.7	
Lumley-Tokeh	21	2013
Bottom-mango-king street-Congo cross		
Limkokwing-regent	3.2	2018
Lumley-juba bridge		2019
Regent road-Hill Station-hill cot-king Harman-Jomo Kenyatta road	6.71	2017–2020
Hillside bypass road	3.7	2005–2020
Wilkinson road	5	2009–2012
Regent-Grafton-Kosso town road	11.26	2012–2014
Lungi-Freetown bridge (proposed)	18 km	–

Source: Authors compiled data

jointly implemented by the MoTA and the FCC. This project seeks to improve urban accessibility in Freetown through actions in some pilot areas.

A few reports (World Bank, 2018; Government of Sierra Leone, 2013) have demonstrated that the above-listed projects have significantly impacted the city in many ways. These include:

- The significant opportunity created for projects to enhance the city's infrastructure through strengthening the institutional capacity of the SLRA to undertake transport infrastructure improvements. Through the IRUMP, actions are already underway to integrate the transport and land use policies.
- The hillside bypass road will improve mobility in Freetown from the city centre to the eastern parts of the city, thereby decreasing traffic congestions in the CBD. The population will benefit from cheaper, safer and reduced travel.
- Build institutional capacity in the transport sectors, for example, the Sierra Leone.
- Roads Authority is now better able to plan and manage road network maintenance.
- Increased private sector participation in road maintenance which now provides employment to 50 small and medium contractors as well as over 500 single-person contractors, thereby helping to reduce youth unemployment.
- Decline in perpetual traffic, particularly in places between Lumley and Juba. The IRUM will make commuting and access to jobs, businesses, social and health services easier, faster and safer.
- Improved access to quality transport through IRUMP which will also address issues associated with climate resilience. The project in turn will support capacity building within government and of Sierra Leonean younger generation by funding multiple initiatives to boost innovation in the transport sector.
- The proposed Lungi-Freetown Bridge will significantly improve the regional traffic conditions and investment environment between Freetown and Lungi International Airport. This will offer convenient, comfortable and safe transportation environment for the healthy and rapid development of the Freetown economy, bringing about more social benefits to people in the region.

11.7.1 Transportation Infrastructure Gaps

Transport infrastructure gap is reflected in the adequacy of transport infrastructure compared with access to electricity, water and sanitation and telecommunications. Deficient urban transport services and infrastructure are major constraints to growth and economic transformation in Freetown, especially as it relates to the manufacturing sector. This is because poor transport infrastructure raises the cost of input to manufacturing and other activities. However, in Freetown, a number of gaps still remain unaddressed. These are discussed under the following headings.

Road Capacity and Infrastructure

The road capacity in Freetown is unable to match the ever-growing demand and contributes to economic stagnation. Due to historical reasons, most of the roads in Freetown are narrow and do not have adequate traffic capacity. As a consequence, the few roads become highly congested as private users of cars ignore the negative externalities they generate. Research (GoSL, 2019; MLCPE/FCC, 2014) however identifies the following gaps in the road infrastructure:

- The pedestrian infrastructure is weak, and in many ways narrow, and many roads do not have pedestrian ways at the two sides. A lot of hawkers occupied the pedestrian ways, which caused serious mixes of pedestrians and vehicles on the roads. This would bring about not only congestion of the roads but also traffic safety hazard.
- The road maintenance is seriously insufficient. The surfaces of some of the recently constructed roads are intact, but the earlier roads are often severely damaged and in poor conditions. The vehicle speed is slower with bumpy rides, not only causing a serious decline in traffic capacity but also leading to poor driving comfort and more deterioration of the vehicles.
- The junctions do not have lighting signal control, and most of the policemen regulating the traffic at the busy junctions lacked the necessary training. The confusions at the junctions lead to unnecessary congestions and delays.
- Freetown's lack of car parking facilities and the management of the roadside car parking are inadequate. This resulted in a large number of roads being occupied by the disorderly roadside parking.
- Many drivers are conceptually weak in the traffic regulations, and the violation of regulations during the driving of vehicles disturbs the normal road traffic.

Density of Paved Roads

Recent estimates suggest that only 5% of total land in Freetown is allocated to roads, of which only 24% are paved compared with regional benchmarks of 10% and 50%, respectively (DFID, 2018: World Bank, 2019; Koroma et al., 2021). It is

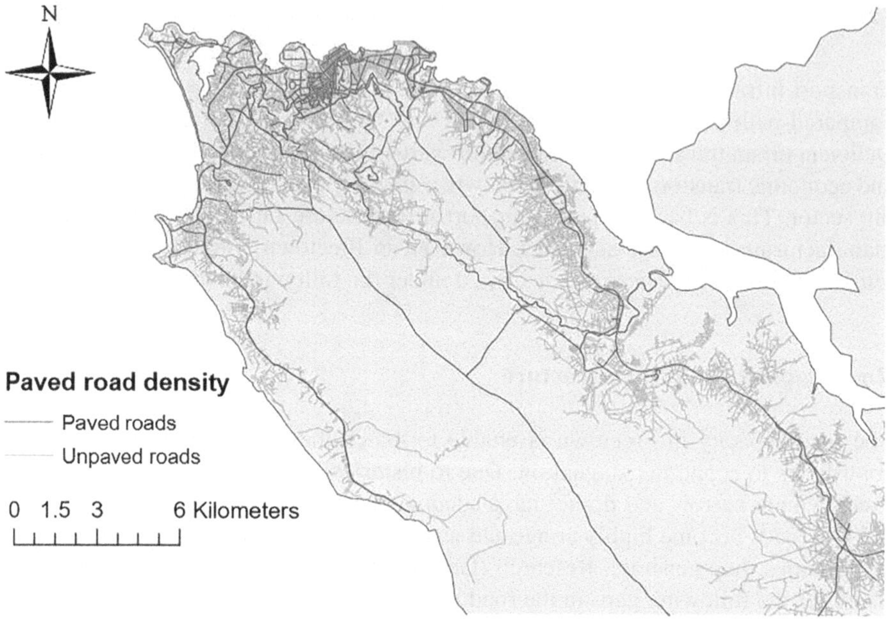

Fig. 11.4 Road density (paved and unpaved roads) in Freetown. (Source: World Bank, 2018b)

particularly so in informal settlements that are located on the hillside where it is mostly inaccessible due to either the lack of connectivity, unaffordability or the lack of transport options (Oviedo et al., 2022).

In the western area, road density per capita is about 165 metres of paved road per 1,000 citizens in the Greater Freetown Area, which is around half of the average in low-income African countries (318m/1000 people) (African Development Bank, 2014[2]; World Bank Group, 2018a). In the 2019 report, paved roads in Freetown have been modelled at an average replacement value of US$341,700km, in contrast to the reported unpaved roads of US$52800/km (World Bank, 2019; Fig. 11.4).

Coordination and Institutional Capacity Gaps

The institutional set-up of the road management had witnessed several coordination, transparency and capacity challenges over the past decades. For instance, the SLRA sits under the Ministry of Works and Public Assets and is tasked with the construction and maintenance of road network, but its operations appear somewhat misguided, with a seemingly greater focus on improving traffic flow in Freetown

[2] Tracking Africa's development in figures (African Development Bank, 2014).

through plans for flyovers and bridges than on the broader road network. The SLRSA who collects driving license and vehicle registration fees disburses to the RMFA, which reports to the MoF. A lack of integrated database at the SLRSA makes both revenue tracking and forecasting challenging. The RMF has been increasingly drawn into road rehabilitation projects, diverting attention from much needed maintenance. Many road construction and rehabilitation projects (ongoing and completed projects over the past 20 years) have been unsustainable, as without funding for maintenance, most roads have fallen into disrepair.

On the other hand, consistently all reports and evaluations done within the transport sector have highlighted the limited domestic capacity in the local construction industry. In general, there is a very limited expertise in urban planning, including transport planning. Human capital in ministries, departments and agencies, local government and academia in the transport sector lacks capacity, resources and proper institutional coordination. Different levels of government have limited capacity in terms of manpower specialised skills, and expertise, which limits their productive capacity. Local universities lack specialised training in transportation and engineering, which limits the availability of local professionals with specialised skills. The only engineering university, Fourah Bay College, University of Sierra Leone, lacks formal curriculum in transport planning, transport economics, programming, use of Geographical Information Systems (GIS) tools or use of new technologies to enhance transport planning.

Financial Investment

Local capital investment in road infrastructure is weak. There is a high dependency on external donor funding, which means initiatives are often externally driven. This will continue to foster a culture of paternalism and is generally considered problematic for sustainable development.

Land Use and Land Markets

Land is at the core of any urban development. Ensuring secure and marketable land rights is therefore of vital importance, also for the transport sector. There is a clear lack of records and clarity on ownership of land. This makes it difficult to understand where development can take place. Expanding transportation infrastructure in the city requires repurposing land from private or commercial use to public use. Land is however insecure and lacks clear regulation, which is a priority for the government to address, as this affects not only transport but also all urban development in the city.

11.8 Recommendations

11.8.1 To Government and Policymakers

1. Implement and enforce relevant land use planning rules and regulations to ensure the planned growth of the city. This should involve regularising the land tenure status of the urban poor and making land and housing affordable for them. Also, matching the enforcement of development control rules in Freetown with the decentralisation of services and other opportunities throughout the country may help to reduce current migratory pressures on the city.
2. Curative planning and redevelopment of blighted areas using appropriate Neighbourhood/Action Area Plans. Using land pooling approaches to create mixed-use neighbourhoods will ensure not only an efficient use of the land but also a lessening of pressure on the existing transport infrastructure.
3. Integrate knowledge in urban development, climate change, disaster risks and the spatial configuration of the city in infrastructure planning and design. Ensuring cross-sectoral collaboration mechanisms involving the transport and land ministries and relevant other agencies (public and private) and professional bodies and transport users and operators will be vital to the process.
4. The government should consider investing more in road infrastructure repair and reconstruction to improve efficiency of the network and to ensure connectivity. This should involve clarifying the different road hierarchies, designing and constructing urban roads based on appropriate standards and specifications and ensuring an efficient traffic management system. Using information and communication technology for monitoring and enforcement of transport rules will enhance the safe and efficient movement of people and vehicles.
5. The government should additionally increase and diversify revenue and funding streams for urban infrastructure development (e.g. public-private partnership, congestion fees, etc.) to reduce the overreliance on donors.
6. Build the technical skills and competence of the various transport and road infrastructure institutions through providing specialised training, mentorship and knowledge/skills transfer. Clarifying and agreeing on the roles and responsibilities of the actors will be critical.
7. Agencies responsible for traffic management systems should adopt more innovative approaches for monitoring and enforcement. Coordination between the different actors and continuous professional training in urban traffic management will be critical to this process.
8. Commitment to routine professional development and to foster organisational and institutional capacity would go a long way to ensure that transport planning is strategic and well informed.
9. Support the government to build the capacity of other urban transport actors and stakeholders.

11.8.2 To Investors

1. To consider investing in the informal transport sector. This should involve a reform of the different transport modes within the sector and aligning it with the public transport system. Attention should be paid mainly on fitting them across the city road hierarchies but also on issues of safety and efficiency.
2. To direct some funds to routine road maintenance since the benefits of upgrading, expanding and constructing new roads may only last for a few years/decades. Poor roads not only limit accessibility and connectivity but also increase the risk of travels including the cost of vehicle repair. Similar to a few African countries, anecdotal evidence suggests a severe lack of road maintenance culture within governments in Sierra Leone.
3. To expand the space for discussing urban transport and infrastructure problems beyond public sector institutions to include transport operators, users and other non-state stakeholders. Such engagement can allow the identification of the more germane issues that should be prioritised in projects.
4. To support governments on local revenue generation and enhancement strategies. This should include building the institutional capacity with skills in viable approaches.

11.8.3 To the Community

1. Transport operators and users to observe key messages arising from awareness-raising campaigns on the use of road space, parking, road safety and traffic management.
2. Local residents, not just government ministries and agencies, should actively participate in area renewal/redevelopment projects including taking self-help actions to improve the existing conditions in their neighbourhoods.
3. Push for more sustainable road infrastructure and mobility options while demanding for greater accountability in the urban road infrastructure sector.

11.9 Lessons for Other African Cities

This study has generated a few lessons for consideration by other African cities. Like in Lagos, Monrovia and a few other cities, this study has shown that sprawling, due to rapid urbanisation, can lead to spatial dysfunction with implications for economic development, infrastructural provision/expansion and mobility. The growth of cities creates new needs for transport with additional spaces required for the construction of roads similar to the innovative urban infrastructure initiatives that is currently being implemented in the city of Johannesburg in South Africa. However, despite the focus of infrastructural development in Freetown, the role of urban planning has been limited. This is particularly critical given the pressure already exerted by unplanned growth on land availability for transport and infrastructure development/expansion in the city and the accompanying problems (congestion, accessibility, etc.) already faced.

The report further shows that the high cost of transport infrastructure makes it almost impossible for some countries like Sierra Leone to fully undertake the cost. Therefore, the inability of the government to invest in infrastructure and transport has had undesirable effects on urban dwellers and their well-being. But while the financing mechanism in Sierra Leone is mainly donor driven, there is fear that the agenda is often set externally with less consideration of locally driven initiatives. From a sustainable point of view, this system does not bode well for the country given the dependency culture it has steered on the part of the central government. The need to find alternative funding sources is not only urgent but also critical. A clear example is the fuel levy charged in Sierra Leone for road maintenance and rehabilitation. This can be expanded to include betterment taxes to capture land value for areas covered by road infrastructures and public transport systems, taxes on businesses and employers and parking charges. Since these roles may correspond to different transport agencies/departments given the generally fragmented institutional arrangement in most countries, the need for the effective coordination of actors is vital.

Additionally, it is shown that transport in Freetown is provided by a combination of modes, which vary widely and are mostly conducted in a system that is not only informal but also ineffective. In particular, the central role which informal service providers and transport organisations play in the transport sector is illustrative. Regarding this, key lessons will need to be learned from efforts to transform a similar dysfunctional transport system in Cape Town, South Africa, into an 'integrated rapid transit' system which was initially fraught with some institutional difficulties (Wilkinson, 2010). In Freetown, it will be critical to leverage on the current dysfunctional set-up with clear imparting of skills and capacity (funds and resources) to improve and reorganise the system to be able to meet the needs of passengers. Besides, actions are needed to address the specific needs of people with disabilities (especially design considerations), including meeting the safety and security needs of women and other vulnerable groups (girls, the aged, etc.).

Annex 11.1: Summary of Transport Infrastructure Projects Last 20 Years in Freetown

Projects	Implementing agencies	Objectives	Project implementation priority	Amounts	Donor	Location	Duration
Integrated Resilient Urban Mobility Project	Ministry of Transport and Aviation	Aims to foster changes in the way the city is growing: from a congested, vehicle-oriented city to a resilient, people-oriented city, where walking and public transport are at the core of the metropolitan vision, and with a focus on vulnerable groups such as low-income residents, women and PWD	Construct transit markets and terminals at Lumley, pedestrian footbridges, formalized okada/keke stops, bus stops, bus only streets, modern buses, clear roads and drainages	US$ 50 million	Grants from the World Bank through the Ministry of Finance	Western Urban District (East and West)	2019-2023
Cable Car Initiative	Freetown City Council	To build a cable car as to complement the mass transit facilitated by the IRUMP project.	To conduct feasibility study and assess station locations	tbd	Tbd	Western Urban District	Tbd

(continued)

Projects	Implementing agencies	Objectives	Project implementation priority	Amounts	Donor	Location	Duration
Building transport resilience in Freetown		The overall objective of the World Bank project was:		300,000 USD	GFDRR/World Bank		2018-2019
		to improve climate resilience of transport infrastructure planning and investments in Sierra Leone;					
		To enhance local capacity to design and implement resilient					
		Transport infrastructure.					
Hillside bypass road	Sierra Leone Roads Authority	To address inefficient traffic management and deteriorated roads causing constant congestion	Payed and provided with culverts and drainage works. Safety measures, such as fencing, guard rails, traffic signs and pavement markings	US$6m	OPEC Fund for International Development	East-west bypass, four-lane, 3.7km-long road	2005-

Construction of the Freetown Ring Roads	Sierra Leone Roads Authority				African Development Bank	The outer ring will include Spur Road, Wilkinson Road, and Coastal Road, Grafton to Cline Town Road, and Regent to Grafton Road. The inner ring road will include Hillside By pass Road, Jomo Kenyatta Road, Hill Cot Road, Spur Road, Wilkinson Road and Freetown Coastal Road	
Road Rehabilitation and Maintenance Project	Sierra Leone Roads Authority and Ministry of Transport and Aviation	Aimed to remove physical bottlenecks in the country's transport network through an extensive road reconstruction and rehabilitation program, build capacity to better plan and manage road maintenance, and encourage private sector participation in road maintenance	Trunk road maintenance and rehabilitation, strengthen road management institutions, support transport sector institutions, promote non-motorised transport and technical assistance	US$45m	World Bank	City roads, particularly the Central Business District	1992-2001

(continued)

Projects	Implementing agencies	Objectives	Project implementation priority	Amounts	Donor	Location	Duration
Freetown Infrastructure Rehabilitation Project	Guma Valley Water Company	Aimed to improve infrastructure in Greater Freetown, with particular regard to the needs of the urban poor, and strengthen the capacity of responsible agencies	Urban water supply, urban upgrading, institutional support to SLRA and FCC, technical assistance	US$26	World Bank	Kennedy street market, fisher street, smyth street and Bombay street. 450 kilometres	1993-2001
Regent Road (Hill Cot Intersection) to Choithram Hospital; Regent Road from Limkokwing to Jomo Kenyatta Road, King Harman, Hill Cot and Hill Station Road	6.71 kilometres (hill Cot)		Design and construction of a dual carriage way, 6.71km long with a newly constructed bridge				
Lumley-Juba Bridge	SLRA/ MoWPA	Ease the problem of traffic congestion and unlock the tremendous economic potentials of the city – tourism, fisheries and other sector	Two-lane bridge, pedestrian walkway complete with accessible mobility ramps, enhanced safety railings, and solar lighting	US2.4m	GoSL/Chinese Government (1.5m)	Lumley-Juba	2019

Increasing Revenue for Freetown City Council through Property Tax collection	FCC together with WARC and international academics						
Limkokwing-Regent Road Project	SLRA/ MoWPA	To rehabilitate and upgrade the 3.2km Ring Road from Limkokwing University to Regent Village; ease traffic congestion and make those areas more accessible for citizens and for businesses	2-way 4-lane road for just over 2.6km of the road and a 2 way 2-lane road for the rest.		GoSL/Chinese Government	Limkokwing-Regent Road	5 months (2018)

References

African Development Bank. (2014). Tracking Africa's Progress in Figures. https://www.afdb.org/fileadmin/uploads/afdb/Documents/Publications/Tracking_Africa%E2%80%99s_Progress_in_Figures.pdf

Allen, A., Osuteye, E., Koroma, B., & Lambert, R. (2020). Chapter 7: Unlocking urban risk trajectories in Freetown's informal settlements. In UN-Habitat (2020). Breaking cycles of risk accumulation in African Cities. United Nations Human Settlements Programme (UN-Habitat) Nairobi.

DFID. (2018). *Infrastructure and cities for economic development – Cities and infrastructure for growth (CIG): High-level scoping in Sierra Leone.*

GoSL. (2019). *Sierra Leone's Medium-Term National Development Plan 2019–2023: A new direction for improving people's lives through education, inclusive growth, and building a resilient economy.* Government of Sierra Leone.

Government of Sierra Leone. (2013). *Sierra Leone integrated transport policy, Strategy and Investment Plan.*

Koroma, B., Rigon, A., Walker, J., & Sellu, S. A. (2018). *Urban livelihoods in Freetown's informal settlements.* Sierra Leone Urban Research Centre (Issue November).

Koroma, B., Oviedo, D., Yusuf, Y., Macarthy, J. M., Cavoli, C., Jones, P. M., Levy, C., & Sellu, S. A. (2021). *City profile: Freetown, base conditions of mobility, accessibility and land use.* T-SUM, University College London.

Ménard, A. (2017). Interpreting conflict and integration through the reciprocity lens: Mobility and settlement in a historical perspective on the Sierra Leonean coast. *Social Identities, 23*(4), 413–429.

MLCPE/FCC. (2014). *Sierra Leone preparatory components and studies for the Freetown development plan. 'The urban planning project'. Freetown structure plan 2013–2028.* European Union.

Oviedo, D., Okyere, S. A., Nieto, M., Kita, M., Kusi, L. F., Yusuf, Y., & Koroma, B. (2021). Walking off the beaten path: Everyday walking environment and practices in informal settlements in Freetown. *Research in Transportation Business & Management, 40*, 100630.

Oviedo, D., Clemence, C., Levy, C., Koroma, B., Macarthy, J., Sabogal, O., Oroyo, F., & Jones, P. (2022). Accessibility and sustainable mobility transitions in Africa: Insights from Freetown. *Journal of Transport Geography., 105*, 103464. https://doi.org/10.1016/j.jtrangeo.2022.103464

Padrosa, S. B. (2009). *Infrastructures in Africa: Analysis and evolution of the road network in Sierra Leone* (issue July). https://upcommons.upc.edu/bitstream/handle/2099.1/8183/01.pdf

Rigon, A., Walker, J., & Koroma, B. (2020). Beyond formal and informal: Understanding urban informalities from Freetown. *Cities, 105*, 102848.

Savage, I. A. B. (2019). Data collection on resilient urban transport in Freetown, Sierra Leone.

Statistics Sierra Leone. (2016). *2015 population and housing census: Summary of final results.* Statistics Sierra Leone.

Statistics Sierra Leone. (2017). *Sierra Leone 2015 population and housing census: National Analytical Report.* Statistics Sierra Leone.

Tarawally, M., Wenbo, X., Dagadu, J. C. (2017). Assessing land-use/land cover change in response to population dynamics in Freetown, Sierra Leone. *Studies in surveying and mapping sciences (SSMS), Vol. 5.*

UN. (2015). *Transforming our world: The 2030 agenda for sustainable development.* United Nations.

Wilkinson, P. (2010). Incorporating informal operations in public transport system transformation: The case of Cape Town, South Africa. *Urbe: Revista Brasileira de Gestão Urbana (Brazilian Journal of Urban Management), 2*(1), 85–95.

World Bank. (2018a, February). *Republic of Sierra Leone priorities for sustainable growth and poverty reduction, systematic country diagnostic (SCD).* World Bank.

World Bank. (2018b). *Integrated and resilient urban mobility project.* World Bank Group. https://projects.worldbank.org/en/projects-operations/project-detail/P164353

World Bank. (2019). *Freetown urban sector review: Options for growth and resilience*. World Bank Group.

World Bank. (2021, June). *Sierra Leone economic update*. World Bank. https://doi.org/10.1596/978-1-4648-1408-2

World Bank Group. (2017). *Sierra Leone rapid damage and loss assessment of august 14th, 2017 – Landslides and floods in the Western area*. The World Bank http://documents.worldbank.org/curated/en/523671510297364577/text/Sierra-Leone-Rapid-damage-and-loss-assessment-of-August-14th-2017-landslides-and-floods-in-the-western-area.txt

World Bank Group. (2018a). *Reviving urban development: The importance of Freetown for the national development*. World Bank Group.

World Bank Group. (2018b). Sierra Leone multi-city hazard review and risk assessment. *Final Report, Vol. 2 of 5*. Freetown City Hazard and Risk Assessment. World Bank Group.

Chapter 12
Urban Infrastructure and Transit-Oriented Development in Abuja, Nigeria

Muhammad Bashar Nuhu, Suleiman Yakubu, Ikpeme A. Ankeli, Nasiru Salihu, and Fatima Bashar Nuhu

12.1 Introduction

12.1.1 An Overview of Abuja City

Abuja was selected as the new capital of Nigeria when it was obvious that population pressures and political and ethnic divisions necessitated a move from Lagos. On February 4, 1976, a decree was signed, establishing the Federal Capital Territory of Abuja and setting up the Federal Capital Development Authority (FCDA). It was named Abuja after a nearby village in Niger State, which was later renamed Suleja. The new capital which was to be a symbol of Nigeria's aspirations for unity and greatness was excised from Niger, Kogi and Plateau, Nasarawa states. It has a total land area of 8000 square kilometres, which is more than twice the land area of Lagos State, and is centrally located as shown in Fig. 12.1.

In order to realise its objective of developing Abuja, the new capital city, into a masterpiece on the African continent, the Federal Capital Development Authority

M. B. Nuhu (✉)
Department of Estate Management and Valuation, Federal University of Technology, Minna, Nigeria
e-mail: mbnuhu@futminna.edu.ng

S. Yakubu
Department of Estate Management, Baze University, Abuja, Nigeria

I. A. Ankeli
Department of Estate Management & Valuation, Federal Polytechnic, Ede, Nigeria

N. Salihu
Department of Estate Management, Bayero University Kano, Kano, Nigeria

F. B. Nuhu
Niger State Urban Development Board, Minna, Nigeria

T. Nubi et al. (eds.), *Transit Oriented Development in West African Cities*, Urban Perspectives from the Global South,
https://doi.org/10.1007/978-3-031-58726-9_12

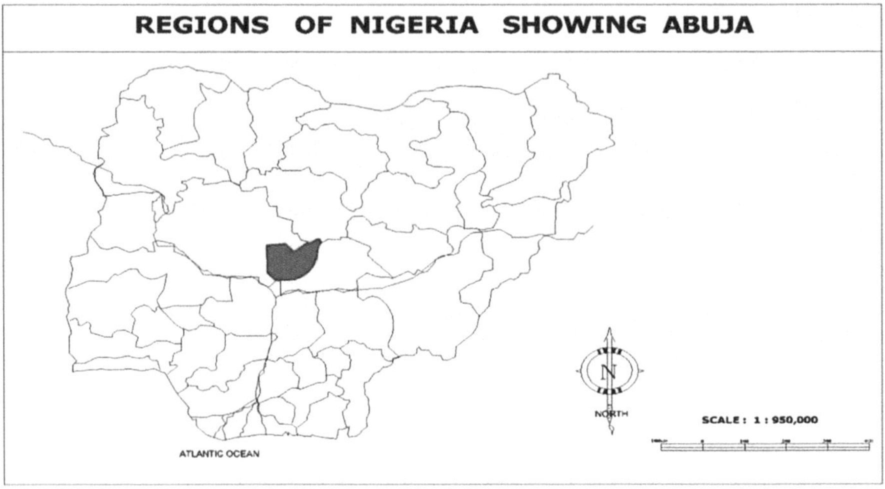

Fig. 12.1 Map showing the centre of Abuja. (Source: AGIS (2019))

(FCDA) became the sole agency vested with the responsibility of planning, designing and developing the city. This brief even extended to constructing and maintaining infrastructure such as 'roads, railways, sidings, tramways, bridges, reservoirs, water courses, buildings, plants and machinery and other works as may be necessary for, or conducive to' the discharge of its functions in light of Decree No. 6 of 1976.

The city was to be built in 4 phases with about 12 districts (Fig. 12.2), with the first scheduled to commence in 1979 to provide space for 230,000 inhabitants across 4000 hectares of land. The capital was scaled to have a population of over 1, 600, 000 and expandable at its sides to 3, 200, 000 with the 'Central Area' as the city centre. The latter was structured with two wings at its sides that can enclose a total of 8 essentially self-contained 'development sector', as the Master Plan calls them, with between 120, 000 and 190, 000 inhabitants each (fcda.gov.ng). The territory is currently made up of six local councils, comprising the City of Abuja, and five local government areas, namely, Abaji, Gwagwalada, Kuje, Bwari and Kwali.

12.2 Transit-Oriented Development: Its Values and Functions

Transit-oriented development (TOD) is a '3C' model for development—compact, connected and coordinated—that has a direct impact on urban mobility and the economy. It reduces travel times, congestion and emissions—expanding access to different areas of the city and thus stimulating economic activity (Priscila & Lara, 2015).

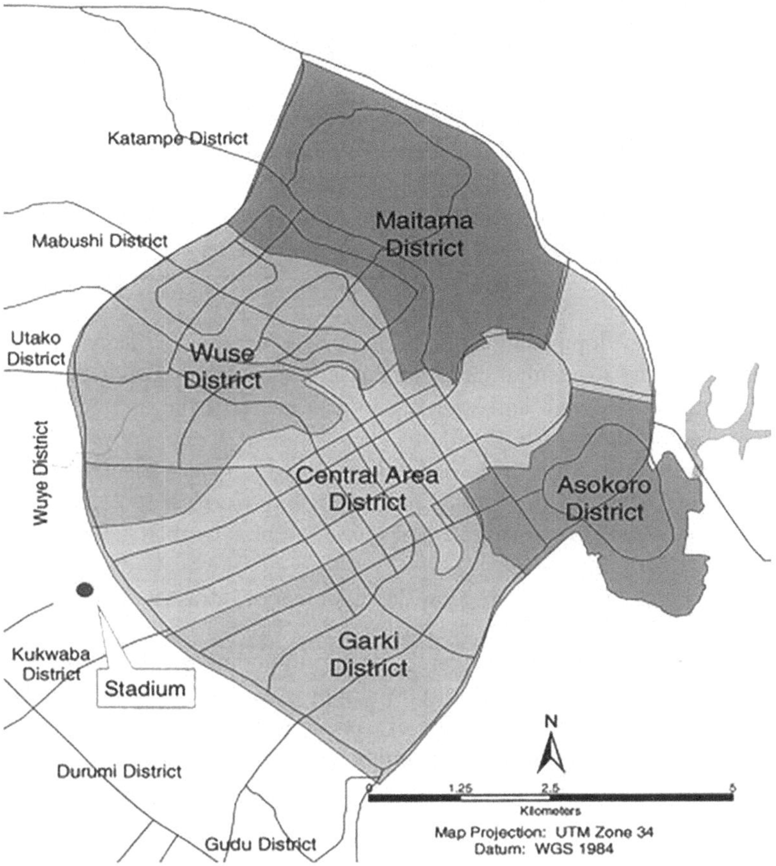

Fig. 12.2 Map showing FCC districts. (Source: AGIS (2019))

The most important function of transit-oriented development is to reduce traffic congestion and increase public transit use in areas experiencing accelerated urban development (Galelo et al., 2014; Loo et al., 2010; Curtis et al., 2009).

The values and functions of transit-oriented development include:

- Promoting higher densities and the concentration of jobs within relatively small areas.
- Proximity of mass transit improves access to transit-oriented development neighbourhoods, boosting their attractiveness and increasing real estate value.
- Compact urban development and high-quality public transit also mutually reinforce each other: mass transit can support the large passenger flows that come with high density development, while the concentration of jobs and housing around stations helps make public transport financially viable.

12.3 Review of the Relationship Between Transportation Challenges and the Thematic Areas of Interest

This is one of urbanisation, health, housing, urban infrastructure, land management, construction sector, governance and spatial data infrastructure in African cities:

12.3.1 Transit-Oriented Development

Transit-oriented development (TOD) is an urban planning strategy that directs mixed-use retail and residential development around public transportation lines. It emerged in the 1990s from anti-sprawl critiques of the social and environmental damage of auto-dependent fringe suburban development. It is a type of urban development that maximises the amount of residential, business and leisure space within walking distance of public transport (Calthorpe, 1993; Cervero et al., 2004). It promotes a symbiotic relationship between dense, compact urban form and public transport use (Caves, 2004).

This rapid urban growth versus slow development of infrastructure for the teeming population has led to scarcity of infrastructures. The rapid urbanisation of Abuja has led to the losses of green land, scarcity of accommodation, shortage of buses for mass transportation, incessant lack of power supply, absence of intra-light rail system, high traffic congestion at the close of work in Abuja (Fig. 12.3), etc.

Transit-oriented development is a planning concept that promotes the creation of a network of well-designed, human-scale urban communities focused around transit

Fig. 12.3 High traffic congestion at the close of work in Abuja

stations. The term refers to communities which comprise high-quality, medium- to high-density mixed-use residential and employment areas, with high levels of access to public transport and pedestrian and cyclist-friendly street environments (QG, 2010).

12.3.2 Abuja Infrastructure Planning Concept

The basic concept of the transportation infrastructure layout in the Federal Capital City is set as follows in compliance with the upper-level concept stipulated in the Abuja Master Plan.

- Highway grid system: Every development district (Phases I–III) and/or an area defined as sector centre in the AMP shall be bound by a highway network. The closed network system of a district/sector centre area is adopted to minimise car trips and secure safety by avoiding traffic since the majority of traffic generated/attracted by a sector centre is limited to the inside as long as the neighbourhood concept works better.
- Dual-transit ways: Each two-transit way both north and south with connection to all sector centres from the CBD is planned along the development wings. Trips between sector centres shall be made by the transit ways.
- Trunk-feeder network system: The feeder lines shall be connected to the trunk lines at appropriate intervals so that almost all residents could access public transport within 5 min. The demand that may arise from the expansion of the city with the increase in population may be addressed efficiently and flexibly by extending the trunk lines.
- Stage-wised public transport upgrading: A transit corridor shall be upgraded depending on urban development and demand. Buses may serve in mixed traffic at the initial stage, but on exclusive lane at the next stage. It may be later on changed to rail transit such as LRT or ordinary rail according to demand increase.

12.3.3 Normative Infrastructure: Spatial Justice and Social Diversity

Because it is argued that it is impossible to have a neutral perspective on development and planning, two normative lenses are to be adopted from which to explore development and planning issues in African cities. These lenses are spatial justice and social diversity (Andrea, 2018).

Spatial justice is a concept that has gained some conceptual traction since the early 1970s and is viewed by European policy-makers as a viable alternative or even substance to territorial cohesion policies. Spatial justice involves 'the fair and equitable distribution in space of socially valued resources and opportunities to use them' (Soja, 2009). For Soja, the idea of socio-spatial relations is fundamental, that is, that space shapes social relations as much as social relations shape space.

Social diversity refers to a successful community in which individuals of different race, ethnicity, religious beliefs, socioeconomic status, languages, geographical origin, gender and/or sexual orientation bring their different knowledge, background, experience and interest for the benefit of their diverse community (Santana, 2017).

12.4 Abuja Transit-Oriented Development Types/Changes and Its Evolution

12.4.1 Abuja Light Rail System

Abuja Rail Mass Transit commonly known as Abuja Light Rail is a light rail transport system in Abuja, FCT, Nigeria. The Abuja Light Rail System is an integral part of the Abuja Rail Mass Transit System and FCT Transportation Master Plan. The development of the light rail mass transport system is the greatest highlight of the Federal Capital transit revolution. The Abuja Metro project was provided for in the Abuja Master Plan but was not implemented until President Obasanjo's administration started working on it in 2007. Six light rail routes, totalling 300 km, are provided in the Master Plan. The construction of two of the routes (1 and 3) has begun and was expected to be completed in 2015. These two routes link the CBD to the Airport (route 3), Idu-Kubwa (route 1A, Fig. 12.4) and Garki-Gwarinpa-Kubwa (route 1B) (TELL, 2014). The remaining four routes are to be financed through the public/private partnership arrangement.

12.4.2 International Airports

Nnamdi Azikiwe International Airport is the main airport serving Abuja and the surrounding capital region. It was named after Nigeria's first president, Nnamdi Azikiwe. The airport has international and domestic terminals. The most ambitious

Fig. 12.4 Route 1a

of the road projects are the Airport Road (renamed Umaru Yar'Adua Expressway), 37.5 km in length; the Kubwa/Zuba Expressway (Outer Northern Expressway), 39.76 km; and the Southern end, the Nnamdi Azikiwe Road. These are ten-lane super multiple carriage highways with full accompaniments of street lights, drainages, pedestrian bridges and flyovers, lay-bys, culverts, telecommunication ducts and interchanges, etc. The Airport Road Kubwa/Zuba Expressways have been able to reduce travel time to the city centre from 3 hours to 30 minutes (over 100% reduction) and billed for completion in 2014.

The Southern Parkway from the Inner Southern Expressway to Ring Road completes the ten-lane ring, and it was completed in 2014. There is also the reconstruction of the 72 km Lower Usuma Dam-Gurara Dam-Jere Road aimed at creating an alternative route for vehicular movement out of the city (Abuja) to Kaduna through the Bwari axis. Another internal road under the influence of the ten-lane model is the extension of the 3.8 km Inner Southern Expressway to Southern Parkway (more than 80% completed (TELL, 2014) aimed at reducing traffic congestion induced by the North-eastern vehicular traffic at the AYA roundabout. Internal roads to complete roads B6, B12 and Circle Road in the CBD were being constructed and billed for completion in 2014 with the object of reducing travel time within the city. Many principal interchanges are also being built. Those constructed and commissioned include the AYA (Asokoro), Banix Junction, Jahi/Mabushi Link, Karmo/Utako and Gwarinkpa II-Kado/Life Camp Junctions. These highways have been completed.

Bus Rapid Transport The first mass transit buses were introduced in 2005 but collapsed thereafter. In 2010, the Operators' License Scheme was introduced and, with it, the feeder route programme and route delineation. High-capacity buses (80–150 passengers each) have been introduced and ply designated routes in the city centre (TELL, 2014). Minibuses ply the feeder roads from satellite towns to defined terminals where passengers disembark and take taxis or the high-capacity buses to the city.

12.5 Abuja Transit-Oriented Development and Formalities

Transit-oriented development (TOD) is a '3C' model for development—compact, connected and coordinated—that has a direct impact on urban mobility and the economy. It reduces travel times, congestion and emissions—expanding access to different areas of the city and thus stimulating economic activity (Priscila & Lara, 2015).

A common definition or criterion remains to be established for transit-oriented development despite the substantial body of research that has discussed the concept. Cerbero et al. stressed the contributions of the 3Ds to the transit- oriented development concept; furthermore, they explained the 3D concept as density in the form of residence and jobs, diversity in the form of mixed land- use development and design in the form of good street connectivity for pedestrians (Galelo et al., 2014).

Calthorpe (1993) defines transit-oriented development as a mixed-use community within a comfortable walking distance of the transit stops and core commercial area that encourages people to live near a transit service and to use public transportation such as bicycling for short trips and transit for long trips. Other scholars interpreted transit-oriented development by borrowing words from its original explanation and providing new empirical additions. The various definitions for transit-oriented development generally state that it should promote compact and highly mixed-use development around transit hubs and should include accessible and walkable neighbourhoods (Cervero, 1995; Knowles, 2012).

Definitions of transit-oriented development often focus on built form. For example, Bernick and Cervero (1996) emphasise the role of the 'three Ds' (density, diversity and design) in the success of transit-oriented development. The term transit-oriented development is often used incorrectly to describe a single development adjacent to, or above, a transit station. Transit-oriented development refers to the set of principles applying to the broader precinct surrounding the station rather than any individual development within it (QG, 2010).

12.6 Values and Functions of Transit-Oriented Development

The most important function of transit-oriented development is to reduce traffic congestion and increase public transit use in areas experiencing accelerated urban development (Galelo et al., 2014; Loo et al., 2010; Curtis et al., 2009).

The values and functions of transit-oriented development include:

- Promoting higher densities and the concentration of jobs within relatively small areas.
- Proximity of mass transit improves access to transit-oriented development neighbourhoods, boosting their attractiveness and increasing real estate value.
- Compact urban development and high-quality public transit also mutually reinforce each other: mass transit can support the large passenger flows that come with high density development, while the concentration of jobs and housing around stations helps make public transport financially viable.
- When located in areas that are less exposed to natural hazards, the transit-oriented development approach can enhance resilience to disaster risks by supporting high-density housing and activities in lower-risk zones.

Other values and functions are (Papa & Bertolini, 2015):

- Facilitating polycentric cities and regions.
- Mitigating urban sprawl.
- Boosting public transport ridership.
- Increasing bike usage and walkability.
- Accommodating economic growth.

12.7 The Key Role of Government, Transport Operators, Civil Society and Others

The 1999 Constitution of the Federal Republic of Nigeria provided certain functions that can be carried out exclusively only by the federal, state or local government even where the provision of public transport is left entirely to the private sector. The government has an important role to play, if only to ensure through appropriate policy measures that the operating environment is conducive to the development of a suitable transport industry. A fundamental requirement is full government commitment to the success of the efficient transport system. The government's key role is the regulatory function and provision of basic infrastructure or enabling environment for operations to function effectively at various levels.

It is pertinent to note that transport is that part of economic activity which is concerned with increasing human satisfaction by changing the geographic position of goods or people. It may bring raw materials to where they can be manufactured more easily or finished goods to places where consumer can make best use of them. Similarly, it may bring the consumer to places where he or she can enjoy services which are being made available. There are five types of transportation mode: road, rail, sea, air and pipeline. Each of it has its own characteristics as well as advantages and disadvantages.

The transportation industry can be broken down into three major groups of companies: shipping, passenger transport and equipment manufacturers. In some cases, particularly within shipping and passenger transport, companies provide services in multiple areas of the industry. Shipping companies are responsible for the transportation of supplies and products to businesses, governments and individual consumers and operate on a global basis. The passenger transport segment provides people with the means to get anywhere on the planet, whether it is by air, sea or land. Finally, the manufacturing segment produces the trucks, planes, ships and railcars along with all the technology that allow transportation to exist in its current form. These manufacturers are just as essential to the transportation of materials and people as are the companies that transport them. Slow economic activity results in lower demand for freight and passenger transportation. According to the IATA, the air transport industry lost more than $4 billion amidst the 2008 global recession.

The underlisted are the stakeholders in FCT—Abuja TOD.

1. Nigerian Institute of Transport Technology (NITT).
2. National Association of Road Transport Owners (NARTO).
3. Independent Petroleum Marketers Association of Nigeria (IPMAN).
4. National Union of Petroleum and Natural Gas Employees (PTDA/NUPENG).
5. Road Transport Employers Association of Nigeria (RTEAN).
6. National Union of Road Transport Workers (NURTW).
7. FCT Traffic Authority.
8. Federal Ministry of Transportation (FMoT).
9. Federal Road Safety Commission (FRSC).

12.8 Formal and Informal Transit-Oriented Development in Abuja

12.8.1 Formal Transit-Oriented Development

These are public transport operators that are registered with the Corporate Affairs Commission to provide commercial transport services to the public. They are called organised, because they are orderly on road and have proper schedule and definite stopping areas to pick and drop passengers. They operate as corporate entities and so have standard workshops, terminal and well-trained staff to run the affairs of the organisation. They mostly operate as a company (Yakubu, 2006).

There are so many organised private transport operators in FCT. But only two of them presently provide intercity and intracity transport service. The rest provide interstate transport services. This includes the following:

1. Peace Mass Transit.
2. NDDC Transport System.
3. Abia Line.
4. ITC (Imo Transport Corporation).
5. Eagle Line.
6. Edo City Transport Service.
7. Cross Country.
8. ABC Transport.
9. Ekesons Transport.
10. Chisco Transport.
11. Abuja Municipal Transport Services.
12. Abuja Urban Mass Transit Company (AUMTCO).
13. Labour City Mass Transit System (LCMTS).

Abuja Urban Mass Transit Company (AUMTCO) and Labour City Mass Transit System (LCMTS) provide intercity transport services, while AUMTCO provide both inter- and intracity services respectively. AUMTCO, which is owned and operated by FCDA, has only about 29 buses, 15 of which are roadworthy and 14 awaiting some repairs at the time of investigation. AUMTCO currently provides services along intra- as well as intercity routes. These fleets are grossly inadequate when compared with the PCU volume in the FCT. The Labour City Mass Transit System (LCMTS) is owned and operated by the Nigerian Labour Congress. It has about 30 buses, all of which have 50 seated carrying capacity, 6 of its fleet are not on the road at the time of investigation, and only 24 are roadworthy, which, if put together with those of AUMTCO, are very inadequate to meet the mobility needs of Abuja residents (Yakubu, 2006).

12.8.2 Understanding Abuja Transit-Oriented Development Risk and Coping/Adaptation Strategies

The risk associated with Abuja transit-oriented development (TOD) includes:

- Smoking engines emitting poisonous gases such as carbon monoxide, nitrogen oxide, lead and volatile compounds into the environment, which pose health risks to motorists and commuters.
- Recklessness, excessive speeding and inexperience of motorists result in accident.
- The scarcity of fleet often causes overloading of the few available vehicles with some of the passengers standing.
- Hindrance of operation and delays.

12.8.3 Informal Transit-Oriented Development

Poverty and unemployment have given rise to entrepreneurship within the informal economic sector. The unmet demands in public transport system in developing countries have given leverage to a variety of informal modes of transport characterised by small vehicles, low-performance services, mostly owner operated with low capital outlay, self-regulated with easy market entry.

and exit jobs for operators deprived of other jobs opportunities (Cervero, 2000; Kumar, 2011).

12.8.4 What Are Urban Informalities?

In the simplest sense, informality is the absence of government regulation. Informality is often used in terms of economics, where the informal sector is not taxed or regulated by the State. Attempts at defining urban informality are often linked to specific epistemological frameworks and more broadly to debates prevailing in specific disciplines, times and places (Nicola et al., 2020). The spatial impact has been massive as these populations compete for scarce land near crowded urban cores. De Soto (2001) argues that the poor engage in informal activity because ‘their only alternative is to live and work outside the official law’ to cope and survive. This sector and its workers are separate and distinct from a formal governmental sector.

Roy’s understanding of urban informality as ‘an organising logic, a system of norms that governs the process of urban transformation itself’ is helpful, moving debates forward by presenting informality as ‘a series of transactions that connect different economies and spaces’ (Roy, 2005). McFarlane’s (2012) re-conceptualisation of informality as critique opens up new ways of understanding how informal and formal actions relate to each other, allowing a recognition that informal practices extend beyond the urban poor to encompass the actions of different sectors including middle- and high-income urban residents, the state and business interests. The significance of this perspective requires greater attention, especially from spatial and economic disciplines, which have tended to overlook the social and institutional aspects of urban informality (Obeng-Odoom, 2016).

For Ranganathan (2014), informality is 'the flexible and uneven suspension of regulation and law in the production of urban space and materiality'. It is a negotiated process in which different actors have (and negotiate) different forms of authority or subservience. While informality can occur through multiple channels, what differs—and particularly what determines whether these channels, experiences and strategies can be used as part of survival or accumulation networks—is the differentiated protections and legitimacies offered by public authorities in their classifications of legal and illegal and even to the different 'types' of illegality (Ranganathan, 2014).

12.8.5 Economic and Spatial Dimensions of Informality

Economic informality and spatial informality are both seen as reasons to repress, informally tax and otherwise further disadvantage the already disadvantaged groups (Varley, 2013). Such realities are described by Watson (2014) when she highlights the reasons why urban planning rules cannot meet the needs of the urban poor, since the elitist nature of planning and building standards means that the urban poor must break the law to secure land and shelter.

The concept of urban informality has been a central focus within studies of the urban since it was first observed. Although theoretical debates have heavily critiqued and attempted to move beyond the formal/informal dichotomy, in policy and practice this retains relevance. We have drawn upon recent research and the new empirical contributions in this issue to argue for a revised conceptual understanding of urban informality, building upon more recent scholarship (McFarlane, 2012; Roy, 2005, 2009; Simone, 2011) that argues for new ways of understanding how informal and formal actions relate to each other across different urban spaces and groups. While the political economy of urban informality has not previously been ignored, its significance has perhaps been obscured by the narrow focus that studies of urban informality have often taken, across particular sectors, settings or outcomes. This is especially the case from economic and spatial perspectives.

The vast majority of informal economy activities provides goods and services whose production and distribution are perfectly legal. In addition, informal economy activities are not necessarily performed with the deliberate intention of evading the payment of taxes or social security contributions or infringing labour legislation or other regulations. The informal economy can however include restricted illegal and restricted legal operations or legal and irregular operators, but no criminal operators (ILO, 2002). The informal economy should therefore not be confused with the criminal economy. The spatial impact becomes massive as the populations compete for scarce land near crowded urban cores.

12.9 Review of the Relationship of Transportation Challenges and Urbanisation

The relationship between transportation and urbanisation cannot be overemphasised, as the two are interwoven together for the functionality and existence of all urban centres. Urban transportation can be described as the complex forms of transportation of people and cargo (goods) within the territory of a city. Transportation is regarded as the main core of urban development; it is like the blood vessel of the urban body in which without it there would be no urbanisation.

The term urbanisation is described as the movement or transition from a rural to an urban society. Urbanisation reflects the proportion of the population living in the urban settlement (Jean-Paul, 2020).

In essence, for urbanisation to persist there has to be a continuous and sustained 'movement' of people into the urban centres, and these movements can only be aided by transportation. Besides, rural-urban migration which is the movement of people and goods from a rural settlement to an urban settlement is regarded as a strong factor of urbanisation. Hence, there is a strong relationship between transportation and urbanisation.

In urban planning, parcels of land and land uses are being demarcated by transportation infrastructure (roads, rails, walkways, etc.), the urban land is given its value based on how accessible it is and accessibility is aided by efficient transportation system. In other words, the efficiency of a city's transportation system is reflected in the city's economic growth. The imprint of transportation networks gives the urban settlement a beautiful pattern, hence its form. In addition, the urban form is regarded as the spatial imprint of an urban transport system as well as the adjacent physical infrastructures. They jointly confer the level of spatial arrangement in cities (Jean-Paul, 2020).

12.9.1 Transportation Challenges and Urbanisation

Urban centres continue to grow proportionately with transportation infrastructural development. So also have the urban mobility problems increased;this could be attributed to the fact that spatial growth of urbanised areas increases the complexity of transport systems, thereby making it difficult to manage, especially in developing countries. Besides, the United Nations predicts that by 2050, about 70% of the global population will be living in urban centres with about 90% of urban growth to occur in developing economies, which will translate to intense pressure on urban infrastructure especially transportation systems (United Nations, 2020).

Within just a few decades, the urban areas globally have become increasingly automobile dominated. This has posed great transportation challenges especially in the developing countries in which the cities have experienced amplified growth of transport-related challenges. Some common transportation challenges faced in developing countries include traffic congestion, pollution, environmental degradation, climate change, public transport decline, environmental degradation, visual intrusion and lack of accessibility for the urban poor (Pojani & Stead, 2015).

Similarly, Breuil and Enskat (2017) in their presentation to the European Commission highlighted the following as the major urban mobility challenges faced within developing countries: explosive urbanisation, rapid private motorisation, growing risk of city dysfunction (i.e. lack of access to jobs and services, high pollution, congestion, etc.), the fastest and increasing source of greenhouse gas emission and lack of managerial capacity for cities and transportation infrastructures.

Besides, Abuhamoud et al. (2011) highlighted lack of proper investment in transportation infrastructure, civil wars, low-quality transport systems, unhealthy transport solutions, improper planning and increasing number of private cars as the major problems owing to transportation in contemporary African city. These were also affirmed by O'Neil (2020) who also included lack of resources and lack of hierarchical highways to the list as part of the urban transportation challenges.

From the Nigerian perspective, Solanke (2013) explained that there is evidence of skewed modal development which is tilted in favour of road transportation to the disadvantage of other means of transportation. He stressed that road transportation in Nigeria accounts for 90% of both freight and passenger transport. And the major challenges accrued from this mode of transport include traffic congestion, road parking by lorries and slow flow of traffic within peak periods which leads to loss of time and energy.

From the above reviews, it is very evident that urban mobility/transportation challenges pose great threat to both the health and economic well-being of the urban populace. Tan et al. (2008) argued that 97% of the global transport systems depend on petroleum which is a finite resource that contributes immensely to the emission of greenhouse gases, leading to global warming and climate change. This poses a great threat to the health of the urban populace. This fact is in agreement with the research of Black and Black (2009) who listed urban development and transportation as an environmental stressor that can impair public health. Urban development leads to change and alteration of the natural environment, thereby reducing environmental quality. This becomes exacerbated with inefficient transportation system that leads to congestion and pollution of the environment. All of these increase the health risk of the urban populace. More so, bad roads due to lack of infrastructural maintenance and standard construction endanger the lives of the urban populace.

According to the World Health Organisation (WHO), road traffic accidents in Africa have increased over the last three decades, and at the same time, they have the highest rate of fatalities from road traffic injuries owing to bad roads which have not been improved upon (Adeloye et al., 2016).

The deteriorating public transportation system and the lack of non-motorable transport option have led to the increase of private motorisation in African cities. The repercussion of this is traffic congestion, stress on infrastructure whereby its life span reduces which in turn leads to high cost of maintenance and, finally, increased carbon emission leading to climate change. Climate change which transportation contributes to affects the production of food for the urban populace due to its effect on the agricultural calendar and rainfall patterns.

To cap it all up, the developing African countries mostly lack the capacity (human and technological) to explore and manage new transportation options that are safer and greener. This has made the current transportation systems stagnant for years without improvements, thus adding to the urban transportation challenges.

12.9.2 Practical Solutions to Curb the Transportation Challenges in Urbanised Regions of Africa

There is no unanimous solution to curbing transportation challenges in various African countries because of the fact that the transportation challenges they face will be diverse and also accruing to the fact that each country will respond to these transportation challenges based on their capacity (by financial and technical expertise). There are however pragmatic suggestions that could curb or reduce transportation challenges in various African countries, some of which includes:

(i) **Adoption of electronic vehicles**: African countries largely run on a fuel economy, and this dominates the transportation option in African countries. This is the major reason why the transportation sector accounts for one quarter of the global greenhouse gas emissions. Countries such as Rwanda, Kenya and South Africa have started the promotion of clean transportation by mandating fuel efficiency standards and also adopting the use of electric vehicles. These approaches are very efficient in reducing carbon emissions especially in congested cities (Posada, 2018).

(ii) **Prioritise mass urban transit**: There has been neglect in the sector of mass public transportation in African countries, as buses are seen deteriorating and inefficient. Also, the available mass transit options are not sufficient to serve the continuously growing urban population; hence, this has contributed to the increase of private motorisation and expensive private transit options (taxi) in African countries leading to congestion on the road. The fact that the public transport is perceived as inefficient and befits the poorest of the social class makes it unpreferable to those who can afford their comfort on the road. There is a need for sensitisation to appease the mind-set of urban dwellers regarding mass transportation options and also a provision of an efficient mass transportation option. In addition, there should be restriction in the use of private vehicles to increase the patronage of mass transit options.

(iii) **Adoption of non-motorised transportation option**: The use of intercity light rails and tramps can be used to reduce the population of fuel-powered vehicles on the road. This will translate to reduction in carbon emissions, congestions and road accidents which all emanate from the use of road transportation.

(iv) **Improvement of transportation infrastructures and proper planning of new routes**: It has been argued that there is a high stagnation in the development of transportation infrastructures within African countries. This phenomenon leads to rapid deterioration of these infrastructures owing to the rapid increase in population which increases the stress on these infrastructures. This deterioration now leads to the high risk of lives of those who use it. Additionally, managerial inefficiency and the lack of maintenance of the newly developed infrastructures make the expectations short lived. Thus, there should be capacity building in the areas of development and management of these infrastructures.

(v) **Priorities for pedestrians and cyclists**: Consideration should be given to pedestrians and cyclists who take short trips within the city. Supporting infrastructures such as sidewalks and bicycle lanes should be provided to encourage people to walk more. This has health benefits and would go a long way in reducing traffic congestion within the city.

(vi) **Consideration for transportation planning**: In matters of transportation infrastructure development, it would be proper to consult transportation planning experts to bring out the best blueprint which would be sustainable to both present and future transportation needs.

(vii) **Policy formulation**: Policies should be formulated and adopted to regulate the competition for mass transit options, for sustainable urban transport financing and integration of transport and land use plans.

(viii) **Institutional development**: This suggests capacity building for the institutions in charge of urban transportation so as to better improve transportation management within the urban context.

12.10 TOD and Its Connection to SDGS

According to the United Nations (2015), nearly half of the world's 3.9 billion urban dwellers reside in relatively urban settlements with fewer than 500,000 inhabitants, while only around 1 in 8 live in the 28 megacities of ten million inhabitants or more. It is estimated that the urban population in developing countries is set to double from 2010 to 2050 while remaining stable in developed countries (United Nations,

2015). Presently, cities with less than 100,000 inhabitants represent a third of world's urban population, a figure which is predicted to grow to 40% in 2050. Transit Oriented Development (TOD) based on the foregoing is a mixed-use community that inspires communities and residence to live near transit services and the reduction of dependence on individual driving to and fro commuting to work, health centres and other services as the practice is in Abuja. TOD is a novel concept focusing on well-organised modes of transportation other than the automobile. TOD refers to a rising model of mobility in cities which connects to the UN Sustainable Development Goals. It can contribute towards the making of a healthy environment which links to SDG 8 leading to stress-free, vibrant and sustainable communities in a liveable city and a productive economy (Goal 8). As several African cities grow with high population and to achieve sustainable transportation in line with SDG 11, making cities and human settlements inclusive, safe, resilient and sustainable should be made prominent. As worldwide urbanisation and expansion of urban areas keep increasing unprecedently, numerous cities throughout the world are developing transit-oriented development policies to facilitate economic fairness, business development and environmental sustainability in line with the expectations of SDGs.

12.11 Abuja Transit-Oriented Development Governance

Devolution of Power Devolution is the statutory delegation of powers from the central government of a sovereign state to govern at a subnational level such as regional or local level. In most federal states, two or more levels of government are created: one at the national level and others at the regional or subnational level described variously as states, provinces, cantons and so on and at subnational or local units called the local councils, municipal councils, country, districts, province, city township, borough, parish, shire, village, etc. In such manner governmental powers and functions are allocated to these levels of government. This is a form of devolution of power (Amah, 2018).

Devolution of power to local people encourages active political participation which acts as a restraint against tyranny and dictatorship, which may occur as a result of the central government being remote from the local people (Dalhatu, 2010). Devolution of power to local people at the grass root enhances the legitimacy of the government and its programmes. Economically, it enables the government to distribute local public goods and services according to the local needs. It further encourages healthy competition between the various sub-national and local governments which results in innovative social and regulatory policies, and politically, it secures greater participation and opportunities to local minorities in the government and the decisions-making process, thereby preserving their distinctive autonomy and linguistic identities (Amah, 2018).

Under the Federal Republic of Nigeria Constitution, there is demarcation and allocation of legislative functions between the national/federal government and the states; the local governments were made to function based on laws made by the states' (the subnational) legislature. This deliberate state of affair has generated much controversy among Nigerian constitutional and political commentators, as most proponents of local government autonomy favoured the total devolution of autonomous power and functions on the local government councils, while others favoured the status quo, whereby the local government remains the appendage of the state government (Amah, 2018).

12.11.1 Scales: City-Level, Metropolitan and Regional

The Abuja skyline is made up of mostly mid-range and a few tall buildings. Only recently have tall buildings begun to appear. Most of the buildings are modern, reflecting that it is a new city.

Plans were made to build skyscrapers such as the Millennium Tower which is partly complete as work has stalled for the last few years. This structure looms 170 metres (560 ft) above the city. The tower is part of a huge cultural development complex called the Nigeria National Complex including the Nigeria Cultural Centre, a 120,000 m^2 (1,300,000 sq. ft) structure dedicated to the art and culture of Nigeria. The Cultural Centre and the Millennium Tower have been designed by the Italian architect Manfredi Nicoletti.

Landmarks include the Millennium Tower; the Central Bank of Nigeria headquarters; the Nigerian Presidential Complex; the Ship House; the National Stadium, which was the main venue of the 2003 All-Africa Games and some games, including the final of the 2009 FIFA Under-17 World Cup; National Mosque; the National Church; Aso Rock; and Zuma Rock.

12.12 TOD Financing/Funding

12.12.1 Financing and Source of Funds

The estimated investment for the development of Abuja light rail project is $824 million. The Federal Capital Territory Administration (FCTA) used loans from the Exim Bank of China to fund approximately 60% of the project. The Federal Capital Territory Administration also signed an infrastructure soft loan agreement with the Exim Bank of China for the procurement and maintenance of 16 light train sets with 3 rail cars each. The $194 million loan agreement is valid for a term of 3 years.

The bank committed to fund $157 million of the total cost incurred on the procurement and maintenance of train sets for lots one and three. The remaining $37 million is provided by the counterpart fund of the Federal Capital Territory Administration.

12.12.2 Urban Challenges

Urban growth is the increase in the number of people living in towns and cities. The rapid increase in the population of urban centres has resulted into rising cost of living because of the higher demand for urban commodities (Amao, 2012). Urbanisation in developing countries has exerted pressure on infrastructural facilities such as housing which has a significant effect on the other basic needs of the urban residents. Since population keeps increasing, the need for more shelter cannot be neglected. The World Health Organisation (WHO, 1992) describes housing as residential environment which includes the physical structure used for shelter and all necessary services, facilities and utilities required for the physical and mental health and social well-being of the people. Therefore, basic amenities like waste disposal system, water supply, roads and location services are part of housing. Likewise, recreational facilities and health and educational infrastructures are all considered as parts of housing (Aribigbola, 2001).

Abuja, the Federal Capital Territory (FCT), had been at the core of urbanisation since its designation as the nation's capital in 1975 (Abuja Journal, 2017). After the shift of the nation's capital from Lagos to Abuja in 1994, Abuja has seen a drastic rise in population till date (Alkali John, 2005). This has overstretched the fiscal, technical and managerial capacities of the Federal Capital Development Authority (FCDA) to properly manage the usage of land, provisions of housing and basic infrastructure of public services and to keep up with the incessant population growth. This continual growth in population along with the rapidly increasing urban population makes it important to analyse this problem and develop appropriate responses to the growing challenges (NUDP, 2012).

As of 2016, the metropolitan area of Abuja is estimated at six million persons, placing it behind only Lagos as the most populous metro area in Nigeria (Andrews, 2017). A greater part of this population, due to inadequate accommodation and high rents in the city, end up settling in the suburbs of the city, in areas such as Kubwa, Karu, Masaka and Nyanya. The urban growth rate of suburbs such as Karu and Nyanya in 2001 was 66.2% compared to Abuja city which was 40.2% (Jinadu, 2004). This urban growth is worrisome, against the background that urban development and infrastructural plans are not as effectively planned and implemented leading to increased numbers of dwellers with little or no access to inadequate means of transportation, water, shelter, sanitisation, education, health, etc.

There is no doubt that this rapid urbanisation rate has resulted in various economic, cultural and environmental issues. The problems and challenges have created uncontrolled and unplanned cities resulting in millions of urban dwellers living in substantial dwelling, mainly slums and shanty towns. Daramola and Ibem (2010), Ogunleye (2005) and Jiboye (2011) argue that these issues resulting from the uncontrolled urbanism in Nigeria include:

- The loss of biodiversity.
- High population density.
- Risk of epidemic outbreak.

- Security.
- Reduced access to social infrastructure.

12.12.3 Co-production of Urban Infrastructure

Co-production in urban areas usually involves the co-delivery of basic municipal services, with roles for both the government and organised citizens. However, there is the potential to move on from this aspect to a more substantive social purpose. The actual delivery of services can be almost ancillary to the relationship implications of the co-production process. They move from considering co-production as a means to meet essential ends to co- production as a means of altering essential relationships and ongoing practices (Diana & Sheriden, 2018).

A study in Lilongwe (Malawi) focused on the co-production of water services through a partnership between the public utility and water user associations in informal settlements. Here, co-production was compensating for the lack of state capacity around services; the model encourages synergies between communities and service providers and the building of community social capital.

In China, in 2010, a widespread co-production effort was initiated, geared towards 'more responsive service provision, infrastructure improvement, and community building through state-enabled decision-making, self-organisation and self-service delivery on the part of communities'. Research in four cities found that although local officials have limited enthusiasm and community members may be hesitant about their contributions, service provision has become more participatory than it used to be, and this may 'inspire and ignite' more genuine co-production (Bingqin et al., 2019).

Co-production can be seen as an array of strategies to access public services initiated by community-based organisations, often presented as a practical strategy to deliver services in what they refer to as unorthodox contexts. But co-production can also be seen as a tool to advance social justice and to challenge the structures that reproduce inequalities (Simon et al., 2018).

12.13 Discussion on the Assessment of the Potentials for TOD in the City

12.13.1 SWOT Analysis of Abuja Transit-Oriented Development (Table 12.1)

Table 12.1 The SWOT analysis of Abuja transit-oriented development

SWOT	
Strength	Manufacturing efficiency like exclusive access to high-grade natural resources. Skilled workforce. Adequate financial resources. Superior image and reputation such as strong brand names. Insulation from strong competitive pressures. Distinctive competence in key areas.
Weakness	Outdated facilities. No clear strategic direction. Lack of management vision, depth and skills. Inability to raise capital. Weak distribution network. Obsolete technology. Low employee morale. Poor track record in implementing strategy. Poor market image.
Opportunities	Possible new markets and an unfulfilled customer need. Emerging new technologies. Complacency among competing organisations. Expansion of product or service line to meet broader range of customer needs. Removal of international trade barriers.
Threat	Increased trade barriers. Entry of lower-cost foreign competitors. Cheaper technology adopted by rivals. Sales of substitute products. Shortages of resources. Recession in economy. Adverse shifts in trade policies of foreign governments. Adverse demographic changes.

12.13.2 Pilot Report on the Means of the City Transportation System Using Kuje Area of FCT

The affordability of several means of the city transportation system is briefly reflected in the pilot survey conducted in Kuje Area Council. It has revealed the following:

Fig. 12.5 Keke napep

1. Okada and keke napep are the major means of public transport in Kuje settlement of FCT.
2. Commuters are less satisfied with the manner Okada and keke napep (Fig. 12.5) operate.
3. Commuters are not satisfied with the number of passengers carried by the operators (i.e. overcrowdings).
4. The study has shown that regulation on traffic rules compliance is not effective.
5. Commuters are satisfied with the overall operations of the available means of public transport in terms of affordability, safety, security, reliability and scheduling.
6. The government-owned intracity transport is no more in operation due to non-profit making and poor management, thereby leaving the intracity transport to the private operators by default.

The main focus of this study on various areas of Abuja transit-oriented development is to note that there are risks associated with Abuja transit- oriented development which should be carefully considered. The government should formulate policies that would enable informal transit system to play complementary role within an integrated modal transport system but with regulations to streamline its operation by the Federal Road Safety Corps (FRSC). Formal transit system should be made more convenient for consumers. Since Abuja is a highly populated city, more transit systems should be created and managed properly to meet the demands of the citizens.

12.14 Recommendations

From the findings of the Kuje pilot study, the following are recommended:

1. The FCTA should revive their public transport operation by acquiring/buying new vehicles to meet the needs of the people.
2. A bus rapid transport system should be introduced. This will reduce the challenge of overcrowding/overloading, discourage the use of private vehicles and decongest the road.
3. The FCTA should ensure adequate maintenance and proper management to avoid previous experience.
4. The FCTA in conjunction with private operators should be actively involved in public transport policy formulation, regulation of operators and educating the drivers, so as to ensure safe operation and avoid reckless or improper ride and parking on the road.
5. Strong enforcement should be introduced to curtail the excess loading of the public transport, in such a way that the vehicles do not carry more than its capacity design, e.g. tricycle/keke napep should not carry more than three passengers, while minibus/taxi should lift a maximum of four passengers (three behind and one at the front), and the number of standing passengers should be restricted to the number of strap hangers in the vehicle.
6. Lastly, the identified motor parks should be relocated to a better place suitable and spacious for a bus park.

12.14.1 Lessons for African Cities

- The Abuja Light Rail System is an integral part of the Abuja Rail Mass Transit System and FCT Transportation Master Plan.
- The development of the light rail mass transport system is the greatest highlight of the Federal Capital transit revolution.
- Six light rail routes, approximately 300 km, are provided in the Master Plan. The construction of two of the routes (1 & 3) is almost completed.
- The most ambitious of the road projects are the Airport Road (renamed Umaru Yar'Adua Expressway), 37.5 km in length; the Kubwa/Zuba Expressway (Outer Northern Expressway), 39.76 km; and the Southern end, the Nnamdi Azikiwe Road. These are ten-lane super multiple carriage highways with full accompaniments of street lights, drainages, pedestrian bridges and flyovers, lay-bys, culverts, telecommunication ducts and interchanges, etc.
- High-capacity buses (80–150 passengers each) have been introduced and ply designated routes in the city centre.

12.15 Conclusion

Abuja transit-oriented development has been a key focus because of the rapid urbanisation that brings about inadequacy and limitations in the transit system of Federal Capital Territory, Abuja. The demand for transport due to its service- oriented capacity cannot be separated from any part of society. The role transport plays in socioeconomic and political activities is not negligible, to bridge the gap between surplus and deficit resources.

References

Abuhamoud, M., Rahmat, R., & Ismail, A. (2011). Transportations and its concerns in Africa: A review. *The Social Sciences, 6*, 51–63.

Abuja Journal. (2017). *Explore Abuja city: Stay in Abuja.com*

Adeloye, D., Thompson, Y. J., Akanbi, A. M., Azuh, D., Samuel, V., Omeregbe, N., & Ayo, K. C. (2016). *The burden of road traffic crashes, injuries and deaths in Africa: A systematic review and meta-analysis* (Vol. 94, p. 510). World Health Organisation.

AGIS. (2019). Abuja Geographical Information System (Map unit) of Federal Capital Territory (FCT) Abuja, Nigeria.

Alkali John, L. S. (2005). *Planning sustainable urban growth in Nigeria: Challenges and strategies. Federal ministry of housing and urban development, Nigeria at the conference on planning sustainable urban growth and sustainable architecture, held at the ECOSOC chambers*. United Nations Headquarters.

Amah, E. I. (2018). Devolution of power to local government: Appraising local government autonomy under Nigerian federation. *Beijing Law Review, 09*, 275.

Amao, F. L. (2012). Urbanisation, housing quality and environmental degeneration in Nigeria. *Journal of Geography and Regional Planning, 5*(16), 422–429.

Andrea, R. (2018). *What is spatial justice? Development and planning in African cities: Exploring theories, policies and practices from Sierra Leone*. University College London. www.futurelearn.com/courses/african-cities/0/steps/483325.

Andrews, J. (2017, August 6). *FCT Minister harps on development of satellite towns*.

Aribigbola, A. (2001). Housing and Nigerian development: Assessment of policy measure and direction. *African Journal of Environmental Studies*.

Bernick, M., & Cervero, R. (1996). *Transit villages in the 21st century*. McGraw-Hill.

Bingqin, L., Hu, B., Liu, T., & Fang, L. (2019). Can co-production be state-led? Policy pilots in four Chinese cities. Environment and Urbanization, Vol. 31, No. 1.

Black, D., & Black, J. (2009). A review of the urban development and public health with particular reference to Australia: Transdisciplinary research teams and some research gaps. *International Journal of Environmental Research and Public Health, 6*(5), 1557–1596.

Breuil, L., & Enskat, M. (2017). *Urban mobility challenges in Africa joint response framework*. European Commission.

Calthorpe, P. (1993). *The next American metropolis: Ecology, community, and the American dream*. Princeton Architectural Press.

Caves, R. W. (2004). *Encyclopedia of the city* (p. 676). Routledge.

Cervero, R. (1995). Sustainable new towns. *Stockholm's Rail-served Satellites Cities, 12*(1), 41–51.

Cervero, R. (2000). *Informal transport in the developing world*. United Nations Centre for Human Settlement, (Habitat).

Cervero, R. et al. (2004). *Transit-oriented development in America: Experiences, challenges, and prospects*. Transit cooperative research Programme, report 102.
Constitution of the Federal Republic of Nigeria. (1999). Laws of the Federation of Nigeria, West Africa.
Curtis, C., Renne, J. L., & Bertolini, L. (2009). Transit-oriented development. Making it happen. *Journal of Transport Geography, 17*(6), 312.
Dalhatu, S. (2010). *Essays on local government administration. Fostering better service delivery, record keeping, accountability and empowerment at the local government*. Benchmark Publishers, 2006. Cited in K. Asaju, Local government autonomy in Nigeria: Politics and Challenges of the 1999 Constitution. *International Journal of Advanced Legal Studies and Governance, 1*(1), 99.
Daramola, A., & Ibem, O. (2010). Urban environmental problems in Nigeria: Implications for sustainable development. *Journal of Sustainable Development in Africa, 2*(1), 124–145.
De Soto, H. (2001). *The mystery of capital: Why capitalism triumphs in the west and fails everywhere else*. Black Swan. https://books.google.com/books?id=8qUseWwx3-IC
Diana, M., & Sheriden, B. (2018). Co-production in cities: Providing services, empowering communities, changing relationships. *Environment and Urbanisation, Brief, 38*.
Galelo, A., Ribeiro, A., & Martinez, L. M. (2014). Measuring and evaluating the impacts of transit-oriented development measures—Searching for evidence of transit-oriented development characteristics in Abuja train line. *Procedia—Social and Behavioural Sciences, 111*, 899–908.
ILO. (2002). *Decent work and the informal economy* (International labour conference, 90th session, report VI). Author.
Jean-Paul, R. (2020, June 17). *The geography of transport systems. Transportation and the urban form*. https://transportgeography.org/?page_id=4609
Jiboye, A. D. (2011). Urbanisation challenges and housing delivery in Nigeria: The need for an effective policy framework for sustainable development. *International Review of Social Science and Humanities, 2*, 176–185.
Jinadu, A. (2004). *Understanding the basics of housing*. Federal University of Technology Minna.
Knowles, R. D. (2012). Transit-oriented development in Copenhagen, Denmark: From the finger plan to Orestad. *Journal of Transport Geography, 22*, 251–261.
Kumar, A. (2011). *Understanding the converging role of motorcycle in Africa cities. A political economy perspectives* (*SSATP discussion paper* no. 13, urban transport services). International Bank for Reconstruction and Development/World Bank.
Loo, B. P. Y., Chen, C., & Chan, H. (2010). Rail-based transit-oriented development: Lessons from new York City and Hong Kong. *Landscape and Urban Planning, 97*(3), 202–212.
McFarlane, C. (2012). Rethinking informality: Politics, crisis and the city. *Planning Theory & Practice, 13*, 89–108.
National Urban Development Policy. (2012, June 20). *The revised National Urban Development Policy*. Federal Ministry of Land, Housing and Urban Development Mabushi, Abuja.
Nicola, B., Melanie, L., & Diana, M. (2020). Urban informality as a site of critical analysis. *The Journal of Development Studies, 56*(2), 223–238. https://doi.org/10.1080/00220388.2019.1577384
O'Neil, P. (2020). *Urban transport in developing cities*. World Bank.
Obeng-Odoom, F. (2016). *Reconstructing urban economics: Towards a political economy of the built environment* (pp. 1–286). Zed Books.
Ogunleye, B. (2005). Environmental degradation control for sustainable urban growth in Nigeria. In W. Fadare, et al. (edns). Proceedings of the conference on globalisation, culture and the Nigerian built environment. Obafemi Owolowo University, .
Papa, E., & Bertolini, L. (2015). Accessibility and transit-oriented development in European metropolitan areas. *Journal of Transport Geography, 47*, 70–83.
Pojani, D., & Stead, D. (2015). Sustainable urban transport in the developing world: Beyond megacities. *Sustainability, 7*, 7784.

Posada, F. (2018). *Unique challenges and solutions to sustainable transport issues in Africa*. The International Council on Clean Transportation.

Priscila, P., & Lara, C. (2015). *How transit-oriented development benefits local economies*. Portuguese on The City Fix Brasil.

Queensland Government. (2010). *Transit-oriented development: Guide to community diversity*. The State of Queensland. Department of Infrastructure and Planning.

Ranganathan, M. (2014). 'Mafias' in the waterscape: Urban informality and everyday public authority in Bangalore. *Water Alternatives, 7*, 89–105.

Roy, A. (2005). Urban informality: Toward an epistemology of planning. *Journal of the American Planning Association, 71*, 147–158.

Roy, A. (2009). Why India cannot plan its cities: Informality, insurgence and the idiom of urbanisation. *Planning Theory, 8*, 76–87.

Santana, D. (2017). *What is diversity and how I define it in the social context*. EmbracingDiversity. http://embracingdiversity.us/what-is-diversity- define-social-diversity/

Simon, D., Palmer, H., Smit, W., & Valencia, S. (2018). *The challenges of transdisciplinary knowledge production: From unilocal to comparative research*. Journal of Environment and Urbanisation (Vol. 30, p. 481).

Simone, A. M. (2011). The ineligible majority: Urbanising the post-colony in Africa and Southeast Asia. *Geoforum, 42*, 266–270.

Soja, E. (2009). The city and spatial justice. [«La ville et la justice spatiale», traduction: Sophie Didier, FrédéricDufaux], justice spatiale |spatial justice | n° 01 septembre | september 2009 (http://www.jssj.org).

Solanke, M. (2013). Challenges of urban transportation in Nigeria. *International Journal of Development and Sustainability*, 891–901.

Tan, Y., Lawrence, F., & Eddo, C. (2008). Challenges to urban transport sustainability and smart transport in tourist city: The Gold Coast Australia. *The Open Transportation Journal*, 29–46.

TELL. (2014). *The change agents: Unlocking Abuja's potential: Special publication*.

UN-Habitat. (2015). *HABITAT III issue papers 22—Informal settlements*. United Nations.

United Nations. (2020, June 16). *United Nations publications. Revision of world urbanisation prospects*. https://www.un.org/development/desa/publications/2018-revision-of-world-urbanization-prospects.html.

Varley, A. (2013). Post-colonialising informality? *Environment and Planning D, 31*, 4–22.

Watson, V. (2014). Co-production and collaboration in planning—The difference. *Planning Theory & Practice, 15*, 62–76.

World Health Organisation. (1992). *Report of the panel on urbanisation*. World Health Organisation Commission on Health and Environment.

Yakubu, D. (2006). *An appraisal of the proposal for public transport initiative (ABUTRANS)*. Department of Urban and Regional Planning Faculty of Environmental Design Ahmadu Bello University, Zaria.

Chapter 13
Urban Infrastructure and Transit-Oriented Development in Osogbo, Nigeria

Abel Olaleye, Timothy Tunde Oladokun, Olorunfemi Ayodeji Olojede, Gbemiga Bolade Faniran, and Esther Ilori Ebunoluwa

13.1 Introduction and General Overview of Osogbo

This study provides the profile of Osogbo, the capital city of Osun State in the southwestern geopolitical zone of Nigeria. This is with respect to its broad infrastructure challenges and specifically its transportation infrastructure. To achieve this, desk review/literature search, field survey including observation and key informant interviews were conducted.

Osogbo lies approximately between latitude 7°44′45″ and 7°48′45″ north of the equator and longitude 4°30′45″ and 4°35′45″ East of Greenwich Meridian (Fig. 13.1). The city is approximately 232 km to Lagos and 510 km to Abuja. It is 206 km to Abeokuta, 117 km to Ado-Ekiti, 108 km to Akure and 95 km to Ibadan, the respective capital cities of Ogun, Ekiti, Ondo and Oyo States in Southwestern Nigeria. Consequent on the carving out of Osun State from the old Oyo State, Osogbo became a state capital on the 27th of August 1991. The city engulfs predominantly Osogbo and Olorunda Local Government Areas (LGAs) and parts of

A. Olaleye (✉)
Department of Estate Management, Obafemi Awolowo University, Ile-Ife, Nigeria

Department of Finance & Investment Management, University of Johannesburg, Johannesburg, South Africa
e-mail: aolaleye@oauife.edu.ng

T. T. Oladokun
Department of Estate Management, Obafemi Awolowo University, Ile-Ife, Nigeria

O. A. Olojede · G. B. Faniran
Department of Urban and Regional Planning, Obafemi Awolowo University, Ile-Ife, Nigeria

E. I. Ebunoluwa
Department of Quantity Survey, Obafemi Awolowo University, Ile-Ife, Nigeria

T. Nubi et al. (eds.), *Transit Oriented Development in West African Cities*, Urban Perspectives from the Global South,
https://doi.org/10.1007/978-3-031-58726-9_13

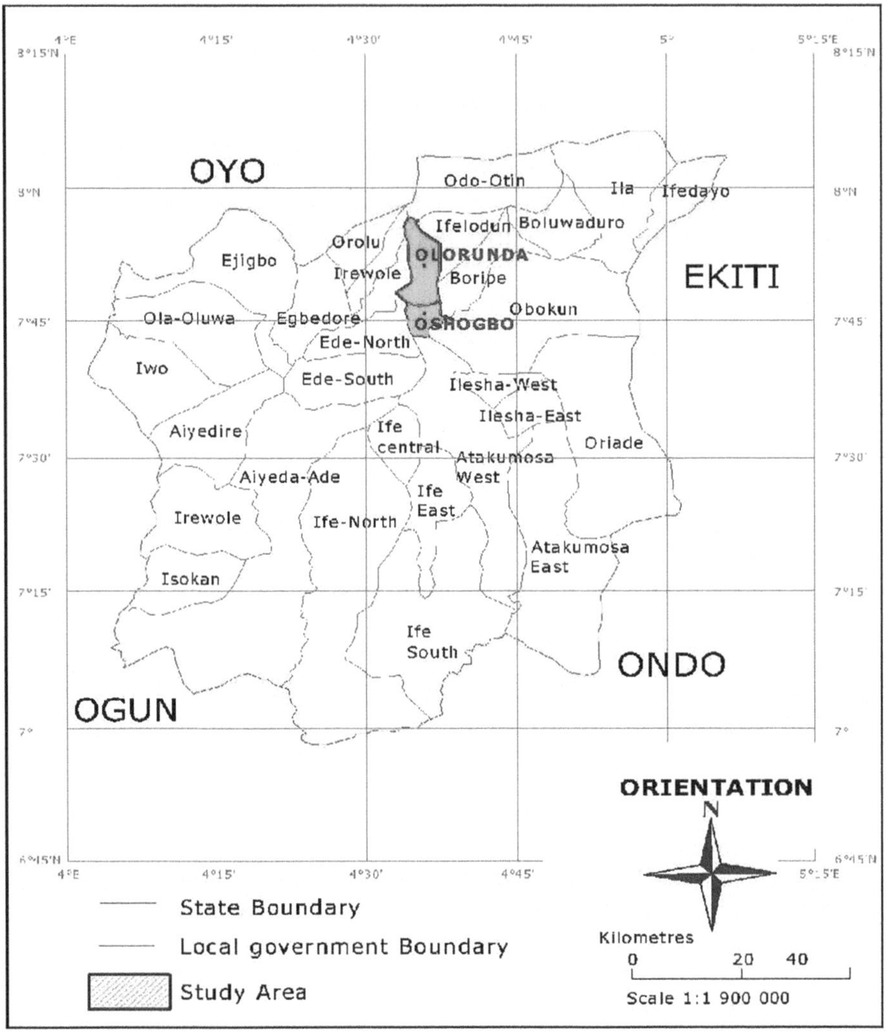

Fig. 13.1 Osogbo City area in the context of Osun State, Nigeria. (Source: National Space Research and Development Agency (2020))

other seven LGAs of Osun State (Fig. 13.1); hence, its extent is often defined as the Osun State Capital Territory (UN-HABITAT, 2014).

Osogbo is often referred to as *Ilu Aro* (a town famous for tying and dyeing). It is famous for its rich arts and cultural heritage, especially the Osun Osogbo Festival with the Sacred Osun Grove, recognised by UNESCO as a World Heritage Site (Gore, 2012; Peter, 2013; Adewale & Adekola, 2019). Osogbo was home to art workshops initiated by Ulli Beier in the early 1960s (Agbola, 1992; Peter, 2011).

Artworks that started then are now available locally and in many parts of the world. Osogbo hosts the moribund Nigeria Machine Tools and Osogbo Steel Rolling Company. It is home to both Osun State University and Fountain University. The city also boasts of natural evergreen forest vegetation of hardwood timbers.

With the arrival of the colonial rulers and the railway line in 1907, Osogbo became a commercial centre. According to the 1963 census, there were 152,424 people in Osogbo. This increased to 460,899 by 1991 and 567,997 in 2006 (National Population Commission, 2010; National Bureau of Statistics [NBS], 2013). At an annual growth rate of 3.2%, the population of Osogbo was projected to be 648,333 in 2010 and 764,925 in 2015 (NBS, 2013, 2016). This is estimated to be 1,077,458 by 2020 and 2,033,000 by 2035 (UN- HABITAT, 2014). These figures are computed on the basis of Osogbo and Olorunda LGAs' total population and 30% of the population of the adjoining seven LGAs.

Apart from farming, the people of Osogbo are known for their commercial activities in pottery and gold-smithing. The people are in the civil service, informal sector and other enterprises such as education and small-scale industry. As the administrative and political headquarters of Osun State, and at just 32 km, 48 km, 46 km, 48 km and 46 km to Ilesha, Ile-Ife, Iwo, Ikire and Ila-.

Orangun respectively, Osogbo is easily connected and accessible by road from other parts of the state. Neighbourhoods within the city are also connected by a network of roads. The city connects with the northern region of Nigeria through the Lagos-Kano rail line. The city's economic and connectivity advantages have in recent time increased the infrastructure challenges, especially that of transportation.

13.2 Overview of Broad Infrastructure Challenges in Osogbo

Cities are complex socio-technical systems governed by a web of ecological, cultural, economic and political relationships (Frank & Engelke, 2005; Muller, 2011; Williams et al., 2010). As places of opportunities and benefits, cities are also faced with a plethora of problems. These range from challenges posed by the phenomenon of urbanisation to those heightened by the actions of city dwellers. The rate at which cities all over the world are growing is alarming. The situation is worrisome in developing countries, where there are no commensurate economic and socio-physical growths. Many cities in developing countries lag behind in the provision of sustainable, resilient and efficient infrastructure fundamental to daily life and central to individuals, groups and households' economic production. No doubt, improved urban infrastructure services enhance welfare and foster economic growth. Contrariwise, inadequate provision of infrastructure impinges on quality of life.

As obtainable in some other cities in developing countries and Nigeria specifically, Osogbo is faced with many challenges. Broadly, there are the demographic, socioeconomic, technological, environmental and financial challenges. These are evident in the state of primary infrastructure such as housing, energy, communication, water and sanitation and transportation. Secondary infrastructure, including education and healthcare, are not spared. For example, neighbourhoods within the core/traditional areas of Osogbo are characterised by old and dilapidated buildings, multi-habitable and overcrowded dwellings and extended family and traditional compounds. Houses in these areas are inaccessible by road and lack organised areas for parks and recreation. The existing infrastructure that supports affordable housing for different groups of citizens is inadequate. According to UN-HABITAT (2014), these areas are described as informal settlements with critical slum features. The few planned areas in Osogbo have suffered deterioration owing to inadequate infrastructure. The general housing environments of neighbourhoods in the city are in deplorable conditions.

Osogbo, though a state capital city in Nigeria, does not have sufficient infrastructure that promotes an integrated provision of water, sanitation and waste management (both liquid and solid) in neighbourhoods. Cursory observation revealed that many households both within the central area and outskirts of the city are not connected to the public water scheme. There is overreliance on water sources provided by individual households. Households' access to improved water and sanitation systems, including solid waste management, is poor. Residents dump waste along the road, on river courses and in uncompleted or abandoned buildings, among other inappropriate places. The nature of urban solid waste management is crude despite the government's acclaimed investments (Afon & Faniran, 2011). This notwithstanding, Osogbo is one of the cities in Southwest Nigeria that enjoy adequate and steady electricity supply. Osogbo hosts the National Control Centre responsible for all Transmission Company of Nigeria grid operations, monitoring and control within the electricity transmission network.

At the heart of any city supply chain, and how it works, is transportation. The Osogbo transportation system is not spared from various challenges, however. Thus, the next section is devoted to the discussion of transportation infrastructure challenges in Osogbo.

13.3 Overview of Transportation Infrastructure Challenges in Osogbo

The growth of cities all over the world, bearing in mind the essentiality of connecting people, goods and services, is highly connected to efficient and effective transportation systems (Adedotun, 2015). When transportation is hindered in an area, development will be retarded; linkages between spatially separated facilities, social contact and access to employments will also suffer. With deficits in transportation infrastructure, a plethora of problems such as congestion, parking demand and auto-crash abound. This is evidently the situation in many African cities, Osogbo included.

The road is the major mode of transportation in Osogbo (Fig. 13.2). As the city grew from a small commercial centre along a number of radial exit roads, many other intra-city road networks (West Bypass, East Bypass, Alekuwodo Road, Odi-Olowo Road, and Oke-Fia-Dada Estate Road) were also constructed and opened for

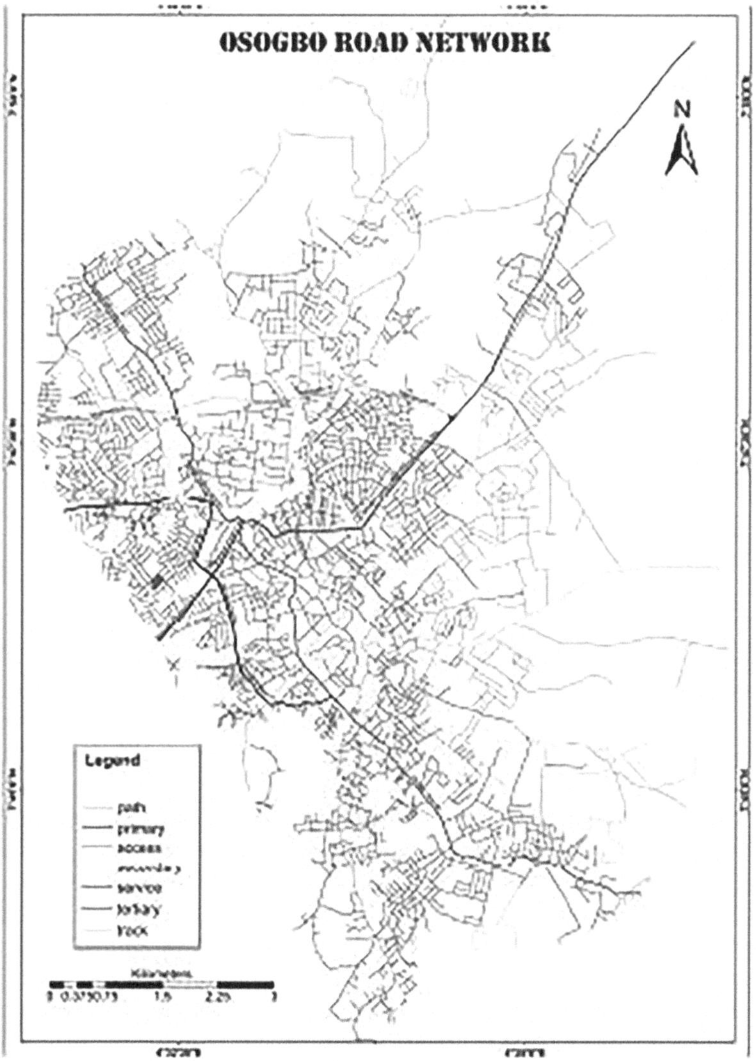

Fig. 13.2 Road network of Osogbo. (Source: National Space Research and Development Agency (2020))

Fig. 13.3 Condition of an intra-city road in Osogbo. (Source: Authors' Field Survey (2020))

use. This affected the city's growth, form and land-use pattern. Most of the roads within Osogbo as well as those connecting neighbouring settlements are in a deplorable condition (Fig. 13.3). For an average of 1 km stretch of a tarred road, there are about 30 potholes with an average depth of 3.8 cm (Adedotun, 2015). Many roads lack adequate and well-maintained drainage systems, walkway, streetlight, road furniture and appropriate calming devices. Indiscriminate blocking of roads within the core residential areas for social functions also contributes immensely to transportation challenges in the city. As many of these roads are single carriageways, the demand for road transport in the city is increasing tremendously.

The Osogbo transportation system largely features poorly regulated informal modes or paratransit. Commercial motorcycles (*okada*) and micro-minibuses (*korope*) (see Fig. 13.4) used for intra-city movement are fast becoming awkward in Osogbo. This is coupled with road users and managers' culture of indiscipline, aggressive behaviour, Highway Code illiteracy, ineffective traffic control, street markets/hawking and indiscriminate installation of traffic-calming devices. These and many other factors contribute to traffic congestion, safety and security issues, road accidents and abandoned wreckage in the city.

Shared mobility systems driven by innovations in technology, such as car-sharing, bike-sharing and ride-sourcing services, are not available in Osogbo. Currently, there is no formal transport or mass transit system in the city. There are

Fig. 13.4 Typical micro-minibuses (korope) used in Osogbo. (Source: Authors' Field Survey (2020))

no well-defined bus stops and terminals in the city either. The existing bus stations in the city, popularly but incorrectly referred to as 'motor parks' or 'garages', are dedicated to inter-city travels. Transportation infrastructure challenges in Osogbo are also associated with on-street parking, loading and unloading of goods (Fig. 13.5). Moreover, vulnerable road users are not adequately considered in the transportation system of Osogbo.

Although the Lagos-Ilorin rail line passes through Osogbo, there are no links to serve the city and its numerous hinterlands. Though efforts have been made by the State Government to revamp an existing aerodrome in Ido-Osun within the Osun State Capital Territory and upgrade it to the status of an airport, these have only translated into the construction of a perimeter fence of the site.

13.4 Governance Framework for Osogbo Transport Sector

Transport governance consists in the set of institutions that provide actors with agency and funding as well as the structure for their actions towards a transport system, with the expected result to attain specific public values (United Nations Economic Commission for Latin America and the Caribbean [UNECLAC], 2018). The governance framework for the Osogbo transport sector is complicated. There is pluralism in the overall governance of the city, as several levels of administration coexist, leading to conflict of interests (Olojede & Daramola, 2015). Besides, Osogbo is technically a mono-modal city, as roads are the only functional transport mode. The moribund railway does not provide commuting services for the

Fig. 13.5 Loading and offloading of goods on the road. (Source: Authors' Field Survey (2020))

residents, and the only river is non-navigable (Olojede, 2019). Furthermore, there are no visible commitments to futuristic travel modes. Consequently, road is the only subsector on which the transport sector governance framework for the city can be reasonably discussed.

The governance framework for the Osogbo transport sector is expected to be provided for by a relevant transport policy. However, there currently exists no transport policy in any city in Osun State, Osogbo included. It is only assumed that the national transport policy could be adopted. Meanwhile, the modus operandi for the adoption and the extent to which it could be adopted have yet to be clearly spelt out. Impliedly, whatever that officially obtains in the governance of the transport sector of the city is at best ad hoc. The Osun State Government only provides muddle-through measures of transport governance as the prevailing situation demands. This it does through such stakeholders as the Osun State Ministry of Works and Transportation, the Osun State Ministry of Regional Planning, the Director of Planning and the Director of Engineering. These stakeholders' roles range from road construction, rehabilitation, maintenance and project funding as well as other roles assigned by the executive arm of the government or the ruling political party.

Another recognised stakeholder of the Osogbo transport sector is the National Union of Road Transport Workers (NURTW). Headquartered in Osogbo, the union's state chapter constitutes a formidable body that regulates the activities of all informal commercial transport services in the state. There are no clear-cut statutory roles assigned to NURTW in the state. However, nationwide, NURTW serves the interests of its members by calling for collective obtaining and pushing for their social stability as defined in its constitution (Agbiboa, 2020). The union has over the years proved to be a formidable force to reckon within the regulation of the activities of informal commercial transport service providers in the state; indeed it owns automobiles (buses) used for the movement of people, goods and services.

Traffic flow in Osogbo is both manually and automatically managed. Traffic lights are installed at Ola-Iya Junction, a major intersection in the city. Traffic in other intersections throughout the city is managed by traffic wardens/police and casual traffic marshal/officers in the employ of the Osun State Government. Occasionally, officers of the Federal Road Service Corps (FRSC) also feature.

13.5 Mobility and Access in Osogbo

Accessibility (or just access) refers to the ability to reach goods, services, activities and destinations (which together are called opportunities). In other words, it refers to the path through which people reach opportunities, which is the ultimate goal of most transport activity. Thus, access can be defined as the potential for interaction and exchange (Engwicht, 1993; Hansen, 1959; Litman, 2020). Mobility, on the other hand, refers to the ease of physical movement, and therefore the quality of travel modes, which, according to Litman (2020), are measured in terms of availability, frequency, speed and comfort, among other indicators. Therefore, all things being equal, increased mobility increases accessibility: the more and faster people can travel, the more destinations they can reach.

Generally, it can be argued that whereas access is measured in spatial terms, mobility is measured in temporal terms. Thus, while land-use planners focus on geographic accessibility (distance between activities), transport planners focus on mobility (or the ease of connecting these activities). However, access and mobility or mobility and access are sometimes used as a single entity, as in this discourse, to describe both the opportunities provided by transport and the ease with which these opportunities are accessed. Impliedly therefore, mobility and accessibility are considered the 'yin and yang' of transportation.

Given the transport situation in Osogbo, mobility and access in the city's urban transport system is a far cry from the ideal. This is because the city's transport system does not make any provision for inclusiveness, especially in the interest of vulnerable travellers. For instance, there is no transport infrastructure investment that caters for the special needs of women, the aged or physically challenged travellers. Few buses made available in time past for the use of schoolchildren are now parked off-road (Fig. 13.6). Meanwhile, there is no known policy in place that addresses this lack of inclusiveness. Similarly, accessibility to residential communities is not consciously provided for in the planning, design and investment of transport infrastructure in the city.

Fig. 13.6 Buses meant for schoolchildren but are now parked off-road. (Source: Authors' Field Survey (2020))

There are serious issues with access and mobility that manifest in huge modal capability challenges in the city. These issues are exacerbated by challenges brought about by such factors as bad and pothole-ridden narrow roads, poor maintenance culture, connectivity issues posed by deterministic topographically restricted geometric designs, poor street lighting, inadequate active transport infrastructure, indiscipline, aggressive behaviour and road rage. Others include Highway Code illiteracy among road users, traffic congestion, safety and security issues, crashes and abandoned wreckage, motor cyclists ubiquity, ineffective traffic control and management, undisciplined law enforcers, street markets and street hawking, indiscriminate and erratic installation of traffic-calming devices (such as bumps, zebra crossing, speed limits, etc.) and natural factors such as the elements of weather and climate, among other issues.

Consequently, even the restricted access to opportunities provided by the inadequate transport infrastructure in the city is further denigrated by sheer immobility. Indeed, temporal potentials are destroyed by spatial realities. The ramifications of this translate to mobility and access issues in the Osogbo transport sector.

13.6 Transportation Investment and Funding Mechanism

The Government of Osun State and the two main local governments in charge of Osogbo Township (Osogbo and Olorunda Local Government Areas) and oftentimes the other seven Local Governments (LGs) are the main funders of transport infrastructure and investments in Osogbo. They provide the financial resources to execute transport infrastructure projects. The state government draws its financial resources from allocations from the federation account and internally generated revenue (IGR). The local governments also source revenue mainly from the federation account, state government and IGR. Additional financial resources, tied to developmental projects, also come from donor agencies such as the World Bank to both the state and local governments. Moreover, at times, external credit facilities in the form of long-term loans are taken, especially by the state government. Usually, road infrastructure projects are given out to private contractors for execution, while the State Ministry of Works and Transportation supervise such projects. Occasionally, direct labour is also employed where both the financial resources and expertise are provided by the respective government.

Ordinarily, the state and local governments fund Trunk B and Trunk C road infrastructure projects respectively, while the federal government is responsible for the funding of all federal roads, otherwise known as Trunk A roads in the state, through the Federal Ministry of Works and Housing. Whenever the state government takes it upon itself to invest in federal road projects, owing to long-time neglect by the federal government, the state government makes claims for requisition from the federal government and recovers the money spent on such investment. Furthermore, whenever Osogbo or other cities and settlements in Osun State benefit from the largesse of international organisations and donor agencies such as the World Bank, the state government usually partners with the donor agencies by determining the communities and areas that would be the beneficiaries of the assisted projects.

There is absolutely nothing like road pricing in Osogbo, as no toll booth is in existence. This is partly because the organised private sector is not in any way involved in the transport sector funding and/or investments in the city. The only items of road infrastructure for which road users are charged are on-street and off-street parking facilities around Station Road, a core area of the city, as well as few private parking facilities and other unbundled parking facilities on some private estates. Moreover, once in a while, a team of emission control workers *invade* the roads and demand that vehicle owners be in possession of emission stickers which they also sell. However, some vehicle owners shun and dismiss them as there is yet no known legislation that backs their activity. Similarly, there is no known transport infrastructure funding and/or management that are executed through the collaboration of the government with the public sector. Thus, currently, nothing is known of any form of public-private partnership (PPP) arrangement(s), as a transportation investment and funding mechanism, for transport infrastructure in Osogbo.

13.7 Actions Taken in the Last 20 Years and Their Impacts

A number of actions have been taken within the last 20 years, and several developments have taken place with a view to improving the state of transportation infrastructure in Osogbo. However, these actions have focused mainly on road infrastructure as the road is the main mode of transportation in the city. For example, owing to the ever-increasing human population in the city and the attendant growing travel demand leading to traffic congestion, a number of new intra-city roads have been constructed, while several existing ones have also been dualised. The dualised roads include Alekuwodo Road, Odi-Olowo Road, Oke-Fia-Dada Estate Road and Ikirun Road. The intra-city roads that have been constructed or rehabilitated and opened up for use include Arubuola Junction-Omigade Road and Igbona-Oke Onitea, among others. In like manner, some roads connecting Osogbo with other cities and neighbouring states, particularly on the Oyo and Kwara States' axes are either being reconstructed or dualised. A number of new roads, notably the East and West Bypasses, and a few other access roads have been added to the existing network over the past two decades.

The state government at one time within the last 20 years provided buses for schoolchildren. However, these buses have since been parked off-road and are currently of no benefit to the initial users and beneficiaries. The dualisation and reconstruction of some of the major intra-city roads, no doubt, have brought about significant enhancements to on-street parking facilities and marked improvement in accessibility and mobility in the city. In a similar vein, traffic flow on some of these newly constructed roads has been automated, while the use of human traffic wardens has been minimised leading to improvement in traffic control and management. There are notable improvements also in the provision and conditions of other road infrastructure such as road drainage system, road furniture and road signs.

Furthermore, it is important to state that the effort of the state government at revamping the existing aerodrome and upgrading it to the status of an airport is, in its own right, a commendable action even if it has not gone beyond the construction of a perimeter fence. However, the investment hitherto can never have any impact until the facility becomes operational. With fund availability, prioritisation and conscientious political will, the dream of making the facility operational can come true someday. When in full operation, the facility will give a significant boost to the socioeconomic development of Osogbo and the entire state.

Generally, it is arguable that the efforts of the government at improving road infrastructure and transport facilities in Osogbo have not yielded much significant success. However, this might be as a result of the constantly evolving trends in the various forms of development that keep affecting the city's growth, form and land-use patterns. In other words, government actions and investments may not have been commensurate with the geometric population growth rate being experienced in the city. Meanwhile, there are still no well-defined bus stops and off-street parking facilities to reduce the impact of the growing population. Moreover, efforts at improving the drainage system have been impeded by recalcitrant elements that keep blocking it with their indiscriminate dumping of refuse into it.

13.8 Osogbo Transportation Infrastructure Gaps

In discussing the transportation infrastructure gaps in Osogbo, the study employed a combination of international benchmarking indicators and comparators that recognise the peculiar characteristics of the city. These include the historical levels of infrastructure provision, coverage and quality; countries with similar socioeconomic characteristics and policy targets and standards (Organisation for Economic Cooperation and Development [OECD], 2017). However, since Osogbo is functionally a mono-modal city (Olojede, 2019), the assessment also employs the adequacy of the components of a comprehensive road transport system. Other related common indicators such as the share of paved roads, road safety and security, road quality, modal split and environmental performance indicators are also used.

The history of road infrastructure provision, coverage and quality in Osogbo is not as should be expected of a city of its status. Granted, a number of new roads (notably the East and West Bypasses and a few other access roads) have been added to the existing network over the past two decades; however, these additions have not been commensurate with the geometric population growth rate being experienced in the city. According to Adedotun et al. (2015), there is an excess of demand over the supply of road infrastructure in Osogbo with about 77% of the roads having potholes, about 88% without walkways and about 92% without road signs. Thus, road infrastructure in Osogbo is suboptimal even when juxtaposed with cities in comparable low-income countries with similar socioeconomic characteristics. Meanwhile, policy targets and standards are largely either non-existent or not communicated. Policymakers only seem to impulsively and arbitrarily take decisions. This situation has been aptly described as planning transport infrastructure investment without data (Fadare & Ogunsanya, 1989).

Osogbo does not fare any better when components of a comprehensive road transport system are used. There is much to be fixed on its highway system, especially in terms of network description, fundamental structures, pavement, drainage system and traffic monitoring and control systems, among other related subcomponents. Moreover, although on-street parking/loading facilities exist, the supply falls short of the demand. Meanwhile, off-street parking facilities are grossly numerically inadequate. Public transport is completely missing, and there is no known transit-oriented development planning attempt. Facilities for cyclists and pedestrians are conspicuously absent, and there is no provision for the physically impaired persons' mobility. Thus, Osogbo fares badly on the scale of urban walkability. In addition, the majority of the roads lack standard streetlights, road furniture and traffic-calming devices. According to Adedotun (2015), most of the roads in Osogbo do not have a good drainage system. This results in storm water eroding the roadways and the emergence of potholes of various sizes, such that for an average of 1 km of road pavement, there are about 30 potholes with an average depth of 3.8 cm. Most of the roads are narrow and in a state of utter disrepair and require very urgent attention.

The share of roads that are paved is often used as an indicator of road infrastructure availability and quality. This is because paved roads can provide faster, safer and less vehicle-damaging links than unpaved roads, especially during extreme weather conditions like heavy rainfall such as is common in Osogbo. Many access roads in Osogbo are either unpaved or have poor road surface quality. Meanwhile, there are no official data on the status of maintenance, road support services, road connectivity to key centres or road safety standards. This might be due to the governance pluralism in the city which often brings about severe infrastructure deficiencies (Daramola & Olojede, 2011). In Osogbo, at least four sources of ownership and control conflicts are possible as there is coexisting of Trunks A, B and C roads owned and maintained by the federal, state and the two local governments respectively. This pluralism in road infrastructure provision is in itself a good example of a bad model of urban road infrastructure provision.

While not an urban road infrastructure indicator per se, road safety and security trends can affect the quality and reliability of the urban road infrastructure and network. Besides, road crashes and transport crimes constitute a significant financial burden on the economy. Empirical studies on the safety and security indicators of intra-urban transport in Osogbo have shown that the Osogbo urban road infrastructure and system are largely unsafe and insecure (Adedotun et al., 2015; Adedotun, 2015; Olojede, 2019).

Another important indicator of efficiency in urban road infrastructure is its relative elasticity and/or inclusiveness, especially in terms of choice flexibility.

With no mass transit in place, and with the notoriety of poorly coordinated commercial motorcycles, the city's mono-modal system is fraught with imperfections. Residents, especially captive riders, are restricted in their modal choice as they can only choose between *korope* and *okada*. This modal choice inelasticity and the collateral nuisance are a true reflection of another embarrassing urban road infrastructure gap in the city.

Environmental and sustainability concerns constitute another important indicator in the assessment of urban transport infrastructure. The performance of the urban road transport infrastructure and network encompasses its ability to contain and/or minimise the negative environmental externalities that are common by-products of traditional urban transport activities. In Osogbo, only lip service is paid to transport-related emission control: there is neither enabling legislation nor physical infrastructure for its enforcement. Thus, even the existing emission control mechanism in the city constitutes both an illegal and impracticable activity.

Furthermore, the best global practices in urban transport infrastructure provision provide for public-private partnership arrangements in both the financing and management of the infrastructure. It has been widely acknowledged that the government alone cannot be expected to provide the huge human and financial resources to meet transport infrastructure gaps in cities. Consequently, cities in developed countries have demonstrated the workability of this arrangement with significant success. However, there is no such arrangement in Osogbo. This constitutes another factor that exacerbates transport infrastructure gaps in the city.

13.9 Recommendations

The need for an efficient and effective transportation system as a means by which cities such as Osogbo can remain relevant as contemporary cities cannot be overemphasised. To be able to function efficiently and overcome the problems associated with rapid urbanisation, cities require the provision of sustainable, resilient and efficient transportation infrastructure and facilities. However, with respect to Osogbo, a city in Southwestern Nigeria and the capital city of Osun State, it has been noted that it is functionally a mono-modal city with the roads as the only operative mode of transport. Despite this, the city does not fare any better when components of a comprehensive road transport system are considered. As road is the only mode on which transport sector governance framework for Osogbo has been discussed, it is important that access be made open to serve most, if not all, of the main neighbourhoods within the core/traditional areas of the city. This will open up the area for private investments. As such, the neighbourhood, which is hitherto characterised by old and dilapidated houses, multi-habitable and overcrowded dwellings as well as extended family and traditional compounds, will be available for redevelopment and modernisation. This will increase investment activities in the area, improve the values of the land and buildings and subsequently enhance income generation in the form of land taxes for the government.

With daily growth in the population of Osogbo and its attendant rising demands for residential and commercial land uses, the existing dilapidated roads connecting Osogbo with other cities and neighbouring states need to be urgently rehabilitated and put in a better condition. Conscientious efforts are required to fill up all potholes on the roads. This should be complemented with the provision of basic infrastructure such as adequate drainage system, walkways, streetlights, appropriate calming devices and other road furniture. It will also be necessary that many of the roads that are single carriageways be upgraded to dual carriageways in order for them to meet the ever-growing transportation needs of the city. It is also imperative that sufficient infrastructure for on-street parking, offloading and loading of goods be provided, at intervals and in appropriate locations. In addition, the provision of well-defined bus stops and terminals across the city is non-negotiable. Also, for the impact of the Lagos-Ilorin rail line that passes through Osogbo to be felt, there is the need for links that will serve the city and its numerous hinterlands. At the same time, there is the need for the concerted efforts of both the public and the organised private sectors to resuscitate and upgrade the existing aerodrome in the state capital territory to the status of an airport.

As earlier observed, inasmuch as the tools for transport sector governance framework consist in transport policies, the governance framework for the Osogbo transport sector is expected to be provided for by a relevant transport policy. There is an urgent need for a transport policy and master plan that will guide transportation planning in the city. This is to ensure coordinated efforts in the quest for better transit-oriented development. While the state transport policy is awaited, the state is encouraged to adopt effectively the national policy on transportation. This should be documented and made available for ease of implementation and use by stakeholders. This policy should define the specific roles or functions expected to be performed in the area of transport governance ranging from road construction, road rehabilitation, road maintenance and road project funding, among others, by each of the stakeholders, namely, the Osun State Ministry of Works and Transportation, the Osun State Ministry of Urban Development and Planning, the Directorate of Planning and Engineering, Osun State Capital Territory Development Authority and the respective local government councils that constitute Osogbo—the Osun State Capital Territory.

It will also be necessary that the automated traffic flow management in Osogbo be extended to all major interchanges, intersections, nodes, junctions and roundabouts in the city. In addition, in newly constructed and/or rehabilitated roads in Osogbo, road pricing, through toll collection, should be a policy consideration. This will guarantee continuous funding for routine maintenance and future rehabilitation of road infrastructure in the city. In another dimension, appropriate and adequate public enlightenment should be promoted by the government. This will help the government to win the confidence of citizens, earn their buy-in and involve them in transport management. With proper enlightenment, mobility challenges/issues such as badly pothole-ridden narrow roads, poor maintenance culture and connectivity issues, among others, will be minimised.

With the consensus, both in theory and practice, the government alone cannot and should not be expected to provide the huge human and financial resources to meet the wide transport infrastructure gaps in cities; there is the need to involve other stakeholders and adopt global best practices. For efficient transport governance, the government should embrace the informal transport sector, especially the union of transporter service providers (NURTW) and the Road Transport Employers' Association of Nigeria (RTEAN). It is recommended that clear-cut statutory roles be assigned to these bodies by the relevant agencies of the government in the state. This will help the agencies to be able to effectively regulate their activities. Similarly, the provision of urban transport infrastructure through PPP arrangements, in both financing and management, is germane as a way out of the transport infrastructural gap in the city. This model is recognised globally and adopted in many known developed and developing nations' cities.

13.10 Lessons for Other African Cities

African countries are encouraged to establish visible commitments to the development of futuristic travel infrastructure. Doing this will require widening the scope of the transport sector beyond the road subsector and the establishment of a governance framework for each city as it is relevant. African leaders will be expected to explore the opportunities of rail and water transport modes to provide alternative commuting services for the denizens. With the visible absence of a state transport policy as a tool for the transport sector governance and coordination, governments in African countries should enable a governance framework that will be applicable to every main city in all the states of their respective federations. State transport policies could be adapted from the national policies to reflect state and local peculiarities. This will help to arrest possible future transportation crises.

In addition, and for effective transit-oriented development, cities in African countries should incorporate the roles of relevant stakeholders in their transport policies. With the roles of each stakeholder well spelt out, responsibilities such as road construction, road rehabilitation, road maintenance and road project funding, among other roles, will be easily and promptly attended to. Also, the governance structure should be developed to accommodate the informal sector, since many individuals and organisations are already key players in the implementation and realisation of better transit-oriented development in African cities. As the city population grows, the management of the traffic flow in major cities should be automatically enhanced. This will require the provision of solar-powered traffic lights that could guarantee efficient management.

Appropriate and adequate enlightenment will help African governments to win the confidence of their citizens and carry them along in transport control and management. With proper enlightenment, mobility challenges/issues, such as bad pothole-ridden narrow roads, poor maintenance culture, connectivity issues, destruction of street lighting, the culture of indiscipline, aggressive behaviour/road rage and Highway Code illiteracy among road users, could be minimised. Other issues such as the challenges of motorcyclists' ubiquity, ineffective traffic control and management, undisciplined law enforcement agents, street markets/hawking and indiscriminate installation of traffic-calming devices could be discussed among relevant stakeholders with appropriate solutions proffered.

The inefficient funding of transport infrastructure, which is notable in many African countries, could be overcome if the governments embrace PPP arrangements for financing and delivery. This model has been adjudged effective in other parts of the world. A contemporary alternative financing arrangement is for African countries to partner with international organisations and donor agencies, such as the World Bank, in terms of road construction and rehabilitation. Effective funding will assist governments to provide solutions to prominent transportation issues which have hitherto hampered city growth, development and mobility. Where necessary and where it might be socially acceptable, African countries should be encouraged to adopt the policy of road pricing in controlling traffic and generating funds for financing the management of existing transport infrastructure and the provision of future ones.

References

Adedotun, S. B. (2015). A study of urban transportation system in Osogbo, Osun State, Nigeria. *European Journal of Sustainable Development, 4*(3), 93–101. https://doi.org/10.14207/ejsd.2015.v4n3p93

Adedotun, S. B., Ogundahunsi, D. S., & Oyeniyi, A. S. (2015). Assessment of road transport infrastructure in Osogbo, Osun State, Nigeria. In *Proceedings of the 22 international conferences on urban transport and the environment. WIT transactions on the built environment* (Vol. 164, pp. 61–72). https://doi.org/10.2495/UT160061

Adewale, M. O., & Adekola, A. O. (2019). Spatial preference of urban residential location in Osogbo, Nigeria. *Ghana Journal of Geography, 11*(1), 140–158. https://doi.org/10.4314/gjg.v11i1.9

Afon, A. O., & Faniran, G. B. (2011). Governance and urban solid waste management: The case of local government budgetary allocations in Osun State, Nigeria. In S. O. Fadare, B. T. Aluko, & A. O. Afon (Eds.), *Urban Governance and Environmental Challenges in a West African Sub-region* (pp. 27–44).

Agbiboa, D. E. (2020). Between cooperation and conflict: The National Union of road transport Workers in Lagos, Nigeria. *Crime, Law and Social Change, 73*, 605. https://doi.org/10.1007/s10611-019-09878-x

Agbola, T. (1992). Osogbo. *Cities, Butterworth Publication, 9*(4), 249–256.

Daramola, O. P., & Olojede, O. A. (2011). Pluralism in urban governance: Implications for urban planning in Nigeria. In S. O. Fadare, B. T. Aluko, & A. O. Afon (Eds.), *Urban governance and environmental challenges in a west African sub-region* (pp. 168–181).

Engwicht, D. (1993). *Reclaiming our cities and towns: Better living with less traffic*. New Society Publishers. www.newsociety.com

Fadare, S. O., & Ogunsanya, A. A. (1989). Planning without data: The example of the Nigerian urban transport planning situation. *Studies in Environmental Design in West Africa, 8*, 57–66.

Frank, L., & Engelke, P. (2005). Multiple impacts of the built environment on public health: Walkable places and the exposure to air pollution. *International Regional Science Review, 20*, 193–216.

Gore, C. (2012). Review of Osogbo and the art of heritage: Monuments, deities and money by Peter Probst. *Bulletin of the School of Oriental and African Studies, 75*(2), 431–433.

Hansen, W. G. (1959). How accessibility shapes land use. *Journal of the American Institute of Planners, 5*(2), 73–76.

Litman, T. (2020). *Evaluating accessibility for transport planning: Measuring people's ability to reach desired goods and activities*. Victoria Transport Policy Institute. www.vtpi.org.https://www.vtpi.org/access.pdf

Muller, B. (2011). *Urban regional resilience: How do cities and regions deal with change?* Springer.

National Bureau of Statistics. (2013). *Annual abstract of statistics 2013, Federal Republic of Nigeria*. http://www.nigerianstat.gov.ng/

National Bureau of Statistics. (2016). *Annual abstract of statistics 2016, Federal Republic of Nigeria*. http://www.nigerianstat.gov.ng/

National Population Commission. (2010). *2006 population and housing census of the Federal Republic of Nigeria; housing characteristics and amenities tables, priority tables (LGA)*. Vol. II, Abuja, Nigeria.

Olojede, O. A. (2019). Urban transport security: Analysis of transit crime in Osogbo, Nigeria. *Analele Universității din Oradea, Seria Geografie, 29*(1), 9–18.

Olojede, O., & Daramola, O. (2015). Urban and regional planning administration in Nigeria. *Journal of Environmental Design and Management, 7*(2), 61–70.

Organisation for Economic Cooperation and Development [OECD]. (2017). Bridging the gap: Inclusive growth. *2017 Update Report*. http://www.oecd.org/inclusive-growth/Bridging_the_Gap.pdf

Peter, P. (2011). Osogbo and the art of heritage. Monuments, deities and money. Indiana University Press.

Peter, P. (2013). From iconoclasm to heritage. The Osogbo art movement and the dynamics of modernism in Nigeria. In G. Salami & M. B. Visona (Eds.), *A companion to modern African art* (pp. 294–310). Malden, MA. https://doi.org/10.1002/9781118515105.ch15

UN-HABITAT. (2014). *Structure plan for Osogbo capital territory and environs (2014–2033).* State of Osun Structure Plans Project.

United Nations Economic Commission for Latin America and the Caribbean [UNECLAC]. (2018). *Transport Governance: Theoretical and Policy Perspectives, Issue No. 365*, No. 5. https://repositorio.cepal.org/bitstream/handle/11362/44246/1/S1800932_en.pdf

Williams, A., Kitchen, P., DeMiglio, L., Eyles, J., Newbold, B., & Streiner, D. (2010). Sense of place in Hamilton, Ontario: Empirical results of a neighbourhood-based survey. *Urban Geography, 31*(7), 905–931.

Chapter 14
Urban Infrastructure and Transit-Oriented Development in Accra, Ghana

Isaac K. Arthur, Ernest Agyemang, Aba Obrumah Crentsil, Charlotte Wrigley-Asante, and George Owusu

14.1 Introduction

Ghana is considered one of the fastest urbanising countries in sub-Saharan Africa, with an estimated urbanised population of about 57% in 2018, far ahead of the averages for West Africa and the rest of Africa (see Fig. 14.1, UNDESA, 2019). Burdened by weak urban planning and governance, and poor investments and infrastructure, increased congestion in inner cities and massive outward expansion of cities characterised by low densities and separated land uses have become a common feature of large Ghanaian cities (Owusu & Oteng-Ababio, 2015; World Bank, 2015). However, nowhere is the process of urbanisation intensive, and its challenges of expansive peri-urbanisation and sprawl are greater than the Greater Accra Metropolitan Area (GAMA). The metropolitan region has an estimated population of five million (see Table 14.1) and is the most densely populated region in Ghana and continues to be the main destination for in-country, regional and international migrants (Owusu & Oteng-Ababio, 2015).

GAMA is a functional economic and geographical area which includes the national capital, Accra and its surrounding municipalities. As the nerve centre of Ghana both economically and politically, this contiguously built area which is made up of 25 municipalities attracts the largest share of both public and private investments (including Foreign Direct Investment [FDI]). Analysing Accra's rapid

I. K. Arthur (✉) · E. Agyemang · C. Wrigley-Asante
Department of Geography and Resource Development, University of Ghana, Legon, Ghana

Centre for Urban Management Studies, University of Ghana, Legon, Ghana
e-mail: ikarthur@ug.edu.gh

A. O. Crentsil · G. Owusu
Centre for Urban Management Studies, University of Ghana, Legon, Ghana

Institute of Statistical, Social and Economic Research, University of Ghana, Legon, Ghana

T. Nubi et al. (eds.), *Transit Oriented Development in West African Cities*, Urban Perspectives from the Global South,
https://doi.org/10.1007/978-3-031-58726-9_14

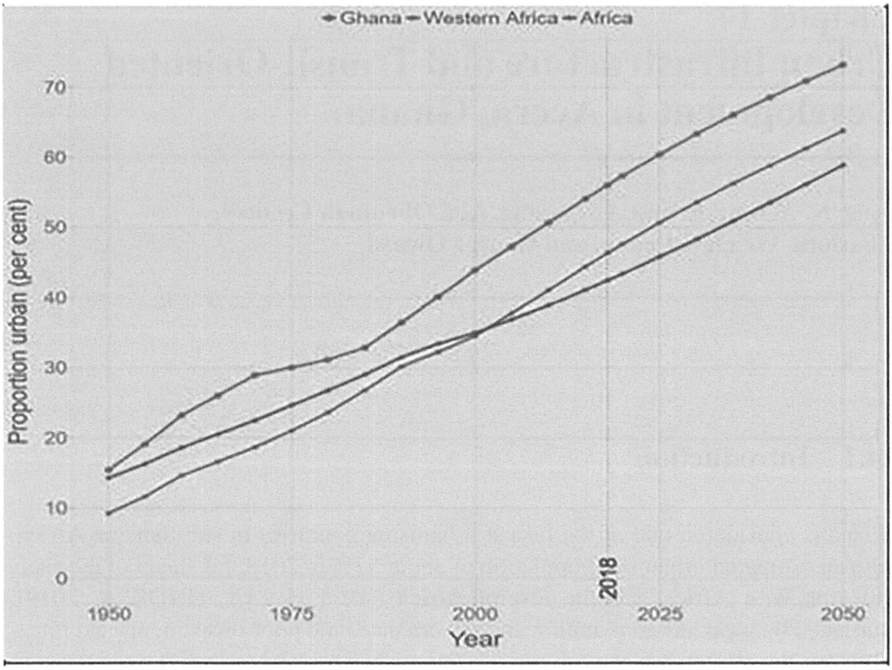

Fig. 14.1 Level of urbanisation: Ghana, West Africa and the rest of Africa

Table 14.1 Population and urban growth rate in GAMA, 1970–2018

	1970	1984		2000		2010		2020[b]	
District	Pop.	G/R[a]	Pop.	G/R	Pop.	G/R	Pop.	G/R	Pop.
Accra Metropolis	636,667	3.0	969,195	3.3	1,658,937	1.1	1,848,614	1.1	2,062,328
Tema Metropolis	102,431	4.4	190,917	2.8	298,432	3.0	402,637	3.0	541,110
Other municipals	88,885	5.2	183,704	8.4	700,780	6.9	1,405,172	6.9	2,738,460
	827,983	3.5	1,343,816	3.9	2,501,196	3.8	3,656,423	3.8	5,341,898

[a]G/R (annual growth rate)
[b]Projection based on 2010 figures
Source: Adapted from Owusu and Oteng-Ababio (2015)

population and real expansion, especially in the last four decades, many analysts have pointed to the impact of globalisation and global capital and the presence of international non-governmental organisations (NGOs), multinational corporations (MNCs), expatriates and international returned migrants, all seeking a foothold in this large metropolitan area of the global south (Briggs & Yeboah, 2001; Owusu, 2013; Grant, 2009; Obeng-Odoom, 2018; Fig. 14.2).

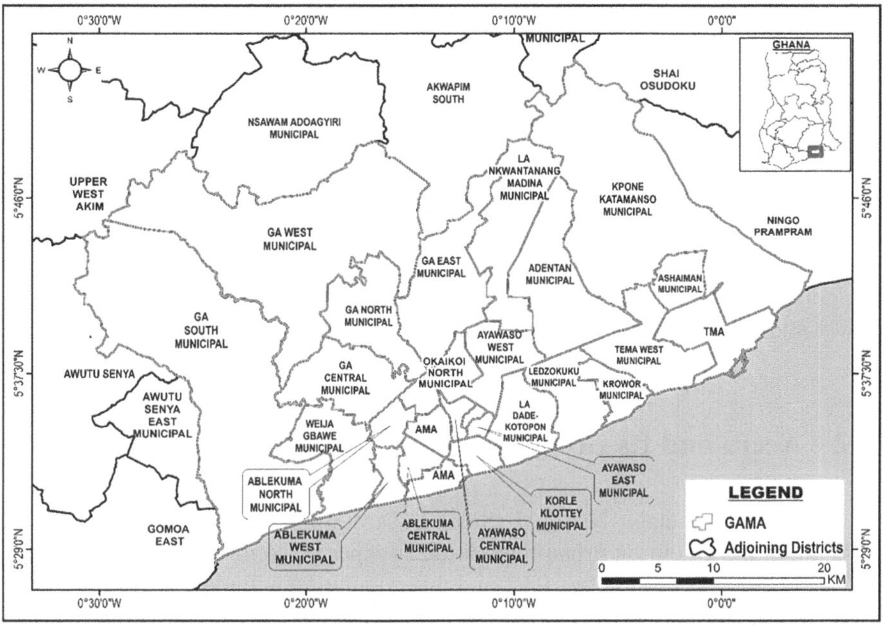

Fig. 14.2 Map of Greater Accra Metropolitan Area. (Source: Authors' construct)

According to the World Bank (2015, 14), 'Ghana's rapid urbanisation, accompanied by unplanned spatial expansion of metropolitan areas, and limited connectivity both within and across Ghana's cities, has presented challenges to economic efficiency'. For GAMA, this challenge has been further exacerbated by the lack of coordination between road transport investments and urban planning, which has allowed roads in and out of the city to be dualised, contributing to the intensification of sprawl amid population growth (Doan & Oduro, 2012; Owusu, 2013). Confronted with the multiple challenges of poor and inequitable accessibility, and the near absence of formal public transport services, transit-oriented development (TOD) has been suggested as a development strategy for cities to enhance accessibility, promote compact development and deal with sprawl and ecological footprints of cities, promote public transport and make cities more efficient and competitive (Hess & Lombardi, 2004; Cooke et al., 2018; Abutaleb et al., 2019). It has been argued that TOD marks a form of land-use and transport planning that makes sustainable transport modes convenient and desirable, resulting in maximising the efficiency of transport services through concentrating urban development around transit stations (Ibraeva et al., 2020).

However, Huang (2017) has argued that despite the success stories of TOD in Asia and Europe, the strategy is still very new in sub-Saharan Africa. He adds that though TOD's concept of leveraging on major transit infrastructure to affect integrated land-use development for greater benefits has gained recognition, there are few examples of successful TOD on the continent with the probable exception of

South African cities, mainly Cape Town and Johannesburg. This scoping study of TOD in Ghana, using Accra as a case study, provides a direct discourse of the specifics of urban transportation in the city and its current challenges, as well as planned and implemented TOD strategy and lessons learnt. Specifically, after the introduction, the paper is structured into seven parts. First, it looked at the city of Accra and its infrastructure challenges followed by an overview of its transportation infrastructure gap. Next is the governance framework for Accra's transportation sector followed by an analysis of the mobility and accessibility in the city (including issues of affordability and inclusivity). The paper then examined planned and implemented transportation investments in Accra and their impacts and ended finally with a conclusion and policy recommendations.

14.2 Accra and Its Broad Infrastructure Challenges

According to Ghana's National Spatial Development Framework (NSDF), 2015–2035, the country's urban population is expected to increase from about six million to about 18 million by 2035, and this population will require a high level of intermodal traffic and transport connectivity, efficient energy and water supply, improved environmental sanitation and the use of modern ICT technology (GoG, 2015, 198). The NSDF adds that reliable and efficient transport links and infrastructure will guarantee the population's access to education and health facilities, business and job opportunities and other public and private services as well as promote agricultural and industrial development (GoG, 2015). However, weak urban policy and institutional focus have impacted negatively on public urban infrastructure growth and expenditure patterns. In the last four decades, government and donor interventions in the area of urban development and poverty reduction have been channelled through the country's medium-term development frameworks (MTDFs). However, these MTDFs have lacked an urban development focus except for the last 5 years following the formulation of Ghana's first-ever National Urban Policy (NUP) and Action Plan.

Consequently, in the face of rapid urbanisation and urban growth, and weak investment in the urban sector, urban infrastructure and services have proven to be inadequate, especially in the areas of transport, utility services provision, sanitation and waste management and housing (Poku-Boansi et al., 2018; Yankson & Bertrand, 2012). These challenges have been compounded by weak capacities of municipal and metropolitan local governments. Experience suggests that most of the investments in infrastructure are publicly funded and are primarily concentrated in sectors like roads, electricity and telecommunications. However, public investments in infrastructure are insignificant and unable to meet effective demand (Shand, 2019). In the absence of formal provision of infrastructure and services, much of the urban population falls on informal channels in obtaining urban services, most at a relatively higher cost or price.

In general, infrastructure provision in urban Ghana has more often than not been provided as a result of development rather than in anticipation of development. As a result, infrastructure provision and improvements have been piecemeal and woefully inadequate. A key problem accounting for this situation is limited funding for infrastructural provision. Further compounding this situation are the effects of rapid population growth, poor maintenance of existing infrastructure and limited creativity by local government in developing innovative strategies for financing urban infrastructure.

While the urban sector has received some level of attention in recent years, the overall investments in the sector fall far short of the needs of the sector (GoG/MLGRD, 2012). In general, past urban development interventions have focused on infrastructure upgrading and environmental sanitation. More importantly, the lack of comprehensive government and donor interventions on urban development has resulted in a situation whereby the impact of these projects has not been effective, especially along the lines of promoting urban centres as engines of growth. It needs to be noted that improving the quantitative and qualitative supply of critical mass of infrastructure and services in urban Ghana must be regarded as, first, a basic need to the population and, second, as an incentive for attraction of investments for growth and development of centres.

14.3 Transportation Infrastructure and Gap: An Overview

Roads, railways, ports and airports are basically the types of hard transportation infrastructure found in Ghana. Ghana's transport infrastructure has improved considerably over the past decades. Compared to other countries in the sub-Saharan continent, the country has a good transport infrastructure base. Public-private partnerships (PPPs) such as the Accra-Takoradi highway dualisation, Boankra Inland Port/Eastern Railway Line, and Takoradi Port Rehabilitation and Expansion, the establishment of a new national airline has helped to attract large private investment in the upgrades of motorways, ports and airports in the country. According to the Oxford Business Group (2019) citing the Ghana Statistical Service (GSS), the transport and storage sector contributed GHS24.6 bn ($5.3 billion) to the economy in 2017. It accounted for 12.8% of GDP, in line with the 12.5% share averaged over the period 2006–2016. The sector grew just 0.4% in 2017, following the expansion of 2.2% in 2016 and 3% in 2015. In addition, a number of initiatives have improved, upgraded or expanded the range of public transport across the country, with major improvements in the urban areas, especially in Accra.

Roads are the dominant mode of transportation in Ghana carrying almost 95% of the passenger traffic and 97% of all movable freight (USG/GOG, 2011). According to Foster and Pushak (2011), 75% of the country's roads are paved and are in good or fair condition, and 74% of the unpaved network is also in good or fair condition. Ghana's transport network accessibility is geographically uneven across the country. Road network is denser in the southern part, while the northern part and especially rural areas lack regular connections with the main transport network, implying an important barrier to the economic development of these areas. Foster and Pushak (2011), using GIS analysis, indicate that only 24% of Ghana's rural population lives within two kilometres of an all-season road which is below the 60% found in Africa's middle-income countries.

Air transport infrastructure in Ghana plays an increasingly important function in connecting the country to the rest of the world and other parts of the country. Ghana's air transport market has seen improvement since 2018 with the construction of a new terminal (Terminal 3) at Kotoka International Airport and the rehabilitation of other regional airports. However, the country lacks a home-based carrier. In relation to railway transport, Ghana has a narrow-gauge railway, totalling 935 km (1.067-meter gauge), and 32-km double-track railroads undergoing major renovation.[2]

Ghana has unrealised potential in shipping, with 560 km of coastline. The Takoradi and Tema ports are the two main ports in Ghana. According to ISSER (2019), continued efforts to modernise these ports for greater operating efficiency through expansion works (new terminals, upgraded IT systems and widened operational partnerships and networks) have resulted in increasing volumes of cargo traffic from about 15.7 million tonnes in 2011 to about 25.5 million tonnes in 2018. A number of other development projects completed in the past years include a new terminal, a four-berth bulk jetty, a one-stop service revenue and data centre, the refurbishment of the transit terminal and the electronic gating system, among many others.

However, gaps in the provision and quality of infrastructure and related services are still present, affecting all modes of transport. Table 14.2 reports the overall standing of Ghana in physical transportation infrastructure indicators vis-à-vis other sub-Saharan countries. Ghana lags behind other sub-Saharan countries such as Kenya and South Africa in almost all indicators.

Table 14.2 Infrastructure facilities in Ghana vis-à-vis other developing countries

Selected countries	Rail lines (km)	Port container traffic ('000TEU)	Air passengers carried ('000)	Air freight (million ton km)
	2017	2017	2018	2018
Benin	..	353	1	0
Burkina Faso	518	..	152	0
Cameroon	1010	387	265	0
Ivory Coast	639	663	779	6
Ethiopia	..	..	11,501	2089
Ghana	..	921	467	0
Kenya	..	1200	5936	295
Nigeria	..	1656	8169	19
Senegal	..	497	21	0
South Africa	20,953	4635	23,922	716
Togo	..	240	566	11

Source: World Development Indicators (World Bank, n.d.)

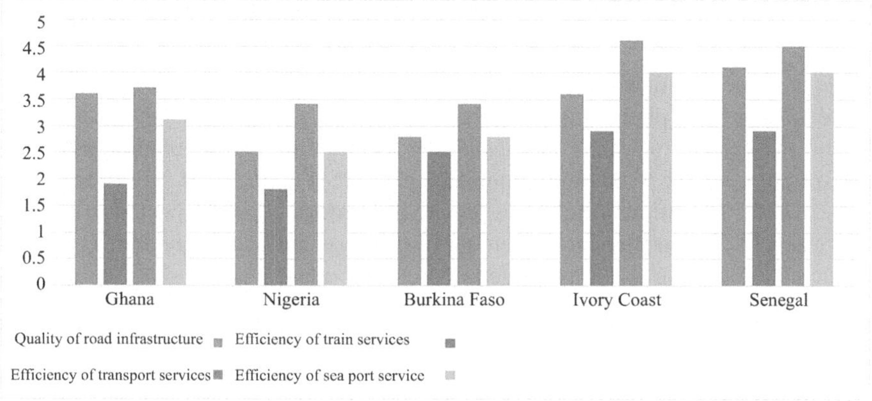

Fig. 14.3 Global Competitiveness Index (1 = worst, 7 = best), 2019 edition. (Source: Global Competitiveness Report 2019 (WEF, 2019))

Similarly, according to the Global Competitiveness Report 2019, Ghana ranks 127th out of 141 countries and is generally poor in overall transportation infrastructure score (WEF, 2019). Figure 14.1 shows infrastructure gaps and their extent in comparison to other sub-Saharan countries. Such international comparisons are helpful, as they offer an indication of the extent to which Ghana needs to improve its transport infrastructure and which sectors have the widest gaps with global comparators (Fig. 14.3).

Train service efficiency is particularly low, according to the World Economic Forum's Global Competitiveness Index (GCI), while Ghana has the highest rating for transportation service efficiency. However, gap emerges when Ghana is compared to selected peered countries. Cote d'Ivoire and Senegal seem to be top in all indicators compared to Ghana. Apart from these gaps, other gaps identified at the national level include the lack of high-quality links between ports and the national highway network resulting in heavy vehicles using inadequate and inappropriate urban streets to access port terminals. Public transport options are not integrated with urban mobility systems; thus, for example, car-based transportation is exclusively used to access all major airports in the country.

Greater investment in transport infrastructure is vital to keep Ghana on the path to greater economic growth. As the country transitions to a high middle-income economy and the population continues to grow, it is likely that today's gaps and negative externalities will worsen. Fortunately, the Government of Ghana has described the year 2020 as the 'Year of Roads'. This is illustrated in the increase in infrastructure sector budgetary allocation from GH¢ 4.63 billion in 2019 to GH¢ 5.11 billion in 2020 with a marginal increase of 10% (PWC, 2020). This boost in investment in the transportation infrastructure sector is expected to close gaps in the sector.

14.4 Governance Framework for City Transportation Sector

The governance framework of Ghana's transport sector under which various interventions have been planned and implemented in the sector consists of institutions defined to perform functions such as policy formulation and regulation, asset and infrastructure management and maintenance services and service provision (GoG, 2008). These institutions (including private and informal) are expected to provide progressively decentralised management of transport infrastructure and services in the country. Working in an integrated and collaborative fashion, these institutions and actors in the transport sector are expected 'to meet broad objectives concerned with enhancing the effectiveness, efficiency and affordability of the urban transport sector, bearing in mind that the overall aim of its development is to contribute to the improvement in the quality of life of the community' (Kwakye & Fouracre, 1998:1), in this case specifically Accra. The governance and institutional arrangements for transport infrastructure provision, management and service are illustrated in Fig. 14.4.

As illustrated in Fig. 14.4, the governance framework, as established in the National Transport Policy, 2008, marks a blueprint for the implementation of urban transport interventions, including the Bus Rapid Transport (BRT). At the top of the governance framework, the Ministries of Transport (MoT), Roads and Highways (MoRH) and Local Government and Rural Development (MLG & RD) are responsible for making policies. The Department of Urban Roads (DUR) is responsible for asset and infrastructure management and maintenance services. The DUR work

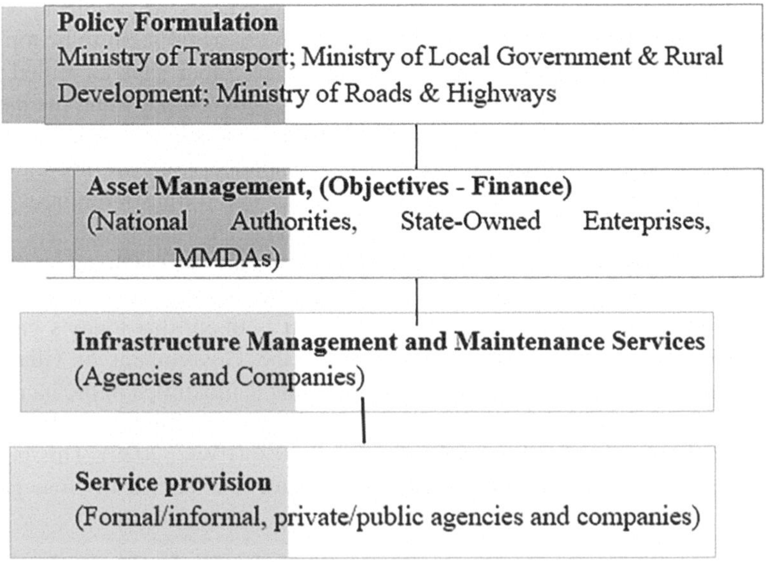

Fig. 14.4 Institutional framework—transport infrastructure. (Source: GoG, 2008)

hand in hand with national institutions such as the Land Use and Spatial Planning Authority (LUPSA) and the Metropolitan, Municipal and District Assemblies (MMDAs) as planning agencies in land-use and environmental planning and supporting the management of public transport terminals. The MMDAs also function as regulators (i.e. enacting bye-laws and issuing permits for transport operators in the paratransit sector). The Motor Transport and Traffic Directorate of the Ghana Police Service (GPS) is responsible for enforcing compliance with route licenses as stipulated in the bye-laws. In addition, the DUR works closely with both national and international private sector road construction firms in the delivery and maintenance of road infrastructure.

In terms of urban transport services to passengers, a number of agencies and actors which can be classified as public and private as well as formal and informal operators can be found in Ghana. The formal operators include the quasi-state-owned entities (State Transport Company [STC] and Metro Mass Transit Limited) and formal private companies (VIP Bus, Pergah Transport Limited, Kingdom Transport Service [KTS], etc). The informal operators are dominated by mini-bus services popularly known as *trotro*[1] and shared taxis which operate under various private transport owners, drivers and unions or associations. These include the Ghana Private Road Transport Union, GPRTU (the largest national union of informal transport operators), Ghana Cooperative Transport Association (GCTA) and Progressive Transport Owners' Association (PROTOA). These associations have local, city and national clusters which organise transport routes for their members. All the unions/associations operate under the Ghana Road Transport Coordination Council (GRTCC), an umbrella body of all transport operators in Ghana that represents the interests of road transport operators in negotiating with the government for transport tariffs and vehicle acquisition (Brookins, 2019).

To address cross-jurisdictional issues of regulation, development and organisation of urban passenger transport, the Greater Accra Passenger Transport Executives (GAPTE) was established. GAPTE is an inter-MMDA body established with the legal mandate to plan, regulate and coordinate urban passenger transport activities in GAMA, as well as to facilitate the implementation of a pilot BRT. Although the BRT is supposed to improve urban transport system and harness sustainable development, its implementation in Accra has been fraught with challenges. Most of the required infrastructure including dedicated lanes to support the efficient running and achievement of the project's objective is lacking. According to the World Bank (2017), the attempt to fully develop a BRT in Ghana was too ambitious as the government overlooked the complexities in carrying out such urban development.

14.5 Mobility and Access in the City

Transport is a critical structure in the city which provides access to essential activities that enables individuals, both males and females and other vulnerable groups, to exercise their rights in urban spaces. Considering the needs of different users in the city and creating equal levels of accessibility to transport could therefore contribute to achieving sustainability in urban mobility (Levy, 2013). Moreover, the expansion of transport network could also facilitate many positive impacts on the lives of urban dwellers especially vulnerable groups. Yet, studies in Ghana have shown that while the majority depends on motorised transport systems, particularly public transport, the poor urban infrastructure and transport services in the country have significant effects on the livelihoods of urban residents (Abane, 2011; Hotor, 2016), with vulnerable groups being more constrained. In spite of the improvements in the transport system in major urban areas, especially Accra, over the years, several challenges such as congestion due to vehicular traffic, poor road networks and frequent mechanical breakdown of vehicles are some factors that affect the mobility of the majority of urban dwellers (Hotor, 2016).

Abane's (2011) study in four major urban centres in Ghana revealed that urban dwellers have a preference for private or public transport system due to its affordability, availability and accessibility. These private transport operators dominate the intra-urban transport services with the majority opting for this mode of transport as the most economical mode of transport. However, workers with higher income levels were more likely to use their private cars as against the public transport system (MoRH et al., 2013). The vulnerable groups have limited options but to depend on sometimes unreliable public transport systems.

Similarly, Hotor's (2016) study among residents of urban Accra revealed that the majority of the population within the city resort to public transport as their main mode of transport due to its low economic cost. Other factors that contribute to the use of this mode of transport included its flexibility and lack of alternative modes of transport. While affordability and lower economic cost were identified as key factors in choosing the 'trotros', the majority intimated that they would opt for improved transport systems such as a taxi or a private car, when their income increased. Interestingly, residents highlighted the fact that they would change their regular mode of transport in case of emergencies. In such instances, the most expensive means could be an option due to its urgency (Hotor, 2016). Andreasen and Moller-Jensen (2017) have also highlighted that resident in the peripheries in many parts of African cities may be severely challenged in terms of accessibility to and within the city due to poor roads and inadequate public transport services. There are also long and unpredictable travel times resulting from the intense traffic congestion within cities.

This may imply that there are different accessibilities to transport among different socio-economic groups of urban dwellers such as high- and low- income persons, men, women and other vulnerable groups (such as the physically challenged) (Abane, 2011; MoRH et al., 2013; Hotor, 2016). Increasingly, an emerging challenge for urban mobility policymakers is the need to adopt more inclusive planning of urban infrastructure including a gender-sensitive perspective. It is noted that while more men than women are likely to have access to motorised and private means of transport, women are less likely than men to make use of other transitional means of transport and may depend more on public means of transport (Tran & Schlyter, 2010). Further, while men do travel more than women regarding more distant locations, women are also inclined to do more frequent short travels in connection with reproductive works (Tran & Schlyter, 2010; Amoako-Sakyi, 2017). This may have implications for their mobility in urban spaces. However, there is limited research in the context of Ghana to ascertain these gender differences.

Madariaga (2013) further notes that gender and transport is a field where new insight from gender analysis can contribute immensely to enhancing policymaking and expanding the quality of life for individuals in society. For instance, the mobility of care, which includes all travels resulting from home and caring responsibilities, escorting others and shopping for daily living, among others, provides a framework for recognising, measuring, making visible and giving value to all the travels associated with care- and home-related tasks (ibid). Thus, the total inclusion of gender aspects at every level and areas of transport research and policy will be very significant to implementing a sustainable transit-oriented development (TOD).

Again, Porter (2008), cited in Amoako-Sakyi (2017), highlights the fact that there are fundamental differences in transportation needs of men and women. This is as a result of the different roles that men and women play within cities. Even though women perform multiple roles as income earners, homemakers and community managers and usually take short and more frequent trips during the day, often accompanied by children, the transport system is not properly directed towards them to meet their needs (Porter, 2008).

In effect, gender and transport in urban development and planning still remains without much discussion within the Ghanaian urban infrastructure discourse. There are therefore critical issues within the urban transport discourse which needs attention and research. The provision of transport infrastructure that addresses the need of persons, men, women and people living in urban peripheries who access the city every day, as well as the vulnerable groups such as the aged, disabled and children, should be explored fully in the context of Ghana. This calls for intensive research in this area to ascertain the dynamics and its contribution to a sustainable transport sector.

14.6 Transportation Investments: Actions Taken and Impacts

Ghana's medium-term development frameworks (MTEDFs) for the past two decades have recognised the critical role of transport infrastructure towards national economic growth and poverty reduction. These MTEDFs have acknowledged that improved transport infrastructure guarantees easy access to economic opportunities for urban dwellers in Accra and elsewhere. Thus, the Road Fund Act (Act 536) has been enacted to support the financing of the construction and rehabilitation of public roads. The Fund derives monies from fuel levies, vehicle registration fees and road use fees, among others. Also, a Consolidated Fund has been set up to support infrastructure provision and maintenance.

Also, transport infrastructure in Ghanaian cities like Accra is funded through international donor organisations and financial institutions. For instance, under the Road Sector Development Programme (RSDP) which began in 2001, some US$ l.2billion has been invested into the road sub-sector. About 75% of this amount was sourced from donors, notably the World Bank, African Development Bank (AfDB), French Development Agency (ADF), Danish International Development Agency (DANIDA), United States Agency for International Development (USAID), UK's Department for International Development (DFID) and the Japan International Cooperation Agency (JICA) (World Bank, 2009). In addition, the Millennium Challenge Corporation (MCC), the Saudi Fund and the European Commission (EC) have been important sources of funds for infrastructure development in Accra, as well as in other parts of Ghana. More recently, the African Development Bank Group (AfDB) in 2016 also granted a loan facility to the tune of US$83.9 million. This facility was to support the Accra Urban Transport Project in the construction of a 3-tier highway interchange at Pokuase and a 10-kilometre stretch of interconnected urban roads which is expected to be operational in 2020.

However, it seems these investments in transport infrastructure have not been at pace with the urbanisation trends of Accra. In addition, transport challenges in Accra have been largely seen by officialdom as an infrastructure challenge rather than management, hence the weak application of TOD. Since 2000, Accra's population size has doubled, and its built-up area has exponentially expanded. For instance, from a mere 260km^2 in 1990, Accra's built-up area in 2015 was 795km^2. The city has sprawled considerably along major road corridors. Meanwhile, most of the important trip destinations for work, shopping and accessing government offices and other social activities have been concentrated at the city core. Thus, most residents of Accra have had to endure long commute distances along severely congested roads. Most of the commutes are undertaken in the *trotro* informal paratransit services (Agyemang, 2017; Amoh-Gyimah & Aidoo, 2013; Abane, 2011). Macroeconomic reforms in the 1980s which saw the scaling down of government's investment in transport infrastructure (i.e. bus operations) partly explain the popularity of *trotros* (Fouracre et al., 1994).

However, recent years have seen attempts to apply TOD strategy in Accra. Key innovations include the formulation of the National Transport Policy (NTP) which seeks to prioritise mass transportation in urban areas, 'aiming to move at least 80% of passengers' (GoG, 2008, 45). Also, the policy seeks to ensure that transport planning is 'fully integrated with development planning and service provision' (GoG, 2008, 49). A key objective of the policy is to make Accra and other major Ghanaian urban centres transit-oriented cities. In 2003, the Metro Mass Transit or the 'Kufuor Bus' service was set up as a joint venture agreement between the state (which owns 45% shares through the assets of the now defunct Omnibus Services Authority) and private investors (55% ownership). High-occupancy buses were imported on a gradual basis from China and Holland or assembled locally. The Italian government also donated some buses to augment the Ghanaian government's laudable efforts. By the end of 2006, the total fleet had reached 677. Despite receiving huge government technical, logistical and financial support, the impact of this project has been minimal as the bus company 'attracts only a small proportion of the travelling public because of operational problems' (Abane, 2011, 321).

As part of an urban regeneration strategy involving TOD, Achimota Transport Terminal (ATT) was started in 2004 and completed in 2009. The project was aimed at reducing congestion on the main road in the eastern part of Accra (Accra-Nsawam Road) which serves as the major corridor connecting Accra to other Ghanaian cities such as Kumasi and Tamale and the West African countries of Burkina Faso, Niger and Mali. It was also to facilitate the development of a new central business district (CBD) on the eastern part of Accra with transit-oriented improvement. However, Abedi et al. (2015) have argued that ATT though conceived as a model project has not met the dimensions of urban regeneration to enhance its usage and sustainability.

By far Accra's BRT has received attention as a TOD intervention. Under the Ghana Urban Transport Project (GUTP), the state attempted to solve Accra's intractable mobility problems using traffic engineering measures, management improvements, regulation of the public transport industry and the implementation of a bus rapid transit (BRT) system (World Bank, 2020). The GUTP received an International Development Association (IDA) financing of $ 45 million on standard credit terms, together with a $ nine million grant financing from the Global Environmental Facility (GEF). Over 95% of the financing both from IDA and GEF was disbursed. Much of the funds was spent on the construction/upgrading of infrastructure along the Accra Central to Kasoa corridor on which a pilot BRT service was to be undertaken. Aside from the construction of a bridge on the Odaw River, not much was heard about the project. Eventually, the Accra Central (Tudu) to Amasaman corridor was selected for piloting the BRT service when in October 2016, the 'Aayalolo' BRT service was launched.

The impact of BRT in making Accra a transit-oriented one is yet to be felt. This may be due to the lack of leadership and proper coordination among state actors responsible for the implementation of the project. For instance, at the national level, the Urban Transport Advisory Committee (UTAC) was set up and accepted in principle among the Ministries of Transport, Local Government and Finance. Also, at the local level, the Greater Accra Passenger Transport Executive (GAPTE) was set up to coordinate the various local city authorities under whose jurisdiction the buses will operate. In practise, though, neither the UTAC nor GAPTE appears to have achieved the desired objective. Dedicated lanes for the BRT buses could not be enforced. The Aayalolo now operates largely in mixed traffic as a 'quality bus system' (QBS), and most commuters still use the *trotros* instead.

14.7 Conclusion and Policy Recommendations

As noted earlier, although TOD in cities has gained global prominence (Cervero & Dai, 2014), its application in Africa in general and Ghana in particular remains very limited. For Ghanaian cities like Accra and the GAMA region, urban growth has largely been shaped by informality as a result of the strong disconnection between physical planning and programming of investments (Owusu & Oteng-Ababio, 2015). These challenges have been further compounded by the weak urban governance system, especially the overbearing control of the central government and the lack of autonomy by city governments, thus, stifling creativity and innovation for addressing urban development challenges (Obeng-Odoom, 2013). Nevertheless, given the rapid pace of urbanisation and urban population as well as the massive sprawl of Ghanaian cities, especially the GAMA region, there is a need for transport planning and investments to go hand in hand with overall city planning and development.

As a result of poor city planning, GAMA's transport sector is characterised by severe traffic congestion, high accident rates, noise and air pollution, overdependence of low-capacity vehicles, weak implementation of traffic management measures, inadequate facilities for pedestrians and cyclists and poor road safety arrangements (UNEP/DTU, 2016; Agyemang, 2015; Kwakye & Fouracre, 1998). There is also the neglect of the transportation needs of different categories of persons in the city. Apart from their effect on the competitiveness and attraction of investments in Accra and other Ghanaian cities and towns, the transport sector contributes significantly to greenhouse gas (GHG) emissions with harmful effects on human health and the environment (UNEP/DTU, 2016). In all, these issues threaten Ghana's effort at achieving Goal 11 of the UN's SDGs, especially Goal 11 of making cities inclusive, safe, resilient and sustainable.

GAMA's TOD strategy of BRT consisting of coordinated improvements in transit system infrastructure was expected to improve efficiency and affordability of mobility, reduce congestion and urban air pollution with related health impacts and reduce GHG emissions (Brookins, 2019; UNEP/DTU, 2016) as well as promote overall urban development. This echoes the vision of Ghana's National Transport Policy (NTP) and National Urban Policy (NUP) of developing an affordable, safe and efficient urban transportation system that supports the overall development and competitiveness of Ghana's urban areas. Apart from promoting inclusiveness in the urban transport system through the provision of accessibility to the disabled and cheaper fares to passengers, the BRT was to provide a key opportunity for reducing GHG emissions in Accra through the use of new, energy-efficient buses that are climate friendly as opposed to obsolete buses, *trotro* and taxis that are currently used on major corridors in the city (UNEP/DTU, 2016).

However, evidence to date indicates that while TOD's application has been limited, the few planned and implemented in GAMA have not been successful. Analysing the failure of GAMA's pilot BRT, Agyemang (2015) provides a myriad of challenges for its collapse including recurring traffic congestion, passengers' inadequate comfort and personal security, the resistance from existing public transport operators, the lack of legal status for BRT system and limited advertising of the intervention. Nevertheless, opportunities remain for exploiting the potentials of TOD in GAMA including enhancing economic activities and competitiveness and improving land utilisation and value, especially along and near transit corridors. Also, transit-oriented development in GAMA may create opportunities for both public and private sector investment in urban development. It is based on enhancing these opportunities and benefits that we propose the following policy recommendations:

- Planning, investments and any envisaged changes expected with respect to TOD intervention need to take into account the appropriate contextual and cultural realities (including gendered realities) of GAMA, particularly the presence of large informal activities and operators.
- Related to the above recommendation is the need to plan TOD taking into account the interest and role of the private sector, both formal and informal. In other words, the need for the private sector to locate and invest in the locations and corridors earmarked as TOD should be of priority consideration.
- Within the context of multiple-stakeholder consultation and participation, a careful balance needs to be maintained to ensure that lower-income communities and the greater public benefit from the resultant economic and social gains over time as a result of TOD interventions.

Besides the above policy recommendations, we accept Hess and Lombardi's (2004: 28) recommendations regarding the general factors that contribute to a successful TOD:

- **Location efficiency**: Good TODs should provide convenient and efficient transportation links that are blended to reduce auto dependency.
- **Value recapture**: Transit-rich regions should pay less for transportation (per capita) than auto-dependent ones, even after public investments in transit are considered.
- **Liveability**: TODs should lead to improvements in air quality, mobility and access to services for all categories of persons of all ages.
- **Financial return**: Mixed-use strategies should allow for greater flexibility in responding to market cycles.
- **Choice**: Good TODs should provide a greater number of options for housing, transportation and lifestyles than standard development.
- **Efficiency**: Efficient regional land use patterns should reduce land consumption and traffic generation.

References

Abane, A. M. (2011). Travel behaviour in Ghana: Empirical observations from four metropolitan areas. *Journal of Transport Geography, 19*, 313–322.

Abedi, L. A., Sasu, A., Ayitey, J. Z., & Boakye-Agyeman, N. A. (2015). The Achimota Transport Terminal in Accra: A model urban regeneration project in Ghana? *Journal of Sustainable Development, 8*(9), 240–253.

Abutaleb, A., McDougall, A., Basson, M., Hassan, R., & Mahmood, M. N. (2019). Towards a conceptual framework for understanding the attractiveness of rail transit-oriented shopping mall developments (TOSMDs). *Urban Rail Transit, 5*(4), 225–239.

Agyemang, E. (2015). The bus rapid transit system in the Greater Accra Metropolitan Area, Ghana: Looking back to look forward. *Norsk Geografisk Tidsskrift-Norwegian Journal of Geography, 69*(1), 28–37.

Agyemang, E. (2017). Mode choice for long distance trips: Evidence from the Greater Accra Metropolitan Area of Ghana. *Journal of Transport Geography, 64*, 150–157.

Amoako-Sakyi, O. R. (2017, September). Scaling up gender mainstreaming in rural transport: Analysis of policies, practices, impacts and monitoring processes. A case study report: Ghana. *International Forum for Rural Transport and Development (AFCAP Project Reference Number RAF)* 2044J.

Amoh-Gyimah, R., & Aidoo, E. N. (2013). Mode of transport to work by government employees in the Kumasi metropolis, Ghana. *Journal of Transport Geography, 31*, 35–43.

Andreasen, M. H., & Moller-Jensen, L. (2017). Access to the city: mobility patterns, transport and accessibility in peripheral settlements of Dar es Salaam. *Journal of Transport Geography, 62*, 20–29.

Briggs, J., & Yeboah, I. (2001). Structural adjustment and the contemporary African city. *Area, 33*(1), 18–26.

Brookins, D. (2019). Transforming urban transport: The role of political leadership. *TUT-POL Sub-Saharan Africa Final Report*. https://static1.squarespace.com/static/5804efd7cd0f68e576ecd423/t/5dfe925d50c530473013db10/1576964717327/Accra%2C+Ghana_Case_Final.pdf

Cervero, R., & Dai, D. (2014). BRT TOD: Leveraging transit-oriented development with bus rapid transit investments. *Transport Policy, 36*, 127–138.

Cooke, S., Behrens, R., & Zuidgeest, M. (2018). The relationship between transit-oriented development, accessibility and public transport viability in South African cities: A literature review and problem framing. In *Proceedings of 37th Annual Southern African Transport Conference (SATC 2018)* (pp. 365–380). Jukwaa Media.

Doan, P., & Oduro, C. Y. (2012). Patterns of population growth in peri-urban Accra, Ghana. *International Journal of Urban and Regional Research, 36*(6), 1306–1325.

Foster, V., & Pushak, N. (2011). Ghana infrastructure: A continental perspective. In *Policy Research Working Paper 5600*. The World Bank.

Fouracre, P. R., Kwakye, E. A., Okyere, J. N., & Silcock, D. T. (1994). Public transport in Ghanaian cities: A case of union power. *Transport Reviews, 14*(1), 45–61.

GoG (Government of Ghana). (2008). *Ghana National Transport Policy, 2008*. GoG. Available at: https://s3.amazonaws.com/ndpc-static/pubication/Transport+Policy_Dec2008.pdf

GoG (Government of Ghana). (2015). *Ghana National Spatial Development Framework (2015–2035)* (Conditions and main issues) (Vol. I). GoG.

GoG (Government of Ghana)/MLGRD (Ministry of Local Government and Rural Development). (2012). *National Urban Policy Framework & Action Plan*. GoG/MLGRD.

Grant, R. (2009). *Globalising City: The urban and economic transformation of Accra*. Syracuse University Press.

Hess, D. A., & Lombardi, P. A. (2004). Policy support for and barriers to transit- oriented development in the inner city literature review. *Transportation Research Record Journal of the Transportation Research Board, 1887*, 26–33.

Hotor, D. E. (2016). *The use of public transport services by residents in the Accra Metropolitan Area*. [Unpublished MPhil thesis], Department of Geography and Resource Development, University of Ghana, Legon.

Huang, C. (2017). *Can Dar es Salaam become the next global model on transit- oriented development?* https://blogs.worldbank.org/sustainablecities/can-dar-es-salaam-become-next-global-model-transit-oriented-development.

Ibraeva, A., Correia, G. H. A., Silva, C., & Antunes, P. A. (2020). Transit- oriented development: A review of research achievements and challenges. *Transportation Research Part A, 132*, 110–130.

ISSER. (2019). *The state of the Ghanaian economy in 2018*. ISSER.

Kwakye, E. A., & Fouracre, P. R. (1998). *Urban transport policy reform in Ghana*. CODATU VIII Conference.

Levy, C. (2013). Travel choice reframed: "Deep distribution" and gender in urban transport. *Environment & Urbanization, 25*(1), 47–63.

Madariaga, I. S. (2013). From women in transport to gender in transport: Challenging conceptual frameworks for improved policymaking. *Journal of International Affairs, 67*(1), 43–65.

MoRH, MoT, & GSS. (2013). *Second National Household Transport Survey Report 2012*. GSS.

Obeng-Odoom, F. (2013). *Governance for pro-poor urban development: Lessons from Ghana*. Routledge.

Obeng-Odoom, F. (2018). Transnational corporations and urban development. *American Journal of Economics and Sociology, 77*(2), 447–510.

Owusu, G. (2013). Coping with urban sprawl: A critical discussion of the urban containment strategy in a developing country city, Accra. *Planum: The Journal of Urbanism, 26*(1), 1–17.

Owusu, G., & Oteng-Ababio, M. (2015). Moving unruly contemporary urbanism toward sustainable urban development in Ghana by 2030. *American Behavioral Scientists (ABS), 59*(3), 311–327.

Oxford Business Group. (2019). *The Report Ghana 2019*. https://oxfordbusinessgroup.com/overview/opportunities-private-engagement-abound-government-renews-its-commitment-infrastructure-investment

Poku-Boansi, M., Asuah, A. Y., & Cobbinah, P. B. (2018). Contextualizing transport infrastructure and services in Ghanaian peri- urbanism. *International Planning Studies, 23*(2), 195–209.

Porter, G. (2008). Transport planning in sub-Saharan Africa II: Putting gender into mobility and transport planning in Africa. *Progress in Development Studies, 8*(3), 281–289.

PWC. (2020). *2020 Budget highlights: Consolidating the gains for growth, jobs and prosperity for all*. https://www.pwc.com/gh/en/assets/pdf/2020-budget-highlights.pdf

Shand, W. (2019). *The finance landscape in Ghana: Mobilising investment in sustainable urban infrastructure*. Paper for the Coalition for Urban Transitions. London and Washington DC. http://newclimateeconomy .net/ content/cities-working-papers

Tran, H. A., & Schlyter, A. (2010). Gender and class in urban transport: The cases of Xian and Hanoi. *Environment & Urbanization, 22*(1), 139–155.

UNEP/DTU. (2016). *Accra bus rapid transit project: Facilitating implementation and readiness for mitigation (FIRM) Project–Ghana*. UNEPDTU Partnership.

UNDESA. (2019). *World urbanization prospects: The 2018 revision*. Retrieved from https://population.un.org/wup/Publications/Files/WUP2018-Report.pdf

USG/GOG. (2011). *Ghana constraints analysis (partnership for growth) final report*. https://2009-2017.state.gov/documents/organization/202533.pdf

WEF (World Economic Forum). (2019). *World economic report: The global competitiveness report 2019*. http://www3.weforum.org/docs/WEF_TheGlobalCompetitivenessReport2019.pdf

World Bank. (n.d.). *World development indicator*. Retrieved from http://wdi.worldbank.org/table/5.10#

World Bank. (2009). *Ghana transport sector project*. Project appraisal document Africa http://documents.worldbank.org/curated/en/514181468254355062/pdf/473240PAD0CORR10BOX338916B01PUBLIC1.pdf

World Bank. (2015). *Rising through cities in Ghana: Ghana urbanization review overview report*. World Bank.

World Bank. (2017). *Implementation, completion and results report (ICR)*. http://documents.worldbank.org/curated/en/933711488510098083/pdf/ ICR-Ghana-Urban-Transport-Project-P100619-FINAL-01262017-01312017.pdf

World Bank. (2020). *Transport global practice working paper – Synopsis of bank funded bus rapid transport projects*.

Yankson, P. W. K., & Bertrand, M. (2012). Introduction: Challenges of urbanisation in Ghana. In E. Ardayfio-Schandorf, P. W. K. Yankson, & M. Bertrand (Eds.), *The mobile city of Accra: Urban families, housing and residential practices* (pp. 25–46). CODESRIA.

Chapter 15
Urban Infrastructure and Transit-Oriented Development in Lagos, Nigeria

Basirat Oyalowo, Olamide Udoma-Ejorh, Oluwaseun Muraina, Taibat Lawanson, and Timothy Nubi

15.1 City Overview

Currently, Lagos is globally recognised as the fourth largest economy in Africa (AfrAsia Bank Africa; New World Wealth, 2019). It encompasses 358,861 hectares or 3577 km^2 (Lagos State Government, 2023). Although the state is the smallest in land mass in Nigeria, it has the largest population. In 1991, the national census estimated the state's population at 5,725,116, and this has grown more than threefold in the last decade. The 2006 Nigerian National Population Commission census puts the population at 17,535,924, and it was estimated to have risen to more than 20 million (Egbejule, 2020) by 2019. This estimation puts the population density at approximately 4907 people per square kilometre. However, it could be as high as 20,000 people per square kilometre in the metropolis' densely populated areas (Lagos Resilience Strategy, 2020).

As the population has grown, so has the spatial spread of Lagos. The metropolitan area is expanding into the neighbouring Ogun State in all directions. Its southern shoreline bordering the Atlantic Ocean is constantly being dredged to support

B. Oyalowo (✉) · T. Nubi
Department of Estate Management, University of Lagos, Akoka, Lagos, Nigeria

Centre for Housing and Sustainable Development, University of Lagos, Akoka, Lagos, Nigeria
e-mail: boyalowo@unilag.edu.ng

O. Udoma-Ejorh
Lagos Urban Development Initiative, Lagos, Nigeria

O. Muraina
Department of Estate Management, University of Lagos, Lagos, Nigeria

T. Lawanson
Centre for Housing and Sustainable Development, University of Lagos, Akoka, Lagos, Nigeria

Department of Urban and Regional Planning, University of Lagos, Akoka, Lagos, Nigeria

T. Nubi et al. (eds.), *Transit Oriented Development in West African Cities*, Urban Perspectives from the Global South,
https://doi.org/10.1007/978-3-031-58726-9_15

real estate development activities. Lagos State has 5 administrative divisions—Ikeja, Badagry, Ikorodu, Lagos Island and Epe (IBILE). It has 20 local government areas (LGAs) and 37 local council development areas (LCDA), making a total of 57 local authorities. Metropolitan Lagos accounts for over 85% of its total population, in an area that is 37% of the State's land area (Lagos State Government, 2023). The state is made up of low-lying terrain of up to 4 m below sea level (Ogunbiyi, 2019), a wetland region of swamp forests, mangroves and freshwater. Generally, the terrain is flat, although there are areas of reasonable altitude in Epe and Ikeja. The maze of lagoons and waterways constitutes about 22% or 787 km^2 of the state's total landmass (Fig. 15.1).[1]

Lagos is Nigeria's economic and commercial capital and therefore continuously attracts people from all over Nigeria and West Africa; thus, it can be described as a multi-cultural metropolis. This description is very different from its humble beginning as a fishing village. As Lagos grew from its origin, Lagos Island, to the west, east and north, different commercial, residential and political zones were created. Unfortunately, land use and transportation services have not kept pace with the scale of development. The majority of the commercial areas are concentrated in the Lagos

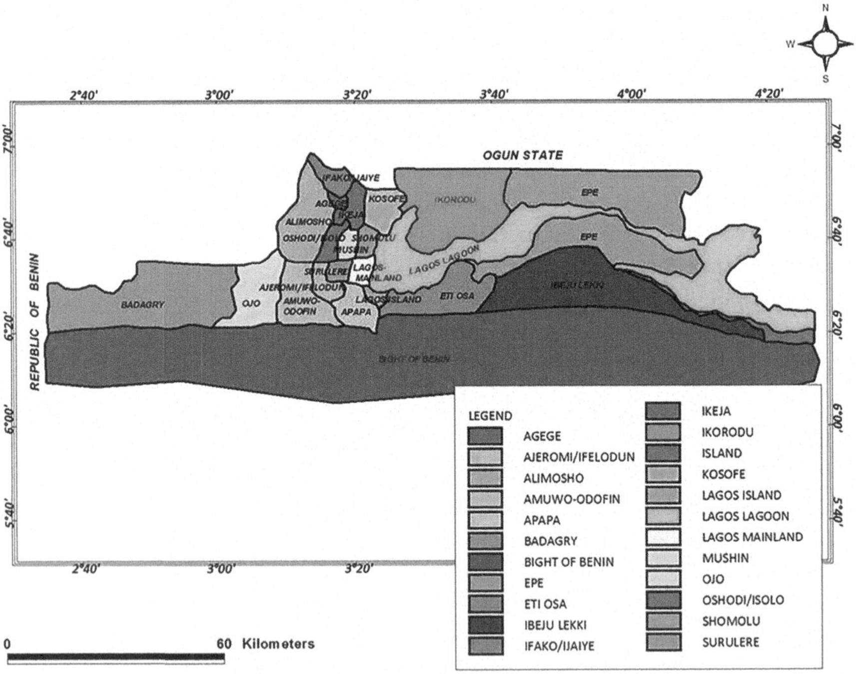

Fig. 15.1 Map of Lagos. (Source: Afolabi et al., 2017)

[1] https://ir.unilag.edu.ng/jspui/bitstream/123456789/4330/2/A%20Study%20of%20Rail%20Mass%20Transit%20in%20Lagos%20and%20its%20Environs.pdf

mainland and Island corridor, while the state's population density clusters around the centre, contributing to significant commuting to and from these axis on a daily basis.

As a coastal city, Lagos hosts one of Africa's largest ports and the fourth busiest in the continent, the Apapa Port., as well as the he Premier Port in Apapa and Tin Can Island Port. 99% of Nigeria's trade is by sea (Stears Business, 2019). In 2017, Premier Port and the largest seaport in Nigeria accounted for 75% of these imports. Also, Lagos houses the busiest international airport in Nigeria: Murtala Muhammed International Airport. The domestic airport is 4.2 km from the international airport with the same name. Within Lagos, people can move through the city using active transport (walking and cycling), motorbike taxis (okada), shared tricycle taxis (keke marwa), minibus taxis (danfo), rail, public buses (Bus Rapid Transport and Lagos Bus Services Limited), water transport, private cars and taxis.

Over the years, Lagos transport system has become inadequate and inefficient for the city's population. The concentration of industries and businesses in a few places, the increase in private car ownership, the lack of an efficient public transport system, poor town and highway planning, poorly maintained and narrow roads and poor driving behaviour have worsened mobility and put pressure on urban public transportation infrastructure.

15.2 City Profile (Including Broad Infrastructure Challenges)

Historically, Lagos's infrastructural development can be traced to its function as a transportation hub and its role in connecting the coast to Nigeria's hinterlands through its internal waterways network. This locational advantage was capitalised upon when Lagos became the colonial's seat and later Nigeria's capital. Its function as a capital city also ensured that it enjoyed significant infrastructural investment and increased its attraction to migrants, resulting in rapid urbanisation that has now outpaced physical and social infrastructure. As a result, city administration daily manages shocks and stresses, borne out of infrastructure deficiency. These deficiencies include disease outbreaks, coastal erosion and flooding, forced evictions, economic downturn, building collapse, high unemployment and underemployment, traffic congestion, inadequate physical and social infrastructure, inadequate transportation system, formal-informal economic contestation, erratic power supply, civil unrest, urban fires, crimes and violence and an inadequate health system (Lagos Resilience Strategy, 2020). According to the Lagos State Development Plan (LSDP 2012–2025), key infrastructure challenges that the city faces include:

Inadequate physical and social infrastructure: Population growth and the lack of strategic infrastructure planning and investment are responsible for the current inadequacy of physical and social infrastructure. These are manifested in massive housing deficits; unequal access to water, education, health and sanitation facilities; insecurity; and poor roads and drainage systems.

Water: There is a water demand of 870mgd in Lagos, and the public water supply system is not responsive to consumers' needs and preferences. With a dearth of technical and managerial expertise and new technology in the sector, there is a need to strategically improve economic efficiency in the public water supply system's operations.

Erratic power supply: Lagos currently suffers a power supply gap of over 10GW due to dependency on electricity supplied from the national grid. The current supply is not adequate to serve small- and large-scale industrial processes and Lagos residents' residential needs. The Lagos Resilience Strategy estimates that more than 40% of the population does not have access to grid power and that 80% of those on the grid have less than 8 h' supply per day. The use of power-generating sets depending on fossil fuel is the general rule for residents and businesses, which exacerbates associated challenges such as pollution and low productivity.

Housing: Over 70% of Lagos residents reside in informal communities bereft of water, sanitation, healthcare and educational facilities. Housing affordability is nearly impossible for the people who require a home, and residents who work in the informal sector are not eligible for mortgages, where available. As a result of macroeconomic factors, the cost of constructing newly built homes in the city is high, and as demand increases, supply has spread to the city's peri-urban areas. Residents purchase land in these far-off communities, and build incrementally as income stream permits, combining with the use of substandard materials to give the settlements a slum outlook. Past and current administrations of Lagos State have attempted to tackle these complex challenges through several policies and interventions. However, the scale of the problem overshadows these efforts.

15.3 Key Transportation and Transport Infrastructure Overview

Transport in Lagos is complex. The various modes can be described as informal and formal public transport and private vehicles. According to the Lagos Resilience Strategy (2020), the transportation demand in Lagos by all modes of transport (including walking) is estimated at 22 million, and the demand is met through danfos (informal minibuses) 45%, walk trips 40%, private cars 11%, BRT 0.41% and Lagbus 1%. The other modes of transport, such as water transport, rail, cycling and taxis, are used by 2.4% of the Lagos population (Heinrich Boll Foundation, 2018; Otunola et al., 2019): water transport 0.34%, rail 1% and others 1%. The number of registered semiformal public transport buses is over 145,000. In 2015, Lagos Metropolitan Area Transport Authority (LAMATA) estimated 2,000,000 private cars with an increasing number of 100,000 cars per annum. Using these numbers, the estimated number of cars on Lagos roads in 2020 would be 2,500,000.

The challenges in the transportation sector are multifaceted and can be summarised in a statement extracted from the Lagos Development Plan 2012–2015:[2]

> *Today, the transport system is inadequate for the growing urban population in the state. All modes of transport have challenges. The bus public transport operation suffers from high levels of fragmentation and inadequate regulation. Rail transport has few existing rail corridors, and the existing corridors are grossly under- utilised. There is no coherence amongst water transport regulatory agencies (LASWA, NIMASA and NIWA). In non-motorised transport, infrastructure facilities are extremely limited throughout the state. Finally, in the paratransit mode of transportation (okadas), there is indiscipline and regulations are not effectively enforced.*

In recent times, BRT Lite was the first push that made a real impact on improving public transport in Lagos State. BRT Lite was opened in 2008 with one line and revamped in 2015. It is managed by Primero, a private transport company, which started its operations with 434 buses. It transports 350,000 commuters daily[3] which is approximately 1.6% of all commuters. A new bus service was inaugurated on April 24, 2019. Lagos Bus Services Limited (LBSL) operates a fleet of 820 buses (520 are HOVs and 300 are midi-buses). However, the dependence on road transportation and the lack of development of alternatives in rail, water and non-motorised transportation have contributed to the myriad of road users' challenges.

15.4 Transport Governance and Investment Framework

The transport sector in Lagos is regulated and managed by two tiers of government: federal and state. Both have important roles to play in air, road, water and rail transportation. At times there is synergy, but there are also cases of overlap and conflict (Fig. 15.2).

Two semiformal institutions that play a part in the Lagos transport system are the Road Transport Employers Association of Nigeria (RTEAN) and the National Union of Road Transport Workers (NURTW). They have registered trade institutions that influence transportation in Lagos. RTEAN are the owners of vehicles (e.g. danfo buses) who engage members of NURTW to operate the vehicles. Lagos State Government understands the immense influence these two institutions have on how public transport works in Lagos and has invited them to be a part of BRT Lite and LBSL through a franchising model.

The policies that govern the transport system in Lagos are:

[2] https://ir.unilag.edu.ng/jspui/bitstream/123456789/4330/2/A%20Study%20of%20Rail%20Mass%20Transit%20in%20Lagos%20and%20its%20Environs.pdf

[3] https://www.theigc.org/wp-content/uploads/2019/10/lagos-BRT-case-study.pdf

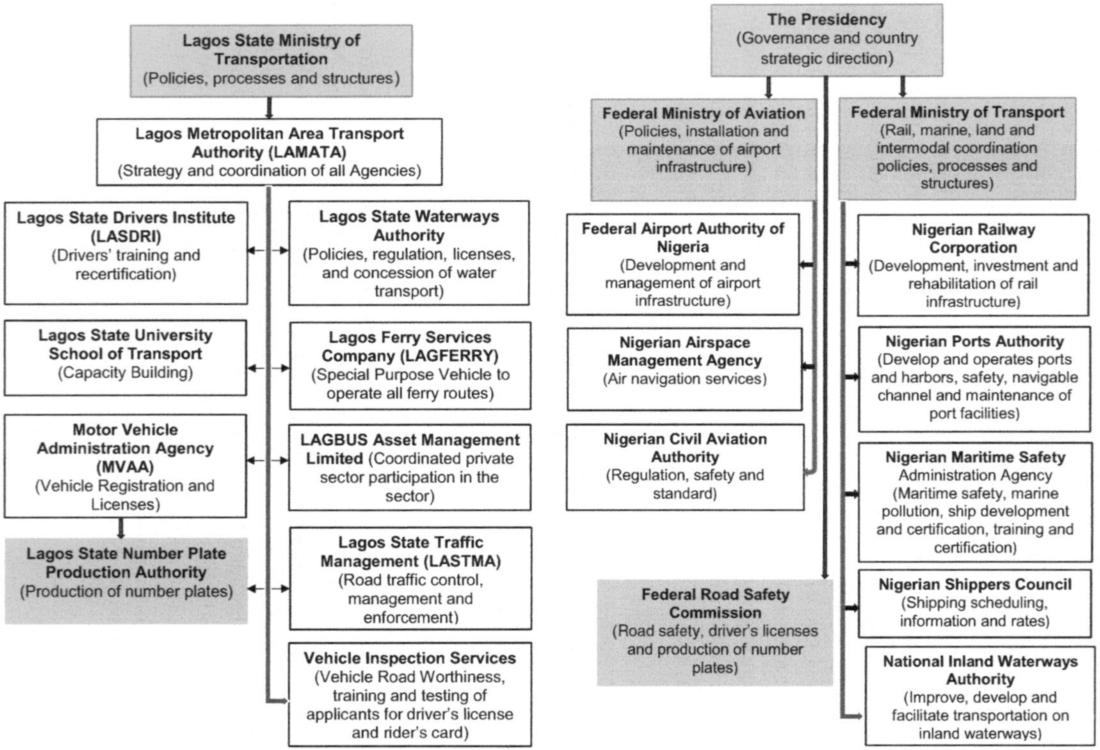

Fig. 15.2 Transport Institutions in Lagos State Via The Lagos State Government and the Federal Government of Nigeria. (Source: Urban Mobility Research Group, Heinrich Böll Stiftung, 2015/2017 cited in Oshodi et al., 2016)

15.5 Lagos State Development Plan 2012–2025

Lagos State Development Plan 2012–2025 provides the direction of the government for the transport sector. Section 8.3 of the plan outlined the state's aim, objectives and targets for transportation. The development plan aims to create a safe, reliable and efficient integrated and multimodal transportation system for sustainable socio-economic development of Lagos State. The strategy will ensure efficient and affordable movement of people and goods, link people to jobs and markets leading to the state's economic competitiveness, lower the cost of production, increase productivity and improve health and time efficiency.

The policy targets focus on access and affordability, efficiency, impact, integration, training, financing and management. It is a clear plan but not a road map; therefore, it does not give life to the goals. The need for up-to-date data is not addressed in the policy, and transportation data is very important for monitoring and evaluation. With 5 years left to achieve the policy goals, it seems unlikely that most of the goals will be achieved. One of the far-reaching goals is cutting down the average travel time anywhere in the state by 50%. More infrastructures are needed for this to happen, and all modes of transport must be integrated, including the formal and semi/informal transport modes. To ensure the reduction of travel time, last-mile transportation needs to be as important as mass transit. Unfortunately, the priority continues to be on large infrastructure projects, leaving last-mile travel to be conducted by the informal sector. In this policy, a transport master plan is called for to coordinate with land use. However, how this will be done is not fully explained, and there is no direct mention of the stakeholders involved in this process.

15.6 Lagos State Strategic Transport Master Plan (2012–2025)

LAMATA developed this plan, and it identifies the transport infrastructure and services required to meet Lagos' demand by 2032. The plan pushes for a modern integrated multimode transportation system and is action-orientated unlike the Lagos State Development Plan 2012–2025. It focuses on infrastructure delivery, the quality of services and funding mechanisms.

The plan hinges on major reforms based on 25 strategic decisions under the following objectives:

- Increasing transport choices for all users.
- Introducing an integrated transport system.
- Making the transit system attractive, convenient, affordable and accessible.
- Reducing urban transportation-induced emissions.
- Optimising usage of the current road network.

- Integration of land use development (urban physical planning) and urban transport planning.
- Securing long-term financing of the investment plan.

The master plan recommends the following transport infrastructure to be able to meet the demand of the population, develop an integrated system and improve passenger and goods flow movement in the city: 6 rail lines, 1 monorail line, 14 BRT routes, 26 water transport routes, 3 cable car lines and several road improvements and traffic management initiatives.

Though it has taken a while, the population is starting to see some of the listed infrastructure being delivered. Three BRT routes are in operation (TBS-Ikorodu, Berger-TBS, Oshodi-Abule Egba). The monorail (Blue Line; Ojo-Marina) is estimated to be completed in 2021. The Red Line's right of way has been determined, and construction should start by the end of 2020, and road improvements are continuous (Mulyungi, 2020). Though some projects are nearing completion, the ambitious infrastructure plan moves slowly and looks unlikely to be achieved by 2032. Like Lagos State Development Plan 2012–2025, integrating land use and transport planning is within this plan, and clear objectives were stated. Transport Oriented Design is hinted at by focusing on future developments along main transit corridors. However, no clear plan calls out the stakeholders needed to develop plans to achieve the objectives.

15.7 Lagos Road Traffic and Administration Law 2012

The Lagos Road Traffic Law guides the Lagos State Transport Management Authority (LASTMA) and the vehicle inspection officers (VIO) on what can be done on Lagos roads. Also, it governs what semiformal and informal transport drivers do on Lagos roads. Unfortunately, the law does not fully consider the nature of transportation and transport operators in Lagos State, and while enforcement agencies such as LASTMA and VIOs are responsible for implementing its provisions, the general public's perception of their efficiency, integrity and approach is often negative. The law demands road users' compliance; however, it does not acknowledge the infrastructure deficit and decline in the traffic control and management problem. The law provided for the administration of road traffic and vehicle inspection in Lagos state, but a major provision was the regulation and ban of okadas (motorcycles) either for private or commercial use as was captured in section 3(1). The law prohibited the operation of commercial motorcycle in about 500 roads out of the 9,200 roads across Lagos (Sunday et al., 2020) causing much dissension. While the government maintained it held consultations before promulgating the law, stakeholders report otherwise.

15.8 Draft Non-motorised Transport (NMT) Policy (2018)

In Lagos, the number of registered motor vehicles has more than tripled from 2005 to 2017 (Lagos State Transportation Policy Draft, 2018), and it will continue to do so unless there is a drastic improvement to the public transport system. The rise in motor vehicles increases traffic accidents and noise and air pollution, reduces safe public spaces and has adverse health impacts. With this in mind and the importance of last-mile travel, the Lagos State Government through LAMATA developed a draft NMT policy. The draft NMT policy focuses on NMT design and management. The policy demonstrates a shift in thinking and focuses on why NMT is important and improving safety, accessibility, congestion and pollution. It also clearly defines design standards and financial mechanisms and gives 5-, 10- and 15-year timelines.

Some of the standards and guidelines are idealistic and stem from best practices; however, they are not in line with other policies or laws that govern transportation. For example, the draft NMT policy encourages street vending, but no other policy supports such. Also, the policy supports the integration of danfo buses. However, there is clear rhetoric from the Lagos State Government to remove them from the public transport system. The policy also needs a well- articulated monitoring and evaluation plan to ensure goals and objectives are met to the standard required. Overall, the policy can improve last-mile travel in Lagos and safety for pedestrians and cyclists but has been a draft for 2 years; the next steps are still unclear.

15.9 Draft Lagos State Transport Policy (2019)

The vision statement of the draft Lagos State Transport Policy is to have 'a transport system that provides reliable and equitable access to all the state's people in an environmentally sustainable and economically, cost-effective manner towards inclusive economic, social, and political growth and development of the state' (Lagos State Transportation Policy Draft, 2018).

The policy thrusts are:

1. To encourage private sector participation.
2. Improved co-ordination in inter-modal transport.
3. Transparent access to subsidies for the provision of transport services for economic and social benefits.
4. Increased access to transport services in rural areas.
5. Integrated transport policy, planning appraisal and implementation across modes (Lagos State Transportation Policy Draft, 2018).

Overall, the draft Lagos State Transport policy fills in the gaps and creates stronger administrative powers for the ministries, departments and agencies that impact Lagos' transportation system. New institutions were recommended, such as the Lagos State Parking Authority. This institution will establish and manage parking spots and lots within each local government. However, two major aspects are absent from the policy. The policy does not put time into the objectives, and it dismissed the informal and semi-informal transport modes, which represent 50–70% of all daily trips in Lagos. Regarding NMT, this draft policy is not as detailed as the draft NMT policy.

15.10 Mobility and Access in the City

A recent study on mobility challenges with Lagos transportation (Lawanson, 2019) indicates poor road conditions and unruly behaviour of paratransit operators, notably drivers, conductors and the ubiquitous enforcers (Agberos) of transportation unions. The study also shows differences in commuters' experience by gender: while men are mostly concerned with infrastructure, women are mostly concerned with safety (Fig. 15.3).

Institutional responses, such as The LSTMP Safety Plan, have also been found to prioritise road accidents rather than mobility concerns such as commuter safety. Thus, transportation in Lagos is conceived and operationalised as infrastructure, not as service, limiting focus on users' needs. In any case, infrastructure and services are not adequately maintained and governed. All of these compound to make the transportation experience stressful for end-users.

For rail transportation, the Lagos Metropolis Mass Transit Service runs from Iddo at the edge of Lagos Island, through to Ijoko/Kajola in Ogun State, a distance of about 26 km, and with 25 train stops. Commuters can use either the Mass Transit Train (MTT) or the Diesel Multiple Unit (DMU) trains. The MTT is cheaper, with a larger capacity than the DMU. However, its coaches are in deplorable condition, unsafe and often unclean and overcrowded. The DMU is newer and air-conditioned

Fig. 15.3 Commuter's mobility challenges and concerns. (Sources: Giditraffic, Twitter (2020); The Guardian Nigeria, Newspaper (https://www.google.com/url?sa=i&url=https%3A%2F%2Fguardian.ng%2Ffeatures%2Fstate-orders-okada-to-vacate-restricted-routes-within-21-days%2F&psig=AOvVaw1girRaC697pamRXqSFT5Qw&ust=1675165737635000&source=images&cd=vfe&ved=0CBEQjhxqFwoTCLjp3eec7_wCFQAAAAAdAAAAAB AE))

and has fewer coaches and is more expensive. The service operates from Monday through Saturday and runs three shifts: morning, afternoon and night. It is hard to find a schedule, and in both cases, additional expenditure has to be made to cover the last mile from the train station to the destinations.

The affordability of public transport depends on the mode. BRT Lite and LBSL have standard fees (regulated by LAMATA), and tickets are usually purchased before entering the bus. Fares are discounted at up to 50% of local public transport providers' fares. This practice is not the same on danfo buses. Here, the fare depends on the passenger's bargaining power, the weather condition in the city, the period of the day, the condition of the vehicle and the operator's mood (Heinrich Boll Foundation, 2018).

Recently, the Lagos State Government, through LAMATA, launched the Lagos Cowry card. It is an electronic card to be loaded with money and used to pay for transportation on the LBSL system and BRT Lite. The aim is that it will also be used on the monorail and water ferries. Inclusion of people living with disabilities is also dependent on the mode. The paratransit (especially minibuses) is not mandated to treat them any differently from other passengers, whereas the BRT Lite and LBSL buses have priority seats for people living with disabilities.

15.11 Transportation Investment

The Lagos State Government has employed a mix of direct funding alongside the financing of transportation infrastructure in recent times. There has been investment in BRT Lite transportation, water transportation and monorail and a significant private sector investment in technology-enhanced transportation models, especially in the paratransit sector with the use of ride-hailing apps. Examples are *Oga Taxi*, *Opay*, *Bolt (Taxify)*, *GidCab*, *Jekalo*, *Uber*, *Max* and *inDriver*.

The Lagos State Government has also collaborated with the Federal Government on the development of the 'Red Line' light rail transport project. Public funds have also been directed towards road expansion and repairs of express roads and routing, expansion and access roads to better link communities. These have been done through either the procurement of civil engineering works or the Public Works Department of the Lagos State Government.

1. Road.

   (a) *Bus Rapid Transit (BRT).*

   One of the major investments in recent times has been Oshodi Transport Interchange, which was planned to convert the popular Oshodi transportation hub (a network of intercity and intracity informal bus garages) into a world-class central business district (CBD) with three bus terminals. The $70-million project was expected to transport one million Lagosians to their destinations daily. It was also conceived as an implementation of multimodal

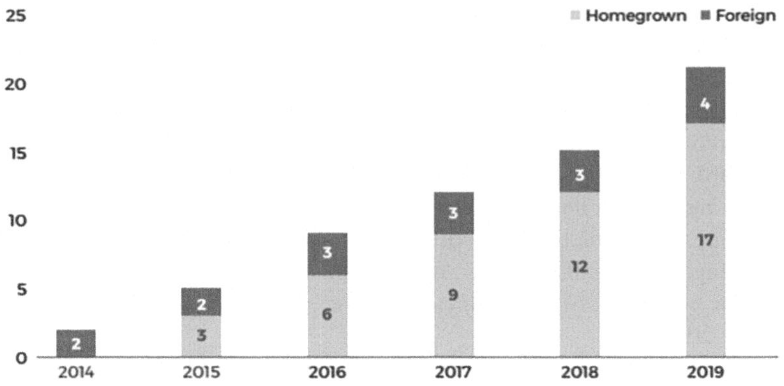

Fig. 15.4 Number of e-hailing platforms operating in Nigeria. (Source: Insight2impact, 2020)

transportation plans, complementing the expansion of existing BRT lanes (Mile 12 to Ikorodu) and linking with the proposed Red Line (31-kilometre railway project).

(b) *Ride-Hailing Providers.*

Since Uber started operations in Nigeria in 2014, digital ride-hailing or e-hailing platforms have quickly become the mainstay in Nigeria, especially in Lagos, reshaping the way we think about personal mobility and work in the transport sector. The revenue generated by Nigeria's e-hailing industry was estimated to reach $292 million in 2020. The 21 e-hailing platforms operating in Nigeria by 2019 are privately owned and financed. They include a few large international players, as well as a range of smaller home-grown platforms (Fig. 15.4).

2. Rail.

The Lagos State Government is constructing a light rail system through the Lagos Metropolitan Area Transport Authority (LAMATA). The urban rapid rail transit system is developed under a public-private partnership (PPP). Although seven lines including Red, Blue, Green, Yellow, Purple, Brown and Orange were proposed for the rail system, only the two phases of the Blue Line are currently under construction and have been since 2008. The Red Line and Blue Line covering a total length of 57 km (Red Line 30 km & Blue Line 27 km) are proposed to cater for a population of over 17 million. The total construction cost is estimated at $1.4 billion, with each line projected at $650 million and $750 million, respectively.

The Lagos State Government entirely funds the project while the system is managed by LAMATA, which is responsible for policy guidelines, legislation and network infrastructure. The private sector supplies supply railway infrastructure, including electric power, signalling, rolling stock and fare collection equipment, under a concession deal. Initially, the Blue Line construction was supposed to be completed by 2011. Due to financial concerns, it was postponed to 2020 (The Punch Newspapers, 2019).

3. Water.

   (c) *New Lagos ferry.*

   Until recently, many Lagos residents were sceptical about the use of water transportation within the state. Recent investments in water transport are increasingly changing their perception of the state's waterways as an alternative natural platform for travel within lagoon areas (Ogah & Adeoye, 2015). They have now become popular and are majorly dominated by private ferry operators and companies. The Lagos State Ferry Services Corporation (LAGFERRY) is a government-owned ferry services provider in the state.

15.12 Actions Taken in the Last 20 Years and Impacts

In 2002, the Lagos Metropolitan Area Transport Authority (LAMATA) was established by the Lagos State Government with the support of the World Bank to execute the Lagos Urban Transport Project (LUTP) 1 and 2. LAMATA is a semi-autonomous authority responsible for the planning and management of the transport sector. LAMATA has successfully achieved its objectives, and there has been a clear improvement in transportation since its advent. As a semi-autonomous authority, they have access to funding, manoeuvre between different government levels and establish partnerships with the private sector. The World Bank explains that they are contributing to the increased awareness of traffic management, transparency and discipline in procurement processes and involving the users in decision-making processes (The World Bank, 2011).

As an important intervention in the last decade, the objectives of the LUTP were to (a) improve the management of the Lagos metropolitan transport sector; (b) enhance the public transport road network in an environmentally, socially and financially sustainable manner; (c) enhance bus services; (d) promote water and non-motorised transport; and (e) prepare future phases of the programme (The World Bank, 2011). The LUTP was done in partnership with the Lagos State Government and the World Bank. The project saw the establishment of LAMATA and the development and implementation of the first route of BRT Lite (completed in 15 months) and the Strategic Transport Master Plan. LUTP 2 and the World Bank plus the French Development Agency saw the BRT Lite corridor's extension (The World Bank, 2011).

The first and most significant project under the LUTP to date is BRT Lite. It was launched in 2008 with high-capacity buses that run on dedicated lanes. The system was developed innovatively with government and private sector financing. Today, the system boasts 2 different lines covering over 35.5 km of track and transporting over 350,000 commuters daily. The construction of lay-bys, bus stops, bus terminals and bridges and reconfiguring of some roundabouts to four-way traffic light junctions have also significantly improved. In addition, the role of LASTMA has improved tremendously, and its image has changed from an extorting arm of law enforcement to a more helpful traffic management agency. In 2012, Lagos Traffic Radio came on air, and Lagosians listened to the traffic situation live on the radio. The traffic radio was an innovative and effective way for commuters already on the road to be guided away from traffic hot spots.

Key to the LUTP and Lagos State Strategic Transport Master Plan is integrating water, rail and bus to create an efficient transport system. A seven-line rail network is to be established for rail, while the Blue Line (CMS-Okokomaiko) is nearly completed and should start operations in 2021. The aim is to transport 400,000 passengers per day and reduce traffic on the route by 30% (Ogunbiyi, 2016). Complementing this for inter-city rail transportation is the Lagos Metropolis Mass Transit Train Service. Although not a Lagos State initiative, the service runs from Lagos onwards to Ogun State carrying many commuters. It is yet to move a significant number of passengers daily, but if more investment is injected into the line and more carriages are bought, and with the advent of the Lagos-Ibadan service, this could be a welcome improvement for commuters in Lagos, Ogun and Osun States (Fig. 15.5).

Water transport has received a lot more attention in the last 2 to 3 years. The Lagos State Government has acquired and is operating through LAGFERRY, ferries on the Lagos waterways. Lagos State Waterways Authority (LASWA) is currently building capacity, aggressively advocating for safety, licensing ferry operators and routes, providing navigation aids and channelling the watercourse all over the state.

Fig. 15.5 Blue Line and Red Line connecting areas (Lagos Rail Mass Transit. (2023, January 26). In *Wikipedia*. https://en.wikipedia.org/wiki/Lagos_Rail_Mass_Transit)

There is also extensive patrol of the waterways, research and removal of wrecks and water hyacinths.

The various state government interventions have recorded a significant impact on travel times and increased public transport capacity. Pedestrian safety has been enhanced due to new pedestrian bridges and educative campaigns. These interventions have enhanced revenue generation, and this can be seen by the revenue expectation of 25 billion naira in 2019 from nine transport agencies.

The construction of access roads through various neighbourhoods that help to defray traffic from the highway and provide alternative routes of travel has also considerably reduced commuting times along these routes.

15.13 Transportation Infrastructure Gaps

Inadequacy of Public Transport Modes

Both the BRT Lite and LBSL bus services, even combined, move a very small percentage of passengers on two routes, which leaves the paratransit sector (semiformal mini buses [danfo]) and informal sector (motorbike taxi [okada], boat operators and tricycles [keke marwa]) moving over 40% of the population daily. Although these operators have transportation unions, they are largely unregulated. They cause traffic chaos due to regular breakdowns, unmarked bus stops and operators being very unruly.

Knowing the large percentage of passengers the semi-informal and informal transport modes carry, there have been efforts to improve the systems to make them safer through regulation and training. The registration efforts taken by the Lagos State Government have not been successful due to lax enforcement. The shortfall of public transport in Lagos and the failure to integrate the semi-informal and informal transport modes have also led to the Bus Reform Initiative (BRI), launched in 2019. The buses are operated under Lagos Bus Service Limited (LBSL), and after 10 months of operation, in May 2020, Dr. Frederic Oladeinde, Lagos State Commissioner for Transportation, mentioned that LBSL had moved 10 million passengers (for more information).[4]

Other modes of transport, such as the monorail, are in the pipeline, and BRT Lite and BRI are part of the Lagos Urban Transport Project 1 and 2. The injection of these new modes of transport should start to meet mobility demand in Lagos State. However, due to the population growth rate at 6% per annum, the LUTP may be a little too late.

[4] https://www.vanguardngr.com/2020/05/lagos-brt-operators-to-roll-out-more-buses-transport-commissioner/

15.14 Inadequacy of Infrastructure

The transport infrastructure is inadequate for the industrial, commercial and residential functions the city performs. Despite the dependence on road transportation, infrastructure is still low. With a road network of about 7598 km of road, Lagos has one of the lowest road provisions per person in West Africa (Heinrich Boll Foundation, 2018). For every 10,000 people, there is 2.2 km of road. Different government levels manage trunk A, B and C roads: state, federal and local government. The local government manages 77% of all roads, while the State Government manages 17% and the Federal Government 6% (Heinrich Boll Foundation, 2018). The majority of primary roads (Federal and State roads) are three-lane and two-lane (Atubi, 2013) and link the major population centres, while secondary roads are singular. One-third of the roads in Lagos State are made of concrete decks[6]. These roads' poor conditions have been of major concern to road users, and maintenance works have not always been proactive enough.

Railway currently plays an insignificant role in urban mass transit in Lagos. The Lagos Metropolis Mass Transit Train Service run by the Lagos district of NRC was launched in 2001 and runs from Iddo terminus in Lagos State to Ijoko/Kajawla (Kajola) in Ogun State, covering 26 km. A similar service, starting in Lagos but terminating in Ibadan was announced to be launched in September 2020. The aim is to have 16 trips a day on a 156-km-long double-track standard gauge. The Lagos Metropolis Mass Transit Train Service comprises 25 train stations. Two types of trains run on this service: the Mass Transit Train (MTT) and the Diesel Multiple Unit (DMU) trains. However, connecting the rail services to other transportation modes can be challenging as there are limited connectivity structures.

This is made even more difficult as a result of a lack of adherence to travel schedules. In 2020 some of the trains in unserviceable conditions were replaced (Aderibigbe, 2020).

The Lagos Ferry Service even has a poor infrastructure to serve any compelling need. For instance, it has a fleet of only 14 boats, complemented by informal boat operators that ply the 40 inland waterway routes. Non-motorised transport infrastructure—walkways, overhead bridges and cycle lanes—are limited across the state.

Safety, Environmental and Social Concerns

In Lagos, in the second quarter of 2017, 2503 road crashes occurred. 44.44% occurred because of speed violations, followed closely by loss of control and dangerous driving. Ninety-four percent of all road accidents affected adults, while 6% affected children (Knoema, 2017). Vehicle conditions and lack of enforcement of transport and traffic laws have also led to environmental concerns like poisonous vehicle emissions, improper waste oil disposal and high traffic noise levels (Heinrich Boll Foundation, 2018). Transport fares continued to stay high due to the fragmentation of the public transport system and forced evictions due to the expansion of transport infrastructure.

Land Use and Transport

Lagos has developed in a haphazard way, where development can happen while disregarding the urban master plans developed by the Lagos State Government. Also, over 70% of Lagos has developed formal and semiformal settlements majorly serviced by other informal public transport such as 'okadas' and 'keke marwa'. This situation has left 52% (in numbers: 2,275,837) of the households in Lagos with a lack of adequate transportation access.

Road traffic congestion is also a major issue due to locational disconnection between the city's commercial and residential zones. There is a large inflow of people from the north, west and east to the south-central areas of Lagos in the morning and a mass exodus after work hours. Commuters on all forms of transport, from BRT buses to private vehicles, can travel anywhere from 2 to 5 h on average on each journey to or from work, depending on the residence-place of work distance.

Modal Integration

At present, all transport modes are fragmented. Danfo buses under the National Union of Road Transport Workers (NURTW) have designated bus parks, but do not always adhere to this, causing serious logjams as they pick up and drop passengers along the highways. They also stop indiscriminately along at places like along street corners, junctions or lay-bys. Other informal public transport modes are hailed and pick up and drop off at their passengers' request. BRT Lite and LBSL have designated bus stops, follow a timetable and stop at bus terminals (e.g. Oshodi, Ikeja, TBS) (Fig. 15.6).

The water transport terminals have their system but do not link with the road public transport services. However, some water transport terminals have a park and ride system, such as the Ikorodu water transport terminal and the Five Cowries Terminal. This lack of integration has made it extremely hard to standardise fees.

Fig. 15.6 Picture showing Lagos bus park. (Source: Guardian Nigeria (https://guardian.ng/features/executive-motoring/lagos-to-decongest-queues-at-brt-busstops/\) & Independent Newspapers(IndependentNewspaper—https://independent.ng/frsc-seeks-support-of-motorcyclists-keke-napep-operators-over-new-driving-licence/)

The public bus service suffers from availability and reliability issues, while existing rail corridors operate marginally. Water transport is also being explored, though with limited private-sector participation.

15.15 Recommendations

Enhanced Multimodal Transportation Solutions
There is a need to continue implementing the Lagos State Strategic Transport Master Plan (LSTMP). New investment in intracity transportation should be focused on introducing first-mile and last-mile solutions to link up existing BRT, water transport and light rail modes. This will better meet the commuters' needs, enhance the efficient movement of persons and goods and potentially reduce city congestion in the long term. The government, therefore, has to appraise the direction of new investment in the transportation sector.

Synergise Transportation Investment and Real Estate Development
Apart from transportation challenges, the city also suffers from massive housing backlogs, although is paradoxically a real estate investment hub. There are existing opportunities to synergise the government's efforts by linking investment in road expansion, BRT routes expansion and light rail projects to existing housing needs, thereby producing clusters of transportation-driven real estate development projects. If financed appropriately, this strategy can provide affordable and social housing to satisfy some of the residents' housing needs. There is a role for real estate developers to initiate, collaborate and find financing for a project of this nature to be realised. At the same time, there is a role for the government in setting directions (and 'enabling environment') for the affordable housing component, possibly through affordability gap financing, and civil society has to be engaged in the project to ensure that rights are not trampled upon, especially where such projects involve the land acquisition by the government. Community groups may also need to be more active participants in a project of this nature if it is considered to have negative externalities such as displacement.

Repositioning Urban Transportation as a Service
Lagos has to improve on promoting inclusiveness and convenience in its transportation strategies, across all sectors. While it currently implements priority seats for pregnant women, people living with disabilities and the elderly on its bus services (the BRT/LSBL), it has limited control over these processes for other non-government-owned services, especially the paratransit modes. As the city seeks more elaborate infrastructure, the government should also pay attention to engendering services and ensuring convenience for all users. Civil society groups that work in this area would be instrumental in amplifying the need for inclusiveness in transportation services and infrastructure provision.

Integrating Existing Transportation Operators into New Projects
The informal transport sector is critical to the running of the state, though their operations often portend a problematic and somewhat chaotic component of the system. The plans enunciated in the Lagos Strategic Transport Master Plan foresees a future where this sector, along with over 145,000 buses that ply routes all over the state, the 50–70% commuter dependency on their operations and the several hundreds of drivers, conductors, *agberos* and unions that embody this system are eliminated and replaced with formal mass transit systems. Although there have been attempts to work with the unions' politically influential leadership, the systems' operators' volatility and restiveness are a very visible threat to achieving the goals. There has to be a conscious effort to understand the operators in this sector, their needs, goals and ambitions and ensure their integration into either new transportation options or other aspects of society. It is only by treating the loss of livelihoods these operators would experience as a potential threat that they can be treated justly in the march towards modernisation of the Lagos transport system, and it is when they are fairly treated that they can be cooperative.

Promoting Non-motorised Transportation
Implementation of NMT systems needs to be prioritised in infrastructure investment in the Lagos State Transport system. While probably not as attractive to finance as motorised transportation because it is a public good, its inclusion in the transport system is critical for achieving the sustainable transport system espoused in the state's master plan.

Increasing civil society advocates seek more seriousness in the provision of the bike and pedestrian lanes to aid this process and the government, as the regulator must ensure more rapid implementation of its proposals for NMT in the state. The international donor organisations can also provide technical support for this policy's achievement, especially given how heavily current transportation investment prioritises motorised transportation systems.

15.16 Lessons for Other African Cities

- Transportation infrastructure provision provides an important opportunity to synergise physical planning, transportation and real estate development if the advantages of transit-oriented development are optimised. There has not been an intentional effort in this direction in the city. While TODs are not the solutions to all city transportation and housing problems, they provide alternative systems for ensuring accessibility to city functions in nuclei and are found to be more sustainable.
- Due to previous dependence on road transportation, the city's other modes of transportation are underdeveloped. The long-awaited light rail project has not been completed since it started in 2008. Road transportation reliance due to limited transportation infrastructure is causing long hours of traffic congestion

and delays, pollution, increased health issues and low productivity. These experiences have become the defined transportation experience in Lagos, but this need not be the case.

- The LUTP is designed with an ambitious intent to eliminate the informal public bus systems, and the road transport system is now clearly bifurcated between the BRT and LSBL systems that currently carry less than 2% of daily commute compared to the paratransit (danfo) upon which over 50% of the commuting public rely. How these ratios will be reversed without massive social unrest remains a very delicate question, even as the bus garages are now shared by both BRT/ LSBL and the paratransit. Paying attention to the public good would be important to the peaceful transition to this system. On the other hand, the ban on Okada or motorcycle taxis needed to be founded on stronger stakeholder involvement and feasible exit plans for those whose daily livelihoods are dependent on this informal activity.
- The Lagos transportation system has witnessed many policy overturns that could be inimical to continued investment in the transportation sector. For instance, the ride-hailing apps supporting transportation through motorbikes were abruptly halted when the government implemented a ban on 'okada' transport in early 2020. The abrupt halt led to the loss of investment to investors (a few of them have now rebranded as delivery companies to support an emerging e-commerce sector), increased dependence on motorised transport (and traffic congestion), increased fares by the informal bus operators (due to a shift in demand) and many other problems. The heavy financial commitments required to improve any city transportation system require a partnership with private sector providers to be fair to the beneficiaries and secure funders' investments.

References

Aderibigbe, A. (2020). *Narrow guage: Iddo-Ijoko train shuttle resumes Tuesday*. https://thenation-onlineng.net/narrow-gauge-iddo-ijoko-train-shuttle-resumes-tuesday/

Afolabi, O., Oluwaji, O., & Fashola, O. (2017, November 1). Socio-economic impact of road traffic congestion on urban mobility: A case study of Ikeja local government area of Lagos state, Nigeria. *Pacific Journal of Science and Technology, 18*, 246–255.

AfrAsia Bank Africa; New World Wealth. (2019). *Africa Wealth Report, 2019*. Afrasia.

Atubi, A. (2013). An evaluation of transport infrastructure in Lagos State, Nigeria. *Journal of Geography and Earth Science*. http://jgesnet.com/journals/jges/Vol_1_No_1_June_2013/2.pdf

Egbejule, E. (2020). Lagos is a country. *The Africa Report*. https://www.theafricareport.com/28089/lagos-is-a-country/

Heinrich Boll Foundation. (2018). *Urban planning processes in Lagos*. Heinrich Boll Foundation.

Insight2impact. (2020). *Africa's digital platforms database*. https://cenfri.org/topic/insight2impact/

Knoema. (2017). *Road accident statistics of Nigeria*. https://knoema.com/knpbiayq/road-accident-statistics-of-nigeria?state=1000250-lagos

Lagos State Government. (2023). *About us*. https://lagosstate.gov.ng/about-lagos/

Lagos State Ministry of Transportation. (2018). *Lagos state transportation policy draft*. Lagos State Ministry of Transportation.

Lagos State Resilience Office. (2020). *Lagos state resilience strategy*. Lagos State Resilience Office.

Lawanson, T. (2019, September). *Transport is not gender blind in Lagos Nigeria* (Paper presentation). Conference on Mobility and Access in African Cities organised by Volvo Research and Education Foundation and GIZ Transformative Urban Mobility Programme, Addis Ababa, Ethopia.

Mulyungi, P. (2020). *Proposed Red Line Rail Project to link Lagos Airport in Nigeria*. https://constructionreviewonline.com/news/nigeria/proposed-red-line-rail-project-to-link-lagos-airport-in-nigeria/

Ogah, D., & Adeoye, T. (2015, October 25). Water transportation: Revving the culture of mass movement. *The Guardian*. https://guardian.ng/business-services/business/water-transportation-revving-the-culture-of-mass-movement/

Ogunbiyi, T. (2016). *Lagos: Need for intermodal transport system*. https://punchng.com/lagos-need-for-intermodal-transport-system/

Ogunbiyi, T. (2019). *Lagos, the rains, the roads and the people*. https://lagosstate.gov.ng/blog/2019/07/17/lagos-the-rains-the-roads-and-thepeople/

Oshodi, L., Salau, T., Seun, O. M., & Richard Emeka, U. (2016). Urban mobility and transportation. In F. Hoelzel (Ed.), *Urban planning processes in Lagos*. Heinrich Böll Stiftung Nigeria and Fabulour Urban.

Otunola, B., Kriticos, S., & Harman, O. (2019). *The BRT and the Danfo: A case study of Lagos' Transport Reforms from 1999–2019*. IGC Cities that Work Case Study. https://www.theigc.org/wp-content/uploads/2019/10/lagos-BRT-case-study.pdf

Stears Business. (2019). *The Nigerian economy revolves around Lagos Ports*. https://www.stearsng.com/article/the-nigerian-economy-revolves-around-lagos-ports

Sunday, E., Olatunji, K., Laba, E., Udonya, B., & Quadri, S. (2020, February 2). Again, Lagos battles commercial motorcyclists, others. Guardian Newspaper. Retrieved from https://guardian.ng/sunday-magazine/again-lagos-battles-commercial-motorcyclists-others/

The Punch Newspapers. (2019, December 18). Lagos Metro Rail will become operational. *Lagos Metro Rail will Become Operational 2021—Sanwo-Olu. The Punch.*

The World Bank (2011). *Implementation completion and results report*. http://documents1.worldbank.org/curated/en/267131468333318985/pdf/ICR18480P074960080501100BOX3618521B.pdf

Index

T. Nubi et al. (eds.), *Transit Oriented Development in West African Cities*, Urban Perspectives from the Global South,
https://doi.org/10.1007/978-3-031-58726-9

SPRINGER NATURE

GPSR Compliance

The European Union's (EU) General Product Safety Regulation (GPSR) is a set of rules that requires consumer products to be safe and our obligations to ensure this.

If you have any concerns about our products, you can contact us on ProductSafety@springernature.com

In case Publisher is established outside the EU, the EU authorized representative is:

Springer Nature Customer Service Center GmbH
Europaplatz 3
69115 Heidelberg, Germany

The manufacturer's authorised representative in the EU is Springer Nature Customer Service Centre GmbH, Europaplatz 3, 69115 Heidelberg, Germany. If you have any concerns regarding our products, please contact ProductSafety@springernature.com

Printed and bound by CPI Group (UK) Ltd, Croydon, CR0 4YY
01/10/2026
02225968-0010